Energy absorption of structures and materials

Energy absorption of structures and materials

Guoxing Lu and Tongxi Yu

CRC Press
Boca Raton Boston New York Washington, DC

WOODHEAD PUBLISHING LIMITED
Cambridge England

Published by Woodhead Publishing Limited, Abington Hall, Abington
Cambridge CB1 6AH, England
www.woodhead-publishing.com

Published in North America by CRC Press LLC, 2000 Corporate Blvd, NW
Boca Raton FL 33431, USA

First published 2003, Woodhead Publishing Ltd and CRC Press LLC

British Library Cataloguing in Publication Data
A catalogue record for this book is available from the British Library.

Library of Congress Cataloging in Publication Data
A catalog record for this book is available from the Library of Congress.

Woodhead Publishing ISBN 1 85573 688 8
CRC Press ISBN 0-8493-1768-1
CRC Press order number: WP1768

Typeset by SNP Best-set Typesetter Ltd., Hong Kong
Printed by TJ International, Padstow, Cornwall, England

Contents

	Preface	ix
	Notation	xiii
1	**Introduction**	**1**
1.1	Vehicle accidents and their consequences	1
1.2	Applications of energy-absorbing structures/materials	10
1.3	Design of energy-absorbing structures and selection of energy-absorbing materials	19
2	**Methodology of analysing energy-absorption capacities**	**25**
2.1	Idealisation of materials' behaviour	25
2.2	Limit analysis and bound theorems	34
2.3	Effects of large deformation	42
2.4	Effects of dynamic loading	47
2.5	Energy method	60
3	**Dimensional analysis and experimental techniques**	**68**
3.1	Dimensional analysis	68
3.2	Small-scale structural models	75
3.3	Experimental techniques	80
4	**Rings and ring systems**	**88**
4.1	Ring compressed by two point loads	88
4.2	Ring pulled by two point loads	89
4.3	Built-in semi-circular arch under point loads	94
4.4	Ring compressed by two flat plates	96
4.5	Laterally constrained tubes	101
4.6	One-dimensional ring system under end impact	106

4.7	Lateral crushing of arrays of circular tubes	109
4.8	Other ring/tube systems	112
4.9	Concluding remarks	113
5	**Thin-walled members under transverse loading**	**114**
5.1	Circular tube under point loading	114
5.2	Indentation of a circular tube by a blunt wedge	117
5.3	Bending collapse of thin-walled members	124
5.4	Other loading systems and comments	143
6	**Axial crushing of thin-walled members**	**144**
6.1	Circular tubes	144
6.2	Square tubes	153
6.3	Top-hat and double-hat sections	164
6.4	Effect of foam filling	167
6.5	Further remarks	172
7	**Impact on structures and inertia-sensitivity**	**174**
7.1	Local deformation of structures due to impact	174
7.2	Inertia-sensitive energy-absorbing structures	197
8	**Plastic deformation with ductile tearing**	**215**
8.1	Measurement of tearing energy	215
8.2	Axial splitting of circular metal tubes	223
8.3	Axial splitting of square metal tubes	228
8.4	Piercing of metal tubes	236
8.5	Cutting of metal plates by a wedge	242
8.6	Concluding remarks	247
9	**Cylindrical and spherical shells**	**249**
9.1	Tube inversion	249
9.2	Tube internal nosing	255
9.3	Inversion of a spherical shell	258
9.4	Propagating collapse of a submarine pipe	261
9.5	Concluding remarks	266
10	**Cellular materials**	**268**
10.1	Honeycombs	268
10.2	Foams	278
10.3	Wood	286

10.4 Impact response of cellular materials 289
10.5 Cellular textile composites 304

11 Composite materials and structures 317

11.1 Factors that influence the energy-absorption characteristics 317
11.2 Axial crushing of circular tubes 318
11.3 Axial crushing of tubes with other geometries 331
11.4 Tubes under bending 333
11.5 Comments on the crushing of composite tubes 336
11.6 Axial crushing of composite wrapped metal tubes 337
11.7 Composite sandwich panels 341

12 Case studies 351

12.1 Rockfall protective net 351
12.2 Packaging using plastic foams 359
12.3 Design of vehicle interior trim 367
12.4 Corrugated guardrail beams 373

References 385
Index 401

Preface

This book is concerned with energy absorption of structures and materials under static and impact loadings. Knowledge of energy absorption behaviour is of importance for material selection and design of energy absorbers, crashworthiness and damage assessment of structures subjected to accidental collision, and packaging design against impact. Investigation of this behaviour requires an understanding of materials engineering, structural mechanics, theory of plasticity and impact dynamics. Over the last few decades, much research attention has been given to this subject. Nevertheless, literature in the field is scattered and it is difficult for someone, especially a beginner, to grasp easily the basic concepts and to apply the principles successfully in other cases. In this book, we attempt to bring together current understanding of the subject area.

Because of the wide variety of complex structures and materials in use, this book mainly focuses on basic concepts and methodologies, and simple structural members and materials. In presenting this material, an emphasis has been placed on physical behaviour and simple analytical treatment. Comprehensive analysis of more practical, complex structures, such as car bodies and aircraft fuselages, is beyond the scope of the present book. Methods of using commercial finite element packages are not included, although results will be given for relevant cases.

The structure of the book is as follows. Chapter 1 gives a brief introduction to the subject. It describes the engineering background of energy absorption of materials and structures and specifies the general requirements of impact energy absorbers. Based on the theory of plasticity and impact dynamics, Chapter 2 presents the fundamentals and methodology for analytical studies. In Chapter 3, dimensional analysis and the closely related concept of small-scale model tests are discussed in the context of energy absorption, whilst conventional experimental methods are introduced.

Chapters 4 to 6 examine energy absorption of several simple structural members under different loading conditions, making use of the fundamentals introduced in Chapters 2 and 3. The structures are circular rings, ring systems, and thin-walled tubes, all experiencing large plastic deformation. The loading conditions include tension, compression (axially or laterally) and indentation. Chapter 7 is concerned with modelling of local deformation under impact and an inertia sensitive structure.

Chapter 8 deals with problems involving plastic deformation and tearing. This class of problems is much more complex. The value of tearing energy is difficult to assess and is problem dependent. Besides, there may be an interaction between tearing and plastic bending/stretching in the far region.

Chapter 9 presents plastic analyses of four problems: tube inversion and nosing, inversion of a spherical cap and buckle propagation in submarine pipelines. These problems illustrate some features of propagating plasticity.

In contrast to the metallic structures discussed in Chapters 4 to 9, Chapter 10 covers energy absorption performance of four categories of cellular materials: honeycombs, foams (including metal foams), wood and cellular textile composites. The analytical approach used here generally invokes a study of a typical cell at a microstructure level, by using the methodology presented in Chapter 2, and then translates its mechanical behaviour to global response of the cellular material as a whole.

Chapter 11 summarises studies on composite structures and materials. Composite tubes and composite sandwich panels are discussed. The detailed energy absorption mechanisms are described in comparison with their counterparts for metallic structures. Available analytical studies are presented.

Finally, Chapter 12 contains four case studies: a rockfall protective net, packaging design using foams, design of a vehicle interior trim and design of roadside guardrail beams. These practical examples are used to illustrate applications of the knowledge presented in the previous chapters.

This book will be very useful to engineering and materials science undergraduate students at advanced stages, postgraduate students, practising mechanical and structural engineers, as well as researchers interested in energy absorption calculations and design of structures and materials against impact.

Many friends and colleagues have helped us in our research and understanding of this subject. In particular, we wish to acknowledge our gratitude to Professors C.R. Calladine and W. Johnson, our respective supervisors as research students, who have been a constant source of inspiration. We wish to thank our colleagues and friends Steve Reid, Norman Jones, Bill Stronge, Yella Reddy, Jason Brown, Majid Sadeghi, Bin Wang, Qing Zhou, Dongwei Shu, Raphael Grzebieta, Xiaoming Tao, Pu Xue and Xi Wang; as well as our students Dong Ruan, Ziyang Gao, Jacquelin Hui and Haihui Ruan for help they provided with various facets of the preparation of the book. Parts of the book were written during the study leave of Guoxing Lu and the sabbatical leave of Tongxi Yu, thanks to the support of the School of Engineering and Science, SUT, and the Department of Mechanical Engineering, HKUST. We thank Stephen Guillow, who carefully read the manuscript and offered many valuable comments, and Xiaodong Huang, who did all the high quality drawings. Mrs Gwen Jones, commissioning editor at Woodhead Publishing Limited, provided us with much valuable advice. Generous permission to reproduce the figures has been given by a number of publishers, who are acknowledged in the text. Lastly, we thank our respective wives, Jue and Shiying, for their understanding and patience over the years.

G. Lu
Melbourne

T.X. Yu
Hong Kong

Notation

A	area
a	acceleration; radius of contact circle
B	material constant in Eq. [2.73]; width
b	width
c	distance; side length of a square tube
c_L	speed of longitudinal elastic stress wave
c_p	speed of longitudinal plastic stress wave
d	diameter
D	energy dissipation; diameter of a circular tube/ring; plastic energy
E	Young's modulus; energy
E_{in}	input energy
E_p	hardening modulus
e	coefficient of restitution
F	force
f	non-dimensional force
G	(concentrated) mass
H	height; half length of a fold; horizontal force component
h	thickness
I	second moment of cross-section
J	moment of inertia
K	kinetic energy
k	spring constant
L	length
l	dimension of cell; rolled length
M	bending moment
M_e	maximum elastic bending moment
M_o	fully plastic bending moment per unit length for plates and shells
M_p	fully plastic bending moment for beams
m	mass; non-dimensional bending moment, M/M_e

m_s	equivalent mass of structure
N	axial force
N_o	fully plastic axial force per unit length for plates and shells
N_p	fully plastic axial force for beams
n	number of cracks
P	load
P_o	initial collapse load
P_s	limit load
p	impulse
p_o	overloading impulse
p^*	porosity
Q	generalised force
q	exponent; distributed load
R	radius
R_{er}	energy ratio, $E_{in}/E^e_{\max}$
R_M	mass ratio
R_p	energy ratio of plastic dissipation to input energy, D/E_{in}
r	radial coordinate; strain-rate index; radius
S	scaling factor
S_l	scaling factor for linear dimension
S_m	scaling factor for material
T	duration of pulse; total kinetic energy of system; total thickness; total time
t	time
u	displacement
V	velocity of projectile; vertical force component
v	velocity
v_y	yield velocity
W	work done
w	flange length; specific work
x, y	coordinates
Y	yield stress
α	angular acceleration
Δ	(representative) displacement
δ	(local) displacement; increment
δ_e	effective length
ε	strain
ε_D	densification strain
ε_f	fracture strain
ε_I	incident strain
ε_R	reflected strain
ε_T	transmitted strain

ε_y	yield strain
η	structural effectiveness; dynamic enhancement factor
θ	rotation angle
κ	curvature
κ_e	maximum elastic curvature
λ	effective length of plastic hinge
μ	coefficient of friction
ν	Poisson's ratio
$\pi_1, \pi_2 \ldots$	dimensionless groups
ρ	density; half angle of hinge rotation
ρ^*	density of cellular materials
ρ_s	density of cell wall solid
σ	stress
τ	shear strength
ϕ	solidity ratio; non-dimensional diameter
φ	non-dimensional curvature, κ/κ_e
ω	angular velocity

superscripts

d	dynamic
e	elastic
ep	elastic-plastic
o	statically admissible
p	plastic
rp	rigid-plastic
$+$	upper bound
$-$	lower bound
$*$	kinematically admissible

subscripts

b	bending
c	compression; composite; characteristic; central
cr	critical
d	densification
e	effective
f	final; flange
fri	friction
$frac$	fracture
g	global
in	input
L	longitudinal

l	local; locking
m	membrane; model; metal
max	maximum
min	minimum
o	initial
p	prototype; peak
R	radius
r	restitution; radial; rebound
s	stretching; static; solid
t	tube; tangential; tearing
tl	total
u	ultimate
w	web
y	yield

1
Introduction

From the crashworthiness of vehicles to the protection of human bodies, the engineering background to studies of energy absorption of structures and materials is reviewed. The general principles involved in designing structures and selecting materials for the purpose of energy absorption are also discussed.

1.1 Vehicle accidents and their consequences

1.1.1 Statistics of vehicle accidents

The modern world relies greatly on various transportation means, and the number of vehicles has been continuously increasing over the last century. In the USA alone, according to the NCSA (National Centre for Statistics & Analysis), in 2000 there were 217 930 000 registered vehicles (an 18 % increase from the figure in 1990), while the VMT (vehicle miles travelled) was estimated as 2 688 312M (a 25 % increase from the figure in 1990). Compared with the population of 275 129 687 for the USA in 2000, the above figures indicate that on average every five persons had four vehicles, and every vehicle travelled 20 000 kilometres per year.

Advances in technology have led not only to increasing numbers of vehicles and VMT, but also to higher speeds and more massive vehicles (e.g. large trucks and aircraft). This means that the vehicles themselves are costly structures and that, if they are involved in traffic accidents, the damage to people and the environment will be more serious.

Motor vehicle related accidents are a major worldwide health problem and constitute a great economic loss to society. For example, vehicular crashes kill more Americans between the ages of 1 and 34 than any other source of injury or type of disease. In the USA, more than 95 % of all transportation deaths are motorway-related, compared with 2 % for rail and 2 % for air.

According to the Traffic Safety Fact Sheets published by the NCSA, in 1999 there were an estimated 6 279 000 police-reported traffic crashes in the

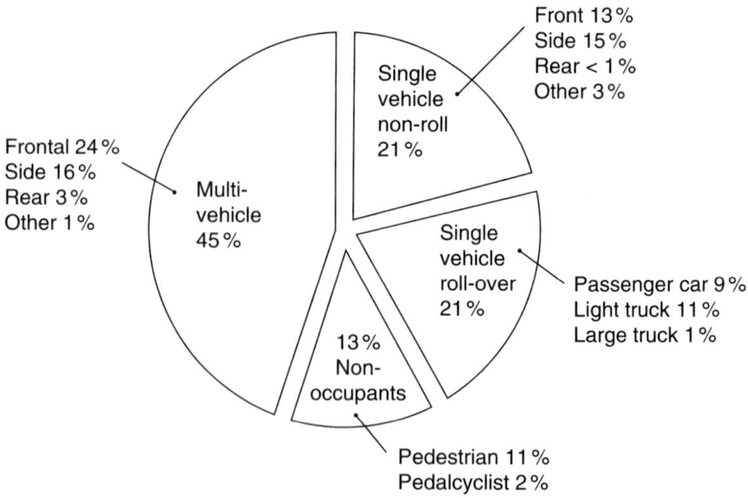

1.1 Persons killed in the USA in 2000 by type of crash (%).

USA, in which 41 611 people were killed and 3 236 000 people were injured. Of these crashes 37 043 were fatal ones, while 4 188 000 crashes involved property damage only. Among the 41 611 persons being killed in 1999's crashes, 25 210 were drivers, 10 596 were passengers and 5805 were non-occupants (pedestrians and pedal-cyclists). Figure 1.1 shows the persons killed in 2000 by type of crash (%).

The trends in the total crash fatality by year and the crash fatality rate per 100M VMT by year are shown in Figs 1.2(a) and (b), respectively. It is seen from Fig. 1.2(b) that the crash fatality rate per 100M VMT gradually reduced from about 2.25 in 1988 to about 1.6 in 2000.

Although road safety issues apply around the world, only the statistics for the OECD (the Organisation for Economic Co-operation and Development) nations are generally available. The OECD has more than 20 member countries, and each OECD member country provides regular road safety statistics for an International Traffic and Accident Database (IRTAD) based in Germany.

The number of deaths for every 10 000 registered vehicles is a method of comparing road fatalities, taking into account the level of motorisation. The median of this number for the OECD nations was 2.0 in 1997, decreased from 6.7 in 1975. Among the OECD nations, this number in 1997 ranged from 0.7 (Iceland) to 11.1 (Korea). Figure 1.3 shows the trend in fatalities per 10 000 registered vehicles for selected OECD countries together with the OECD median.

Another meaningful number is the deaths for every 100 000 of population, which is a measure of the public health risk associated with road use.

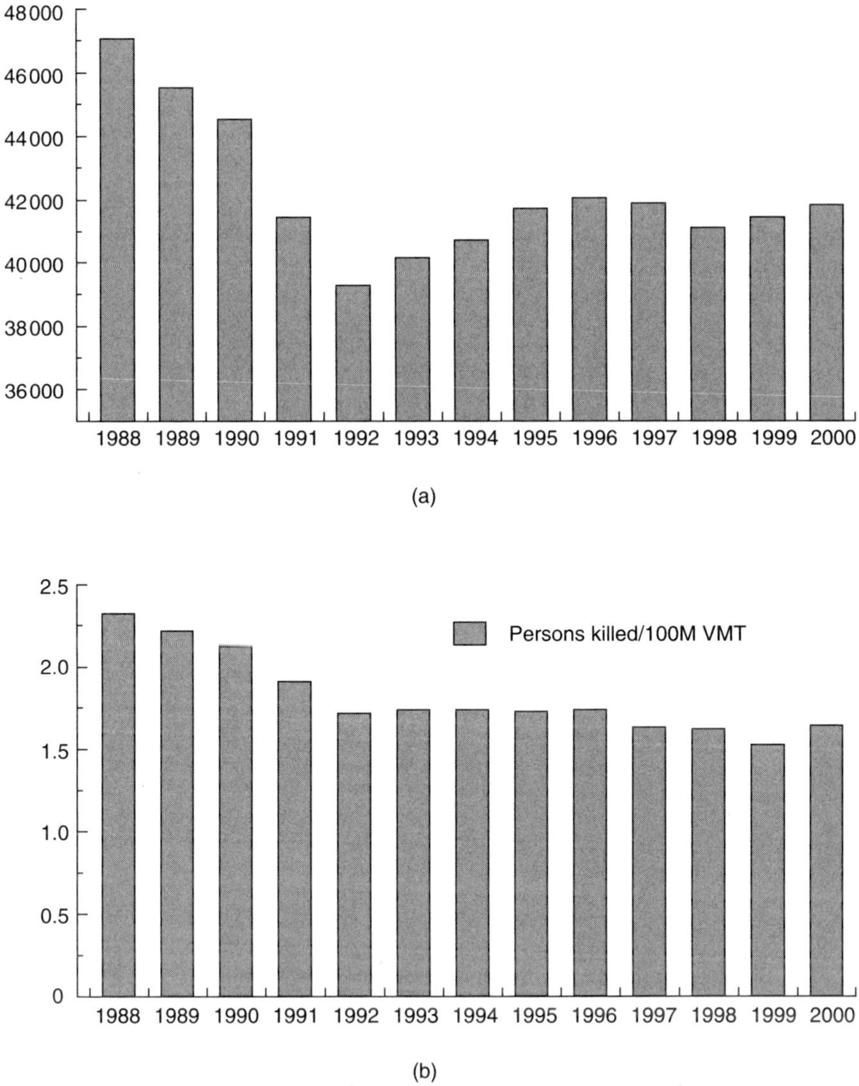

1.2 Crash fatalities in the USA (data from NTSA): (a) total number by year; (b) rate per 100M VMT by year.

The median of this number was 11.7 in 1997, decreased from 18.8 in 1975. Among the OECD nations, this number in 1997 ranged from 5.5 (Iceland) to 29.3 (Korea).

Incomplete statistics can be found for other countries or regions. For example, the Peoples' Republic of China (PRC) reported that about 84 000 people were killed by road accidents in 2000. This number is rather high if

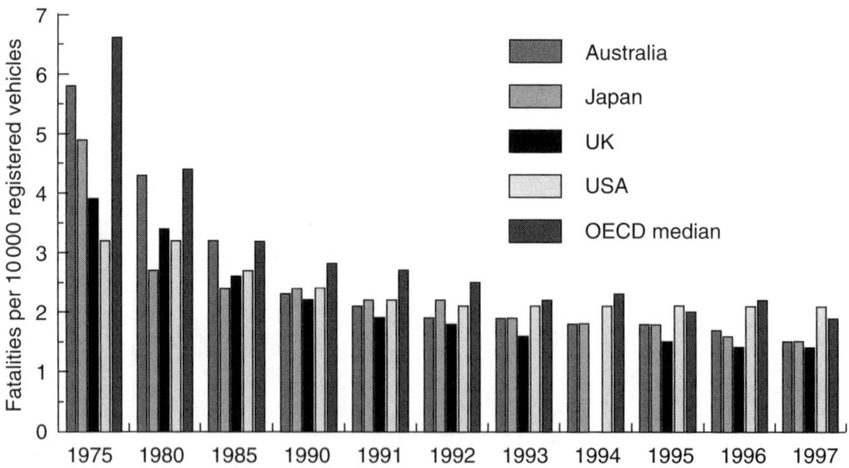

1.3 The trend of the fatalities per 10 000 registered vehicles for selected OECD countries and OECD median.

compared with the number of registered vehicles; but it is not too high if compared with China's 1.3 billion population. As a SAR (Special Administrative Region) of PRC, Hong Kong had about half a million registered vehicles in 2000, while about 200 people were killed by road accidents each year. Some statistics for Hong Kong are plotted in Fig. 1.4.

The economic loss caused by road crashes is also tremendous for every country. For instance, according to Australian Transport Safety Bureau (Annual Review 2000), in 1996 the annual costs of road crashes in Australia by type of crash are (in AUD): $7.15 billion for serious injury crashes; $2.92 billion for fatal crashes; $2.44 billion for property damage only crashes; and $2.47 billion for minor injury crashes. Altogether, the annual cost of road crashes in Australia was about $15 billion in AUD.

1.1.2 Consequences of vehicle accidents

It is well known that, as with other impact events, a crash of rapidly moving vehicle(s) happens in a very brief time period. In the first place, an average force F which prevails over time t and which arises at the impacting interfaces is generated by the need to change momentum mv; it is inversely proportional to t, i.e. $F = mv/t$. Thus, the shorter time t is, the larger is force F (a detailed analysis of collision will be undertaken in Chapter 7 of this book). This large impact force will cause a huge acceleration (indeed, deceleration) to the occupants of the vehicle(s), especially for their heads, which are unrestrained by seat belts. Figure 1.5 displays a typical head impact

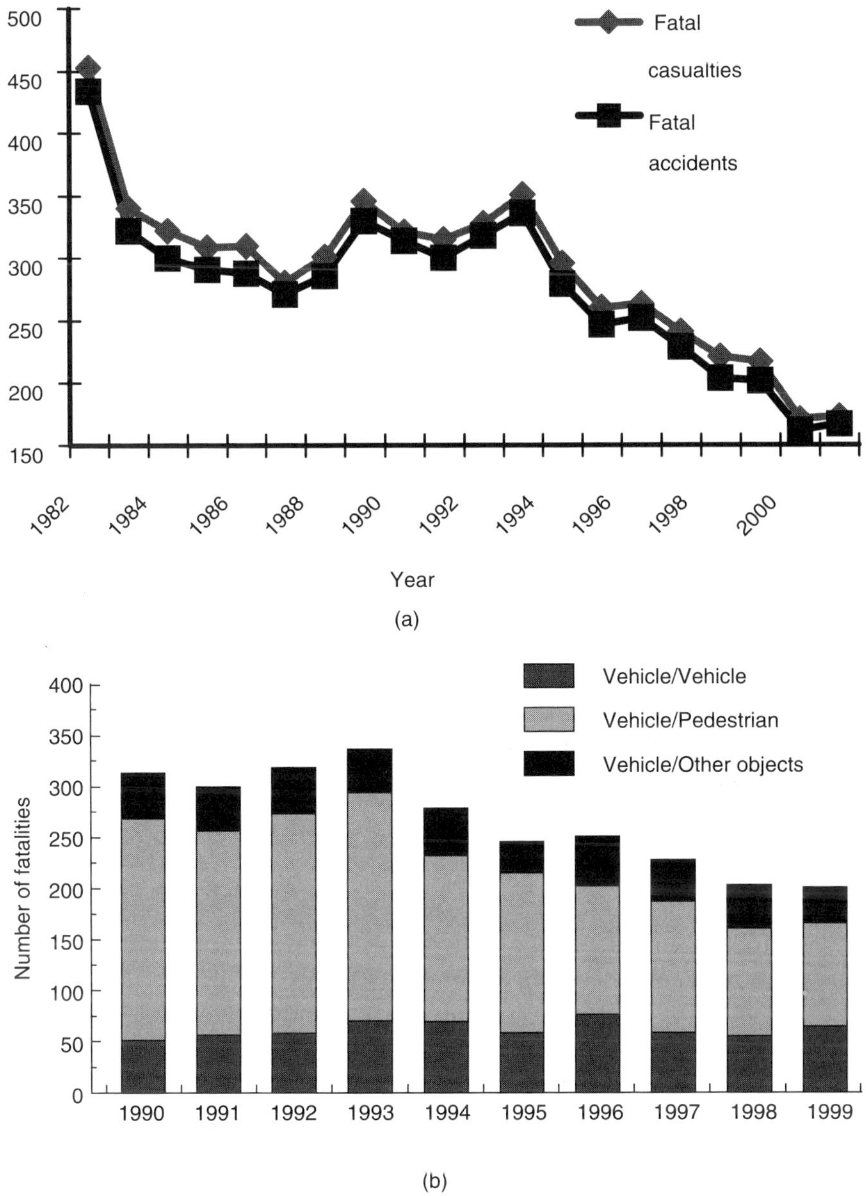

1.1 Crash fatalities in Hong Kong: (a) variation of the fatal accidents with year; and (b) types of fatalities.

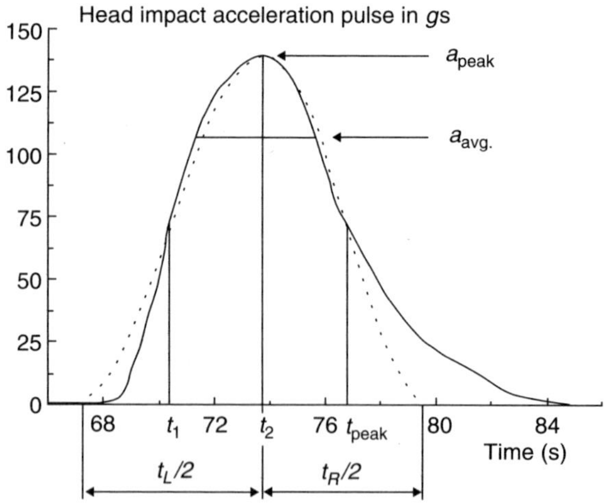

1.5 Typical acceleration pulse in a car crash test (reproduced with kind permission of World Scientific Publishing).

acceleration pulse, indicating a high speed, short duration (usually 3–25 ms) impact with rapid loading and unloading.

This large force or acceleration may lead to serious damage to people and structures. Broadly, the results of a crash of vehicle(s) may be listed as (refer to Johnson, 1990):

- damage or injury to people (or occasionally to other living organisms such as animals), which pertains to the physical and/or psychological injury and trauma caused to the vehicle's occupants and/or people external to the vehicle;
- structural damage, which pertains mostly to the inadvertent plastic deformation and fracture of the vehicle's structure and also includes destruction by fire subsequent to the crash;
- damage to cargo, such as the damage caused by shifting of cargo, oil spills due to tank grounding; and
- damage to the environment, such as the damage to roadside objects (trees, poles, guardrails, etc.)

Regarding the damage and injury to the vehicle's occupants, it is worth noting that in a vehicle crash accident with rapid deceleration, occupants' heads may impact with parts of the vehicle's interior, such as pillars, side rails, roof or windshield. This is called the second impact, and may also be very dangerous.

1.1.3 Human body's tolerance to impact

The occupant of a vehicle can be injured in a crash as a result of the occurrence of one or more of the following four events (refer to Carney III, 1993):

- unacceptably high deceleration
- crushing of the occupant compartment
- impact with part of the vehicle interior
- ejection.

Research on the biomechanics of injury and the associated occupant risk criteria has been conducted in order to determine the severity of crash events. Many of the advances in this general area have been made in the research laboratories of automobile manufacturers and in research supported by government departments of transportation.

Head/brain injury criteria

Head injuries have been recognised for many years as being the most debilitating type of trauma experienced in accidents. Annual estimates of severe head injuries indicate that, of the nearly 5800 cases occurring in accidents involving passenger cars, vans and light-duty trucks in the USA, 67 % are serious injuries. Injuries sustained by the head and brain are difficult to treat, and frequently result in long-term dysfunction. They often involve great cost to society, either because of losses due to an early death or the costs of long-term treatment and loss of productivity. In the USA, it is estimated that approximately 135 000 persons are hospitalised each year for brain injuries as a result of motor vehicle accidents. The in-hospital cost of these injuries is of the order of 370 million US dollars (refer to Carney III, 1993).

Biomechanics studies have revealed that the tolerance of human head (skull and brain) systems can be assessed by a curve first proposed by Lissner *et al.* and modified by Patrick *et al.* (refer to Johnson and Mamalis, 1978). This curve is referred to as the ***Wayne State Tolerance Curve*** (Fig. 1.6) and is widely used in automobile safety research. It claims to define the level at which acceleration, or retardation, of the head causes concussion and skull fracture. It is based on an average acceleration of the skull at the occipital bone for impacts of the forehead against a plane, unyielding surface.

The Wayne State Tolerance Curve provides a basis for several indices of injury severity. The most popular index was developed by Gadd; this gives

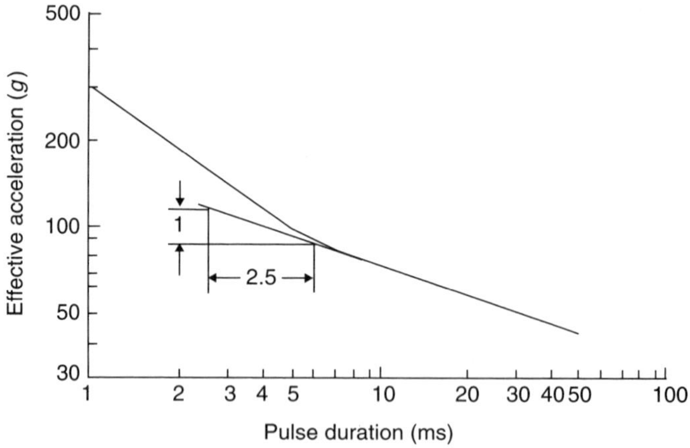

1.6 Wayne State Tolerance Curve.

a single number to represent tolerance for various regions of the human body. For the head, the ***Gadd Severity Index*** (GSI) is defined as (Perrone, 1972; Johnson *et al.*, 1982)

$$GSI \equiv \int_0^T a^{2.5} \mathrm{d}t < 1000 \qquad [1.1]$$

where a denotes the acceleration (or deceleration) in terms of 'g' (gravitational acceleration), t is the time in microseconds and T is the total pulse duration in milliseconds over which the acceleration (or deceleration) is applied. The value of 1000 is taken as the threshold for serious internal head injury for frontal impact. This purports to identify the level that could be tolerated without permanent brain damage or skull fracture being incurred for a normal healthy adult.

The pulse duration T should be in the range $0.25 < T < 50$ ms, because the index 2.5 in Eq. [1.1] represents a straight line approximation to the Wayne State Tolerance Curve in this range of T. In 1972, Gadd proposed that a level of 1500 be adopted for distributed impact.

Subsequently, the *GSI* has been superseded by the Head Injury Criterion (*HIC*), which is considered to be the best available head injury indicator (refer to Chou *et al.*, 1988 and Zhou *et al.*, 1998):

$$HIC \equiv \max(t_2 - t_1)\left(\frac{1}{t_2 - t_1}\int_{t_1}^{t_2} a(t)\mathrm{d}t\right)^{2.5} < 1000 \qquad [1.2]$$

where t_1 and t_2 are the initial and final time during the pulse for which *HIC* attains a maximum value and $a(t)$ is the resultant acceleration. The current requirement is for *HIC* to be lower than 1000. Higher than that, one's head or brain will have great possibility of suffering irrecoverable damage.

For head impact, currently *HIC* is still the criterion used, but the range of searching for the maximum has changed to 36 ms from the whole range. There have been ongoing debates about the validity of *HIC*. Some people argue that *HIC* does not reflect rotational acceleration, which is believed to cause brain injury. *HIC* correlates better with skull fractures but not so well with brain injuries. However, *HIC* is still specified in the standards for certifying vehicles, since no other criterion is so widely accepted.

Regarding the 'second impact', in which occupants of a vehicle collide with its interior, the US motor vehicle safety standard mandates that all upper vehicle interior components be impact tested. This involves a Hybrid III headform of 4.5 kg that travels in free-flight mode at a velocity of 6.7 m/s, corresponding to the average velocity for the onset of severe injuries. The free motion headform is the dummy head detached from the 50th percentile male dummy of the Hybrid III family. The dummy is shown in Fig. 1.7 together with its finite element model (Zhou, 2001). Of primary concern are the magnitudes of the hypothetical occupant impact velocity with the interior of the vehicle and the maximum 10 ms average deceleration of the occupant following this impact. The recommended threshold

1.7 The dummy and its finite element model (Zhou, 2001) (reproduced with kind permission World Scientific Publishing).

values for occupant impact velocity and ridedown deceleration are 12 m/s and 20 g respectively (Carney III, 1993).

Chest injury criterion

For chest injury, the early measures of tolerance were based on acceleration or force and they are still valid. Subsequently it was suggested that chest compression provided a better correlation with injury. For adults, 32–40 % compression is the threshold for severe injury. In the safety standard for the US 50th percentile male, the maximum allowed chest compression is 76 mm. After the compression criterion, a Viscous Criterion was developed by General Motors. The value of VC (chest velocity V multiplied by chest compression C, all being relative parameters) should be smaller than a certain value. The VC criterion is now generally accepted and used in the industry; but it has not entered into the federal vehicle safety codes. Further discussions on chest injuries can be found in Viano and Lau (1988) and Cavanaugh (1993).

Thoracic Trauma Index

Another important measure of occupant risk deals with injuries to organs within the thoracic cage. Damage to the liver, kidneys and/or spleen can be life threatening. Extensive lateral impact tests have been performed on human cadavers and surrogate specimens to determine physiological response and develop an injury index. The result is the Thoracic Trauma Index (TTI), which can be expressed in the form (refer to Hackney *et al.*, 1984)

$$TTI \equiv 0.5(G_r + G_{ls}) < 100 \qquad [1.3]$$

where G_r is the greater of the peak of either the upper or lower rib acceleration in g, and G_{ls} is the lower spine peak acceleration in g. Life-threatening injuries are unlikely when the TTI is less than 100.

1.2 Applications of energy-absorbing structures/materials

The public, now more than ever, is educated and vocal enough to be able to demand a higher degree of personal and public protection and to be able to exact greater legal penalties for mechanical failures. All these factors make up-to-date familiarity with the design of passive safety measures a prerequisite for their application in this field.

The research into and development of energy-absorbing structures and materials, which dissipate kinetic energy during impact or intense dynamic loading, has received attention since the 1970s, especially for the automobile and military industries (Johnson and Reid, 1977 and 1986). In the following, we will summarise their major applications in five aspects.

1.2.1 Energy-absorbing structures used to improve vehicles' crashworthiness

In the design and testing of various types of vehicles, crashworthy protection has become a challenging issue. The term 'crashworthiness' refers to the quality of response of a vehicle when it is involved in or undergoes an impact. The less damaged the vehicle and/or its occupants and contents after the given event, the higher the crashworthiness of the vehicle or the better its crashworthy performance (refer to Johnson, 1990).

Figure 1.8 shows some common terminology used in motor vehicle body structures. Most vehicle body frames are thin-walled steel columns. Upper and lower rails in the frontal part of a vehicle body are the main crash energy-absorbing members. The bumpers of a vehicle may play a role for a minor frontal or rear impact when the vehicle collides with a pole or a tree at a relatively low speed, such as in a car park (e.g. refer to Johnson and Walton, 1983a and 1983b).

Referring to Fig. 1.8, the A-pillar, B-pillar and roof side rails are designed to maintain the structural integrity of the passenger compartment in the event of an impact accident. On the other hand, they are also possible areas with which the occupant's head may impact in a situation of rapid deceleration. Apparently, if the head directly impacts with the sheet metal of a pillar, it would be almost impossible to meet the limited force requirement (see below). Additional cushion on the pillar as an energy absorber has to be provided (refer to Zhou, 2001).

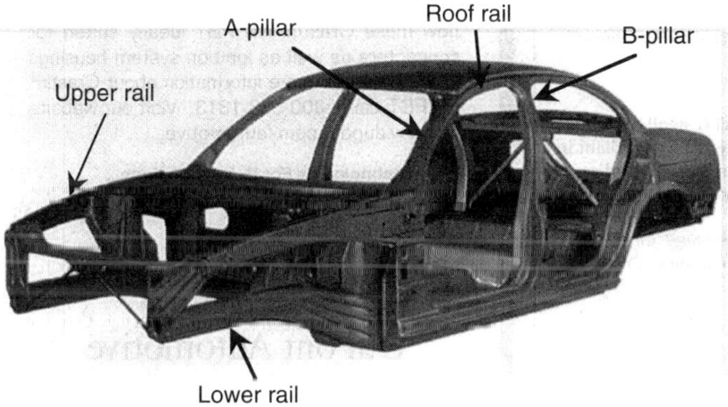

1.8 A vehicle body structure (reproduced with kind permission World Scientific Publishing).

When a dense traffic flow passes over a complicated road system, side impact protection for the driver or passengers (covering about 13 % of all serious car injuries) has become very important, although of course frontal and rear-end impacts continue to be of the greatest importance. It presents a special problem in that, for many cars, in the region of impact there is little distance to accommodate intrusion before a human body is encountered. Several manufacturers have introduced reinforced doors and strong posts to withstand 50 km/h impact, allowing only a modest invasion of the 'safety cell' around the driver and passengers. New designs have also allowed the seats, the steering wheel system and other structural members within vehicles to have a certain energy absorption capacity, so as to increase the total energy absorption of vehicles during collision.

1.2.2 Energy-absorbing structures used for highway safety

To reduce damage caused by vehicles' collisions, various types of hardware have been investigated and installed along highways in the last few decades. A motorway impact attenuation system is usually designed gradually to decelerate a vehicle to a safe stop under the conditions of a head-on crash, or to redirect the vehicle away from a hazard under side impact conditions. Their prudent use has saved numerous lives through the reduction of accident severity.

The most commonly employed guardrail system in the world consists of a galvanised steel beam shaped in the form of a W (W-beam) and supported on steel (tubular or channel section) or wood posts, as shown in Fig. 1.9. The supporting posts are buried in a foundation. When a vehicle collides

1.9 A W-beam guiderail system.

with a guardrail system installed alongside a highway, the kinetic energy carried by the vehicle is dissipated mostly by the deformation of the W-beam and the posts, as well as by the movement and fragmentation of the foundation.

Early installations of W-beam longitudinal barriers were constructed with untreated, blunt ends. This design resulted in many severe accidents, which were sometimes characterised by piercing of the occupant compartment of the errant vehicle by the sharp end of the W-beam section. Many studies have been done on the guardrails end treatments (e.g. the breakaway cable terminal, the guardrail extruder or the brakemaster, see Carney III, 1993 for details) to minimise the spearing and roll-over problems. Now all developed countries have adopted standards or codes to guide the design and installation of these guardrail systems. A recent study was made in the research group led by the second author of this book (T X Yu) at HKUST, see Hui and Yu (2000) and Hui *et al.* (2003). Some of the results are summarised in Section 12.4.

Other protection systems used along highways include concrete parapets and wire rope safety barriers. The function of the former is mainly to redirect the errant vehicle to its original travelling direction, while a part of the kinetic energy of the vehicle is converted into potential energy when the vehicle is raised by the inclined plane at the lower portion of the parapet. The latter can also redirect errant vehicles by wires parallel to the road direction, but little energy will be dissipated because the wires are deformed mainly elastically.

Some spots in highways are identified as 'black spots' for traffic accidents, e.g. at forks and sharp bends where a collision may lead to serious consequences. Engineering experience has indicated that installation of specifically designed energy-absorbers (e.g. the Connecticut impact attenuation system shown in Fig. 1.10) in such 'black spots' can significantly reduce the hazard caused by collisions. With similar motivation, an early study Johnson and Yu (1981) examined the possible use of large elastic–plastic deformation of helical springs in a vehicle arresting system.

1.2.3 Energy-absorbing structures used for protection against industrial accidents

A typical accident reported in the UK in 1973 involved a coal-pit cage carrying 29 men crashing into the coal-pit bottom as a result of an over-wind, see Johnson and Mamalis (1978). Many kinds of energy-absorbing devices would be suitable as pit-cage arresters. Similar safety considerations have to be made for lift-wells and for dead-ends of railway tracks.

In mountainous areas, rolling rocks from steep slopes are hazardous to people and vehicles passing by, especially on rainy days. For most hazardous

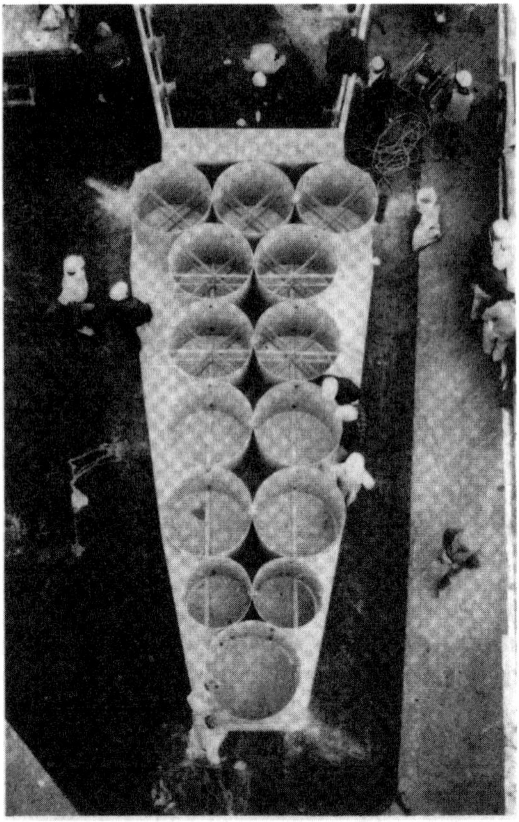

1.10 Connecticut impact attenuation system (Carney III, 1993). Reproduced with kind permission of Kluwer.

spots, protective systems can be installed, which absorb the kinetic energy of the rolling rocks, for instance by plastic deformation of the metal ring-net shown in Fig. 12.1(a).

In the design of mining, construction and agricultural machinery, FOPS (falling object protective structures) and ROPS (roll-over protective structures) are two important concepts, since these machines usually work in a hazardous environment or on sloped ground. For instance, when the roof of the driver's cabin is hit by a falling rock (Fig. 1.11(a)), or the cabin is pushed sideward during a roll-over accident (Fig. 1.11(b)), the deformed cabin has to leave a survival space for the driver; hence the cabin structure should be designed to absorb sufficient energy under these collision conditions.

Structural damage due to rolling stock collisions was illustrated by Johnson (1983). As a consequence of frontal collision, swift turning or

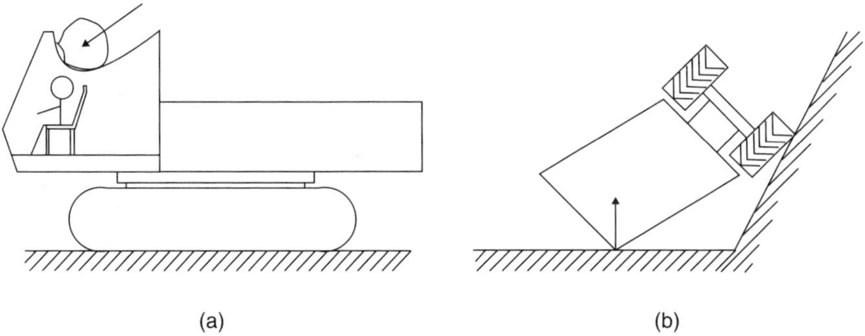

1.11 Schematic illustration of (a) FOPS (falling object protective structures); and (b) ROPS (roll-over protective structures).

damaged road surface, rolling stock collision may occur to trains, coaches, buses, trucks, tankers, etc. Similar to the consideration given to FOPS and ROPS, the ability to maintain a survival space for the occupants is vital. In the case of rolling stock containing hazardous materials, tank punctures as well as subsequent fire and explosion should also be prevented. Therefore, sufficient ductility of the material and sufficient energy-absorbing capacity of the structures must be considered as major design criteria. Concerning the safety of coaches during traffic accidents, particular attention has been paid to the collapse of the roof-pillars structure of coaches during a roll-over accident (e.g. see Lowe *et al.*, 1972).

The study of energy absorption is also essential for protection against pipe-whip, where pipes are often used to transport fluids under high pressure and at high velocities. This is a big safety issue for the nuclear, power and chemical industries. Although such pipes are designed to withstand these high pressures, there is always the danger that more intense pressures and pressure fluctuations could be produced within the system. Failure of the pipe may be result from corrosion, fatigue, creep, earthquake or even the accidental dropping of a heavy tool.

Hence, to satisfy the regulatory authorities with regard to safety, designers must demonstrate that the pipe systems can cope with guillotine breaks in the high-pressure piping run without catastrophic consequences. When a pipe breaks, the jet of high-pressure fluid which escapes from the broken section exerts a lateral reaction force (the blowdown force) on the pipe, causing it to accelerate rapidly and deform (Fig. 1.12). The pipe is, therefore, a potential hazard to the rest of the plant. Normally, to cope with this type of problem, the designer incorporates a pipe-whip restraint system (Reid *et al.*, 1980). Such a system incorporates energy-absorbers that dissipate the kinetic energy of a whipping pipe before it strikes any neighbouring equipment. Clearly, in order to design the restraints, the designer has to

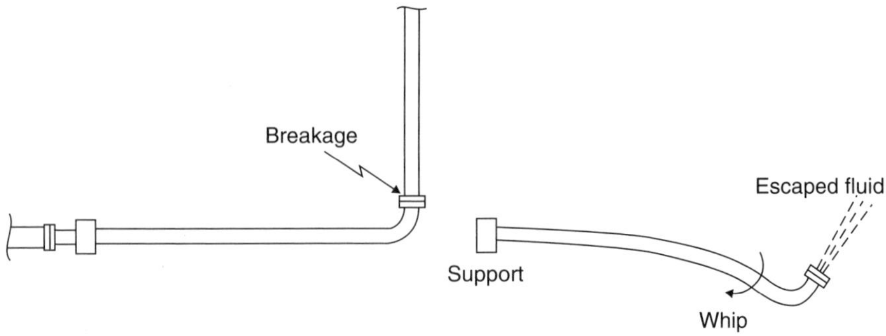

1.12 Schematic illustration of pipe-whip.

estimate the magnitude of the kinetic energy which needs to be absorbed, and this requires an understanding of the motion of the pipe. In the last two decades, a number of research studies have been carried out in relation to testing and modelling pipe-whip phenomena, e.g. see Reid and Prinja (1989), Reid and Wang (1995), Reid *et al.* (1995), Reid *et al.* (1996), Reid *et al.* (1998).

1.2.4 Energy-absorbing structures used for personal safety

Various auxiliary protective devices, like bicycle helmets, hard hats and bullet-proof jackets, are all required to posses high energy-absorption capacity. In construction sites, workers may be hit by falling objects such as small tools or pipes falling from a higher level. For instance, according to the data provided by the Labour Department of the Hong Kong SAR Government, of 67549 construction accidents reported in Hong Kong during the period of 1994–1997, 4037 were caused by falling objects and resulted in head/neck injuries of workers. As an effective protection device, the hard hat has been widely adopted as a personal safety item in most countries. The outer plastic hard shell of a hard hat can sustain and reduce the peak load created by the falling object of energy up to 50 J (i.e. an object of 5 kg in weight dropped from 1 m above the head of a worker), whilst the energy is partly dissipated by the suspension system and the shell itself.

The bicycle helmet is another example; it protects the wearer's head in a crash onto a road surface if the wearer accidentally falls from the bicycle. The international standard requires that when a headform of 5 kg covered by a bicycle helmet drops onto a hard ground from 2 m (carrying a kinetic energy of 100 J, as shown in Fig. 1.13), the maximum acceleration in the headform should be less than 300 *g*. This requires a higher energy-absorption capacity compared with hard hats used in construction sites.

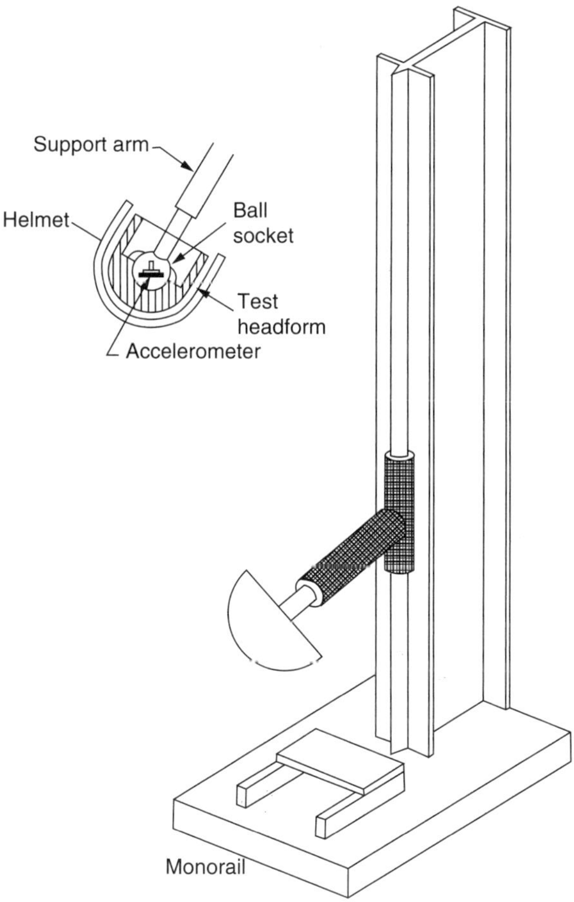

1.13 A test rig for bicycle helmets.

In games and sports – football, boxing, skiing, skating, horse-racing and car-racing for example – knowledge of crashworthiness and energy absorption is an important consideration and has great safety value for those popular activities; exactly the same science and technology applies for the human body as for vehicles.

1.2.5 Energy-absorbing structures/materials used for packaging

Packaging, as an important method of preserving and distributing goods, is an essential part of our way of life. Packaging, along with better transportation, has made it possible to centralise production facilities into areas where raw materials are concentrated and, therefore, take advantage of the economies of large-scale operations. Protective packaging then enables products to be transported to areas of major consumption. The product and the package have become so interdependent that we cannot consider one without the other.

One of the fundamental functions of packaging is, of course, to protect the goods from external damage caused in transportation and/or storage. For example, to transfer a powder product from the place of manufacture to the point of use requires some kind of container not only to carry it, but also to protect it from external damage; the package also serves as a barrier to separate the preserved item from outside contamination and spoilage.

Frequently, the external damage applied to the goods is caused when they are dropped or by the impact of other objects on them during transportation or storage. Because of this people have used various materials as a cushion or wrap to cover the goods. Since ancient times, popularly used cushioning materials include wood shavings (excelsior), straw (bagasse), crumpled or shredded waste paper, cellular wadding and rubberised hair. In its many forms, paper is widely used as a loose-fill type of cushioning. Cellulose wadding is an inexpensive form of crepe paper, available in various thicknesses with different backings, facings and embossings. This material will absorb about 16 times its weight in water and up to 12 times its weight in oil. This is important for shipment of liquid goods. The cellular structure of corrugated paperboard makes it useful for blocking and cushioning. Single-face, single-wall, double-wall and honeycomb corrugations can be die-cut in different shapes and folded to make spring pads and filler blocks. When this kind of packaging material is made of recycled papers, the cushioning becomes more environmentally friendly.

While the traditional materials are still used in various amounts, these have largely been replaced with cushioning created from polymers, which can be tailored for more precise protection. Probably the most popular polymer-based cushioning is in the form of foams. This is a plastic that has had its density substantially decreased by the creation of cellular structures dispersed through its matrix. For packaging purposes, the dispersion process generally distributes a gas throughout the molten resin which, with heat, creates void 'cells' that are allowed to grow to the desired size and are then fixed in place by cooling the material. The process is compatible with a number of thermoplastics, and it can produce rigid and flexible foams.

Rigid foams are engineering structures created by reaction-injection-molding. Packaging applications include pallets, crates and large trays. However, rigid urethane foam is a term also used for foam-in-place cushioning. These materials and techniques have been widely adopted for packaging and cushioning electronic components or fragile goods, the aim being to absorb impact energy during transportation. Also bubble sheets made of

PE (polyethene) or PET (polyethene terephthalate) can be used to handle a broad range of fragile goods. Technical details for the foams used will be further illustrated in Section 12.2, along with an example of selection of packaging material and sizes.

1.3 Design of energy-absorbing structures and selection of energy-absorbing materials

1.3.1 General features of energy-absorbing structures

Under working loads, conventional structures (e.g. those used in civil engineering and for machinery) will undergo only small elastic deformation. These structures are usually required to possess certain strength and stiffness under specified loads, so that the material's selection and the structural design are mainly based on the elastic stress or strain which the structure has to sustain. Failures are mostly attributed to fatigue, corrosion or degradation of material after a long service life.

On the other hand, the design and analysis of energy-absorbing structures are very different from conventional structural design and analysis. The energy-absorbing structures have to sustain intense impact loads, so that their deformation and failure involve large geometry changes, strain-hardening effects, strain-rate effects and various interactions between different deformation modes such as bending and stretching.

For these reasons, most energy-absorbers are made of ductile materials. Low carbon steel and aluminium alloys are the most widely used, whilst non-metallic materials such as fibre-reinforced plastics and polymer foams are also common especially when the weight is critical.

Studying the behaviour of energy-absorbing structures usually starts with quasi-static analysis and testing. Quasi-static characteristics include the predominant geometrical effects, which also occur under dynamic loading. For structural impacts occurring at relatively low speed (of the order of, say, 50 m/s), the effect of strain rate in increasing the yield stress and flow stress can usually be allowed for by using the simple Cowper–Symonds equation based on mean strain rate, see Section 2.4.3.

A number of studies have concluded that particular modes of deformation are more sensitive to dynamic effects than others, which leads to problems in assessing deformation and failure by means of scale model testing. This problem will be addressed in Section 7.2.

To summarise, crashworthiness and impact protection are well-defined subjects but lack sufficient depth of scientific research. On the other hand, engineering plasticity is a highly developed discipline, which will be extensively utilised to analyse and predict the performance of energy-absorbing structures made of ductile materials. Therefore, both the purpose and methodology in analysing energy-absorbing structures are very different from conventional structural analysis. This will be further elaborated in Chapter 2.

1.3.2 General principles

It is evident from the discussion of the engineering background that the design of energy-absorbing structures and the selection of energy-absorbing materials should suit the particular purpose and circumstances under which they are to work. Although the design and selection can vary notably from one application to another, in all cases design and selection aim to dissipate kinetic energy in a controlled manner or at a predetermined rate. Therefore, some fundamental principles are generally valid for all applications and can serve as guidelines. The main ones are illustrated below.

Irreversible energy conversion

The energy conversion by the structures/materials should be irreversible; that is, the structures/materials should be able to convert most of the input kinetic energy into inelastic energy by plastic deformation or other dissipation processes, rather than storing it elastically.

Why must it be inelastic? If the initial kinetic energy (or more generally, the input energy due to dynamic loading) is converted into elastic strain energy of the structure, then, after a maximum elastic deformation is achieved, this elastic strain energy will be completely released and cause subsequent damage to the person and structure to be protected.

For instance, imagine that a car travelling at a high speed collides with a large elastic spring. In the first phase, the spring is compressed elastically; the car is decelerated (Fig. 1.14(a)) and all the initial kinetic energy of the car is converted into elastic strain energy of the spring. After the maximum elastic displacement of the spring is achieved, the second phase takes place in which the spring will gradually recover from its deformation; the car will be accelerated (Fig. 1.14(b)) and eventually all the elastic strain energy will be converted back into kinetic energy of the car. Consequently, the occupants of the car will suffer a severe deceleration followed by a severe acceleration. This could lead to much more serious consequence for the occupants of the car, because the injury to a person in an accident increases with the time duration in which the acceleration or deceleration applies, as indicated by Eq. [1.1] or Eq. [1.2].

Many forms of irreversible energy exist in the large deformation process of structures and materials, such as plastic dissipation, viscous deformation energy, energy dissipated by friction or fracture. Some of them may relate

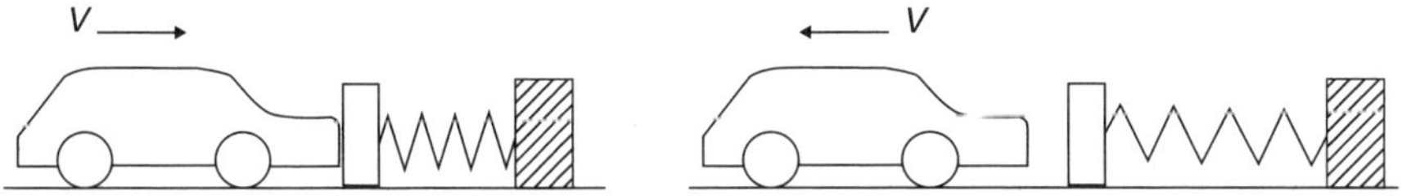

1.14 A car collides with an elastic spring: it is (a) decelerated by the compressed. spring; (b) re-accelerated by the recovery of the spring.

to macroscopic deformation or macroscopic fractures; while some of them come from meso- or micro-scale mechanisms. For instance, the delamination of polymer-matrix laminate composites usually involves not only macroscopic inter-laminar fractures, but also viscous deformation of the matrix, micro-cracks at fibre/matrix interfaces and internal friction in cracked materials. Among these energy-absorption mechanisms, the attention of this book will be mainly focused on energy absorption due to plastic deformation of structures and materials, because this is the most effective mechanism for absorbing energy in ductile materials (e.g. metals and polymers) and has the widest practical applications.

Restricted and constant reactive force

The peak reaction force of an energy absorber should be kept below a threshold; and ideally the reaction force should remain constant during the large deformation process of the energy-absorbing structure.

While providing sufficient total energy-absorption capacity in the large deformation process, the peak force (and thus the peak deceleration) of the energy-absorbing structures/materials under impact must be kept below the threshold that would cause damage or injury; the reactive force should remain constant or almost constant to avoid an excessively high rate of retardation.

When a rigidly seated occupant in a vehicle is restrained by a seat belt, the acceleration to which he is subjected during a collision is approximately equal to the acceleration of the vehicle itself. Based on the variation method, it is not difficult to prove that when the initial velocity of the vehicle is given, the *GSI* expressed in Eq. [1.1] becomes a minimum if the acceleration $a(t)$ in the colliding process remains constant. In other words, the resistant force from the colliding structure should remain constant during a collision in order to minimise the injury and damage caused by it. In this sense, the energy-absorbers act as a special kind of load-limiter, ideally possessing an approximate rectangular force-displacement characteristic.

Long stroke

For an energy-absorbing structure, as pointed out above, the reactive force has to be restricted and almost constant; but the work done by a force is equal to its magnitude times the displacement experienced along the acting line of the force. Therefore, if the structure is to absorb a large amount of input energy, the displacement, i.e. the stroke, should be sufficiently long.

Besides considering values of force F, crashworthy situations also need to be analysed in terms of kinetic energy dissipation. When the initial kinetic energy to be dissipated is specified, the longer is time t, the smaller is force F; and this leads to the notion of 'buying distance with time' – a principle to be followed for reducing impact damage or injury. To decelerate uniformly from speed v to zero requires a distance of $vt/2$, and it is this distance that F acts over to dissipate the damaging kinetic energy. The longer time that force F prevails, the gentler is the arresting force required and the smaller is any injury sustained.

Based on this consideration, the ratio of the stroke (i.e. the maximum deformable distance of the structure along the loading direction) to a characteristic dimension of the structure serves as an important measure of the efficiency of energy-absorbing structures. For instance, when a ring or a tube is loaded in the transverse direction, the stroke is limited by the diameter, so that when a long stroke is required, the designer has either to increase the diameter of the ring or tube, or to pack a few layers of rings or tubes together, see Chapters 4 and 5. When a tube is axially loaded, on the other hand, the admissible stroke could be comparable to the total length of the tube, see Chapter 6. It is observed, therefore, that in the former cases (transversely loaded rings and tubes), the main effort in the design of the energy-absorbing system would be to satisfy the stroke requirement. Conversely in the latter case (axially loaded tubes), the main effort in the design is devoted to satisfying the restriction in the reactive force.

In the selection of energy-absorbing materials, such as in the case of selecting packaging materials for valuable or fragile goods, the ratio of the stroke (i.e. the deformable distance) to the original dimension of the material becomes of vital importance. It is evident that this ratio refers to the maximum compression ratio Δ_{max}/H, where H is the original thickness of the packaging layer and Δ_{max} is the maximum compression distance. It is known that this ratio is restricted by the compressibility (or the durability in compression) of a material; hence ordinary solid metals or solid polymers cannot achieve a high ratio of Δ_{max}/H. However, a much higher ratio of Δ_{max}/H can be achieved by adopting cellular materials, such as honeycombs and foams. This is because the large amount of space within the cells provides for a greater compressibility of the material. In this case the ratio

Δ_{max}/H can be of the same order as the relative density of the cellular material adopted, see Chapter 10.

Stable and repeatable deformation mode

To cope with very uncertain working loads, the deformation mode and energy-absorption capacity of the designed structure should be stable and repeatable, so as to ensure the reliability of the structure in its service.

It should be anticipated that external dynamic loads which would apply to energy-absorbing structures and materials have strong uncertainties in their magnitude, pulse shape, direction and distribution. Hence those structures and materials should possess stable and repeatable deformation modes that are insensitive to the above uncertainties of loading but ensure the required energy-absorption capacity.

For instance, a W-beam guardrail system (Fig. 1.9) or an impact attenuation system (Fig. 1.10) used in an expressway could be subject to impact by a passenger car, or a lorry, approaching at various angles, at any portion of the system. In design and testing, the fundamental requirements for the maximum force and the total energy dissipation have to be satisfied for all the combinations of these uncertainties.

Light weight and high specific energy-absorption capacity

The energy-absorbing component should be light itself, possessing high specific energy-absorption capacity (i.e. energy-absorption capacity per unit weight), which is of vital importance for vehicle-carried energy absorbers (especially for aircraft) and personal safety devices.

When the automobile manufacturers modify their designs to improve the crashworthiness of vehicles, they have to consider carefully the possible increase in vehicle weight, because any increase in weight implies more consumption of fuel and more pollution of the environment.

For various auxiliary protective devices, light weight is also an extremely important design parameter. Much effort has been expended to reduce their weight. For instance, today a typical bicycle helmet available on the market usually weighs 250–300 g, but wearers would welcome a new type of helmet, weighing below 200 g.

Among many candidate materials for energy-absorption purposes, of particular interest is cellular material consisting of a base material (e.g. polymer or aluminium) and air in the formed cells. Thanks to its porosity, it is light in weight compared to a solid bulk made of its base material, while its stiffness, strength or other mechanical properties may have advantages in terms of per unit weight of material. In other words, compared to the base material, the reduction in a mechanical property is often less than the reduction in weight (refer to Gibson and Ashby, 1997; Zhou, 2001). In addition, when it comes to energy absorption, the cellular porosity feature is desirable since it can give a long, nearly constant plateau force level under compression.

Low cost and easy installation

The manufacture, installation and maintenance of such energy-absorption devices should be easy and cost-effective.

In today's competitive world, the design of energy-absorption devices is always constrained by the available budget. Therefore, all protective structures must operate within these economic boundaries. This is particularly true for energy-absorbing devices because they are usually one-shot items, i.e. having once been deformed they are discarded and replaced.

2
Methodology of analysing energy-absorption capacities

To provide basic models and tools for analysing the energy-absorption capacities of materials and structures, this chapter illustrates the idealised models of materials' behaviour, as well as fundamental concepts, principles and methods; the effects of large deformation and dynamic loading are also discussed.

2.1 Idealisation of materials' behaviour

2.1.1 Mechanical properties of materials in tension

To characterise a material's deformation behaviour in response to applied loads, the most straightforward and conventional way is to conduct a ***simple tensile test*** on cylindrical or flat coupons made of the material. This simple tensile test can be applied to various engineering materials. For instance, Figs 2.1(a), (b) and (c) depict sketches of the stress–strain curves for mild steel, aluminium alloy and knitted textile composite, respectively.

As seen from Fig. 2.1, most engineering materials exhibit an elastic deformation stage when the applied tensile force is relatively small. The behaviour of a material in its linear elastic stage can be characterised by two material constants, that is, ***Young's modulus E***, which denotes the constant slope of the stress–strain curve in this stage and ***Poisson's ratio*** ν, which denotes the ratio of the negative strain in the transverse direction to the longitudinal tensile strain.

For metals and polymers, when the applied load reaches a certain level, the material will ***yield***. This not only implies a deviation from the linear stress–strain path, but also marks the onset of non-recoverable plastic deformation. Mild steel (Fig. 2.1(a)) usually undergoes a continuous deformation as the stress remains yield stress Y (also denoted by σ_y in some books). For most of other materials, however, the stress required will increase as the deformation continues. This phenomenon is termed ***strain-hardening***. For some materials (e.g. see Fig. 2.1(b)), strain-hardening can be approximated by a linear or a power law between stress and strain.

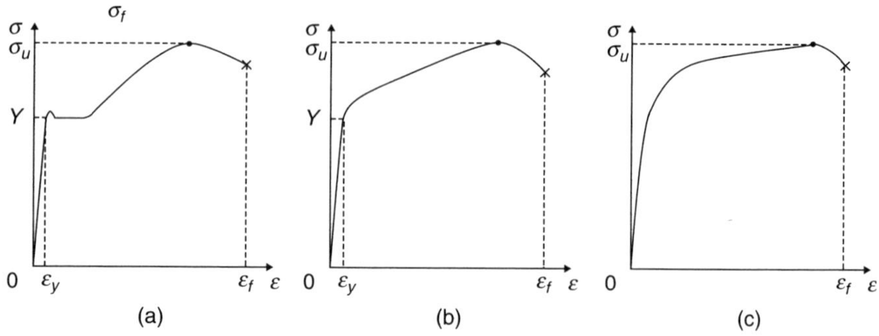

2.1 Stress–strain curves of materials under tension: (a) mild steel; (b) aluminium alloy; and (c) knitted textile composite.

The plastic deformation stage will end when the material coupon is eventually broken under tension. Before fracture occurs the stress will reach a maximum, denoted by the ***ultimate stress*** σ_u. When fracture occurs, the corresponding strain is termed the ***fracture strain*** denoted as ε_f. These two quantities (refer to Fig. 2.1(a)) represent the strength and ductility, respectively, of the material under tension.

When subjected to simple compression or pure shear, most engineering materials display behaviour similar to their behaviour under tension, although the relevant material constants may be different.

2.1.2 Idealised material models

In order to establish reasonably simple theoretical models to analyse the energy-absorption capacities of materials and structures, first of all the materials' mechanical behaviour should be idealised so that their stress–strain relationship can be expressed by simple analytical functions.

When the deformation (strain) is small, a linear elastic material model can be adopted, while Young's modulus E and Poisson ratio ν serve as two specified material parameters. After the material's yielding, if the strain-hardening is insignificant, then an ***elastic, perfectly plastic*** material model can be adopted, as shown in Fig. 2.2(a). Here the term 'perfect plastic' implies that strain-hardening is negligible; in other words, the material will continue to deform plastically from the initial yielding till fracture whilst the stress remains Y.

If the material's strain-hardening after the initial yielding is significant, then either an ***elastic, linear hardening*** model (Fig. 2.2(b)), or an ***elastic, power hardening*** model (Fig. 2.2(c)) can be considered, depending on the material's actual behaviour measured from its simple tensile test.

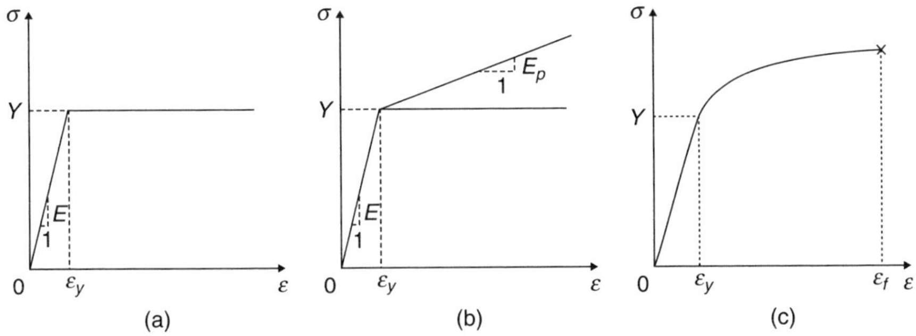

2.2 Idealised stress–strain curves of materials under tension: (a) elastic, perfectly plastic; (b) elastic, linear hardening; and (c) elastic, power hardening.

Corresponding to the models shown in Figs 2.2(a), (b) and (c), the idealised relationship between stress σ and strain ε can be analytically expressed as

$$\sigma = \begin{cases} E\varepsilon & \text{for} \quad \varepsilon \le \varepsilon_y = Y/E \\ Y & \text{for} \quad \varepsilon_y \le \varepsilon < \varepsilon_f \end{cases} \qquad [2.1]$$

$$\sigma = \begin{cases} E\varepsilon & \text{for} \quad \varepsilon \le \varepsilon_y = Y/E \\ Y + E_p(\varepsilon - \varepsilon_y) & \text{for} \quad \varepsilon_y \le \varepsilon < \varepsilon_f \end{cases} \qquad [2.2]$$

and

$$\sigma = \begin{cases} E\varepsilon & \text{for} \quad \varepsilon \le \varepsilon_y = Y/E \\ Y + K(\varepsilon - \varepsilon_y)^q & \text{for} \quad \varepsilon_y \le \varepsilon < \varepsilon_f \end{cases} \qquad [2.3]$$

respectively, where ε_y is the ***yield strain***, E_p denotes the ***hardening modulus***, K and q (hardening exponent) are material constants determined experimentally. It is obvious that if $q = 1$ and $K = E_p$, the power hardening model is identical to the linear hardening model.

When used for an energy-absorption purpose, the materials, structural components and devices will usually undergo a large plastic deformation, as illustrated in Chapter 1. In these cases, the plastic strain will be much larger than the elastic strain, so the latter can be neglected in the analysis. Effectively, Young's modulus can be taken as infinite, so that the material exhibits rigid behaviour before initial yielding. Thus, the idealised material model is called a ***rigid-plastic*** model. Figures 2.3(a), (b) and (c) represent a ***rigid, perfectly plastic*** model, a ***rigid, linear hardening*** model and a ***rigid, power hardening*** model, respectively.

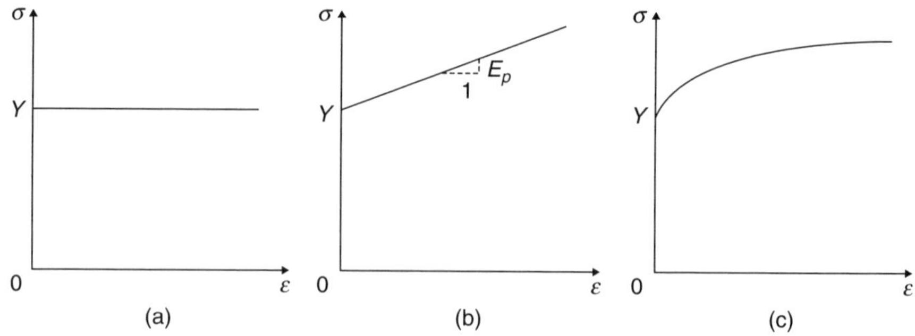

2.3 Idealised stress–strain curves of materials under tension: (a) rigid, perfectly plastic; (b) rigid, linear hardening; and (c) rigid, power hardening.

The idealised stress–strain relationship for rigid, perfectly plastic model (Fig. 2.3(a)) and rigid-hardening models (Fig. 2.3(b) and (c)) can be expressed as

$$\begin{cases} \sigma \le Y & \text{for} \quad \varepsilon = 0 \\ \sigma = Y & \text{for} \quad 0 < \varepsilon < \varepsilon_f \end{cases} \qquad [2.4]$$

$$\begin{cases} \sigma \le Y & \text{for} \quad \varepsilon = 0 \\ \sigma = Y + E_p\varepsilon & \text{for} \quad 0 < \varepsilon < \varepsilon_f \end{cases} \qquad [2.5]$$

and

$$\begin{cases} \sigma \le Y & \text{for} \quad \varepsilon = 0 \\ \sigma = Y + K\varepsilon^q & \text{for} \quad 0 < \varepsilon < \varepsilon_f \end{cases} \qquad [2.6]$$

respectively.

2.1.3 Moment–curvature relationship for plastic beams

For beams (or other 1-D structural components, such as rings and arches) made of elastic-plastic materials, the relationship between applied bending moment M and the curvature κ of its central axis will be linear if the applied moment is small ($M < M_e$, with M_e being the maximum elastic bending moment), or non-linear if $M > M_e$. The actual M–κ relation can be derived from an integration of the appropriate σ–ε relation over the beam's thickness.

Typically, for a rectangular cross-sectional beam (or another 1-D structural component, such as a ring or an arch) made of elastic, perfectly plastic material, the stress profile across the beam's thickness is as shown in Fig. 2.4(a), which contains an elastic 'core' sandwiched by two plastic deforma-

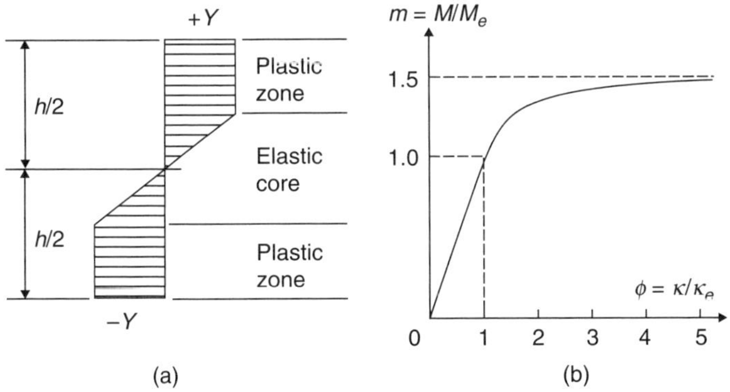

2.4 Bending of an elastic, perfectly plastic beam: (a) stress profile across the thickness; (b) non-dimensional moment–curvature relation.

tion zones. Consequently, the moment–curvature relation is expressed as (e.g. refer to Section 1.4, Yu and Zhang, 1996)

$$\frac{M}{M_e} = \begin{cases} \dfrac{\kappa}{\kappa_e} & \text{for} \quad 0 \le M \le M_e \\ \dfrac{3}{2} - \dfrac{1}{2}\left(\dfrac{\kappa_e}{\kappa}\right)^2 & \text{for} \quad M_e < M < M_p \end{cases} \qquad [2.7]$$

where $M_e = Ybh^2/6$ is the ***maximum elastic bending moment***, $\kappa_e = M_e/EI = M_e/(Ebh^3/12) = 2Y/Eh$ is the ***maximum elastic curvature*** and $M_p = Ybh^2/4$ denotes the ***fully plastic bending moment*** of the beam of rectangular cross-section, with b and h being the width and thickness of the cross-section, respectively.

Equation [2.7] can be recast into a non-dimensional form as

$$\phi = \begin{cases} m & \text{for} \quad 0 \le m \le 1 \\ \dfrac{1}{\sqrt{3-2m}} & \text{for} \quad 1 \le m < \dfrac{3}{2} \end{cases} \qquad [2.8]$$

where $m = M/M_e$ and $\phi = \kappa/\kappa_e$ denote non-dimensional bending moment and non-dimensional curvature, respectively. This relation is sketched in Fig. 2.4(b).

An important special case is when a rigid, perfectly plastic relation between stress and strain (see Fig. 2.3(a) and Eq. [2.4]) is adopted, the stress profile across the beam's thickness will contain plastic zones only, if the beam's section has a non-zero curvature, as shown in Fig. 2.5(a). Consequently, the relation between bending moment and curvature is expressed as a step function, as shown in Fig. 2.5(b). That is

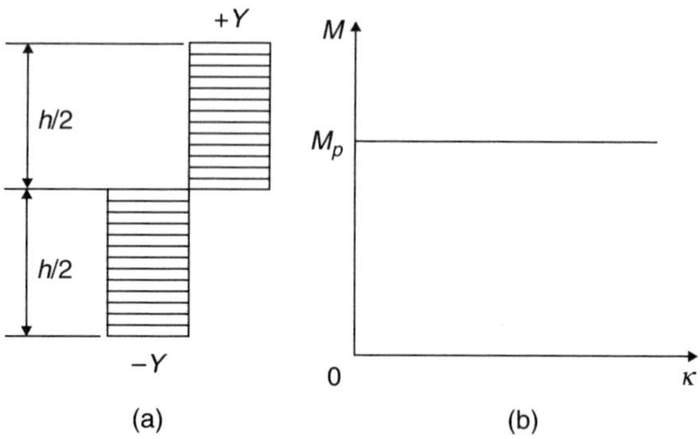

2.5 Bending of a rigid, perfectly plastic beam: (a) stress profile across the thickness; (b) moment–curvature relation.

$$\begin{cases} M \le M_p & \text{for} \quad \kappa = 0 \\ M = M_p & \text{for} \quad \kappa > 0 \end{cases} \qquad [2.9]$$

2.1.4 Plastic hinge and hinge-line

When a beam (or another 1-D structural component, such as a ring or an arch) is idealised as being rigid, perfectly plastic, its plastic deformation will be concentrated at one or a few cross-sections, where the magnitude of the applied bending moment reaches the fully plastic bending moment of the cross-section, M_p (e.g. $M_p = Ybh^2/4$ in the case of rectangular cross-sectional beams). These cross-sections are termed ***plastic hinges***. Any bending moment whose magnitude is larger than M_p, is not ***statically admissible*** for an equilibrium configuration (see Section 2.2 for further illustrations). On the other hand, any bending moment whose magnitude is smaller than M_p, will produce no plastic deformation.

As a result, the deformed configuration of a rigid, perfectly plastic beam (or a ring or an arch) will contain one or more plastic hinges only. As indicated by Eq. [2.9] and Fig. 2.5(b), the curvature at a plastic hinge can take an arbitrary value or infinity. Therefore, a finite relative rotation θ occurs at a plastic hinge as the result of the application of M_p.

Away from those plastic hinges all the segments in the beam (or the ring or arch) will remain rigid and their curvatures will remain unchanged (i.e. $\kappa = 0$). However, those rigid segments are allowed to have translation and/or rotation, provided these motions are ***kinematically admissible*** in a proposed deformation mechanism (see Section 2.2 for further illustrations).

From the viewpoint of energy dissipation, when a rigid, perfectly plastic beam deforms, plastic dissipation takes place only at the discrete plastic hinge(s). The total energy dissipation in the beam, therefore, can be calculated by

$$D = \sum_{i=1}^{n} M_p |\theta_i| \qquad [2.10]$$

where n is the total number of the hinges in the beam, and θ_i denotes the relative rotational angle at the i-th plastic hinge.

In plates or shells, the rigid, perfectly plastic idealisation of materials will result in their plastic deformation being concentrated at discrete ***plastic hinge-lines***. Along those hinge-lines the magnitude of the bending moment per unit length must be equal to M_o, which denotes the ***fully plastic bending moment per unit length***, $M_o = Yh^2/4$. Note that the unit of M_o is N, while the unit of M_p is Nm.

In a similar way to the behaviour of plastic hinges in a beam, relative rotations are allowed along plastic hinge-lines in a plate or a shell. Consequently, the total energy dissipation in the plate or shell can be calculated by

$$D = \sum_{i=1}^{n} M_o |\theta_i| L_i \qquad [2.11]$$

where n is the total number of the hinge-lines in the plate or shell, θ_i denotes the relative rotational angle at the i-th plastic hinge-line, and L_i is the length of the i-th plastic hinge-line.

It is observed that by adopting the rigid, perfectly plastic idealisation of material, the plastic deformation in a structural component of dimension N ($N = 1$ for beams, rings, arches, etc and $N = 2$ for plates, shells, etc.) will be concentrated in discrete regions of $(N - 1)$ dimension. This will greatly simplify the plastic analysis of structures and will be very useful for the theoretical modelling of energy-absorption components, as will be seen in the following sections.

2.1.5 Mechanical models for materials' idealisation

The idealised materials' behaviour can be demonstrated by mechanical models such as those shown in Fig. 2.6. The mechanical model shown in Fig. 2.6(a) contains a linear elastic spring and a friction pair, so that when a force F applies along the axis of the spring, the relationship between force F and axial displacement Δ will be

$$F = \begin{cases} k\Delta & \text{for} \quad \Delta \le \Delta_y = F_y/k \\ F_y & \text{for} \quad \Delta_y \le \Delta < \Delta_f \end{cases} \qquad [2.12]$$

where k is the spring constant, and F_y is the critical friction when relative motion begins. The clear similarity between Eq. [2.12] and Eq. [2.1] indi-

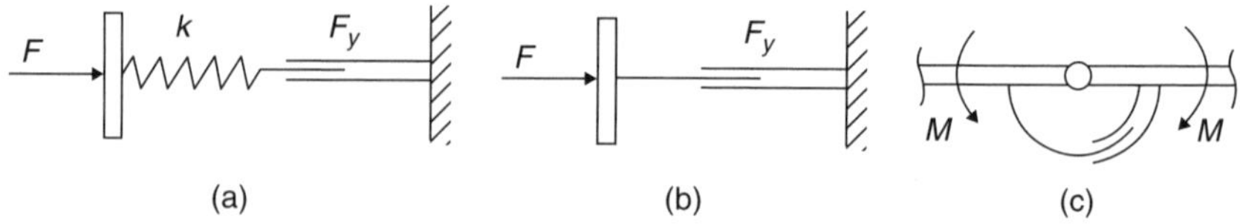

2.6 Mechanical models for materials' idealisation: (a) an elastic, perfectly plastic model; (b) a rigid, perfectly plastic model; and (c) a plastic hinge.

cates that the mechanical model shown in Fig. 2.6(a) appropriately represents the behaviour of an elastic, perfectly plastic material. If the elastic spring is taken away from the model shown in Fig. 2.6(a), which means that the elastic deformation of the material is negligible, then the mechanical model (Fig. 2.6(b)) demonstrates a rigid, perfectly plastic behaviour.

When the bending behaviour of beams is considered, the rigid, perfectly plastic relation between bending moment M and relatively rotational angle θ can be similarly demonstrated by a mechanical model as shown in Fig. 2.6(c). In this model, the two rigid segments are connected by a mechanical hinge that can rotate freely. The angular friction pair in the model requires a critical moment M_p (applied in either the clockwise or the counterclockwise direction) to motivate its non-zero relative rotation.

2.1.6 Validity of rigid-plastic idealisation

Physically, the rigid-plastic idealisation of materials' behaviour is based on the fact that the elastic strain (typically limited by $\varepsilon_y \approx 0.002$ for structural metals) is much smaller than the plastic strain occurring in structural components used for energy-absorption purposes. However, the elastic deformation is always a precursor of subsequent plastic deformation. No matter whether the structure is subjected to a quasi-static load or a dynamic load, the first phase of its deformation is always elastic and after the plastic deformation is completed, the structure must undergo an elastic springback to achieve its final deformed configuration. Owing to these reasons, the validity of rigid-plastic idealisation should be thoroughly examined.

A simple way to conduct this examination is to employ the one-dimensional mechanical models shown in Fig. 2.6, and to compare the response of the elastic-plastic model (Fig. 2.6(a)) with that of the rigid-plastic model (Fig. 2.6(b)).

First, assume that a force F is quasi-statically applied to the left end of the elastic-plastic model shown in Fig. 2.6(a). Consider a deformation process of the model in which the displacement at the left end gradually increases from zero to a total displacement $\Delta_{tl}^{ep} > \Delta_y = F_y/k$, where the subscript tl pertains to the total displacement produced, and the superscript ep pertains to the elastic-plastic model. Correspondingly, force F first increases from zero to F_y (as $0 < \Delta < \Delta_y$) and subsequently remains F_y (as $\Delta > \Delta_y$). In this elastic-plastic deformation process, the total input energy (i.e. the work done) is

$$E_{in} = \frac{1}{2}F_y\Delta_y + F_y\left(\Delta_{tl}^{ep} - \Delta_y\right) = E_{\max}^{e} + F_y\Delta_p^{ep} \qquad [2.13]$$

where $E^e_{\max} = F_y\Delta_y/2$ is the maximum elastic energy which can be stored in the model, and $\Delta_p^{ep} = \Delta_{tl}^{ep} - \Delta_y$ represents the ***plastic (permanent) component*** of the displacement. The total input energy E_{in} given in Eq. [2.13] can be represented by the shadowed area shown in Fig. 2.7(a).

Next, consider that a force F is quasi-statically applied on the left end of the rigid-plastic model shown in Fig. 2.6(b). In this case, the deformation becomes possible only when force F takes the value of F_y. Assume that the final displacement at the left end of the model is Δ_{tl}^{rp}, where the superscript rp pertains to the rigid-plastic model, and then the total input energy (i.e. the work done) in the process is

$$E_{in} = F_y\Delta_{tl}^{rp} = F_y\Delta_p^{rp} \qquad [2.14]$$

Here the plastic component of the displacement, Δ_p^{rp}, is equal to the total displacement Δ_{tl}^{rp} because the elastic component is neglected. The total input energy E_{in} given in Eq. [2.14] can be represented by the shadowed area shown in Fig. 2.7(b).

Suppose the same amount of energy is input into these two models (i.e. the values of E_{in} in Eqs. [2.13] and [2.14] are identical) and define an ***energy ratio***

$$R_{er} \equiv \frac{E_{in}}{E_{\max}^{e}} \qquad [2.15]$$

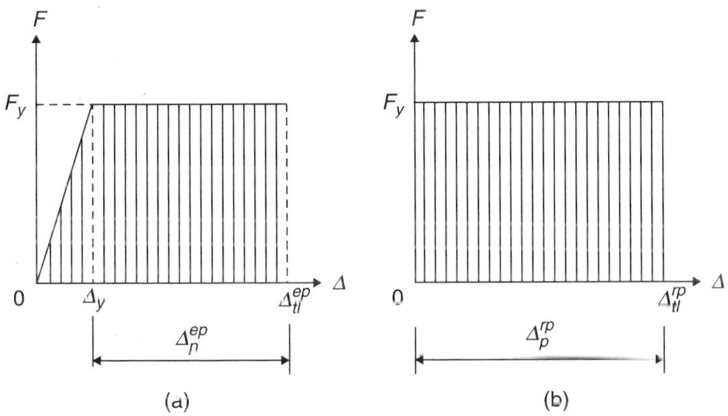

2.7 Force–displacement relationship of perfectly plastic models: (a) elastic-plastic model; (b) rigid-plastic model.

then Eqs [2.13] and [2.14] result in

$$F_y \Delta_p^{rp} = F_y \Delta_p^{ep} + E_{\max}^{e} \quad [2.16]$$

Divided by E_{in}, Eq. [2.16] leads to

$$1 = \frac{\Delta_p^{ep}}{\Delta_p^{rp}} + \frac{E_{\max}^{e}}{E_{in}} = \frac{\Delta_p^{ep}}{\Delta_p^{rp}} + \frac{1}{R_{er}} \quad [2.17]$$

or

$$\frac{\Delta_f^{rp}}{\Delta_f^{ep}} = \frac{R_{er}}{R_{er} - 1} \quad [2.18]$$

Therefore, the relative 'error' of employing a rigid-plastic model in predicting the plastic displacement is found to be

$$\text{'}Error\text{'} \equiv \frac{\Delta_f^{rp} - \Delta_f^{ep}}{\Delta_f^{ep}} = \frac{1}{R_{er} - 1} > 0 \quad [2.19]$$

which indicates that the plastic deformation predicted by the rigid-plastic model is always slightly larger than that obtained from the corresponding elastic-plastic model, whilst the 'error' caused by the rigid-plastic idealisation reduces with increasing energy ratio $R_{er} \equiv E_{in}/E^{e}_{\max}$. For instance, if $R_{er} = 10$ in a particular structural problem, then the 'error' of the rigid-plastic model in predicting the plastic deformation is about 11 %, which is acceptable for most engineering applications.

2.2 Limit analysis and bound theorems

2.2.1 Limit state of perfectly plastic structures

Based on the classical theory of plasticity (e.g. refer to Hill, 1950; Prager and Hodge, 1951; Martin, 1975), if strain-hardening of materials and the geometric change of structures under applied loads are negligible, the following properties can be proven for ***perfectly plastic structures***.

(1) When the distribution of external loads is specified, a ***limit state*** exists for the structure made of either an elastic, perfectly plastic material or a rigid, perfectly plastic material; under this limit state the structure will continue to deform plastically whilst the external loads remain unchanged. The corresponding loads are termed ***limit loads***, and the associated plastic deformation mechanism is called the ***collapse mechanism*** of the structure.

(2) The limit loads are proportional to the material's yield stress Y and independent of its Young's modulus E; so that an elastic, perfectly

plastic structure and a rigid, perfectly plastic structure have identical limit loads, provided they have the same configuration and same yield stress Y. Therefore, the limit load and collapse mechanism of a structure can be conveniently obtained from a rigid-plastic analysis.

(3) The limit loads of a structure are independent of the loading history; so that the limit loads form a ***limit surface*** in the ***load space***. The load space is an n-dimensional space with coordinate axes being the n generalised forces (loads) applied on a structure or a cross-section of the structure, while the limit surface is a $(n - 1)$-dimensional curved surface within the load space, which must be convex and contain the origin of the load space. The limit surface can be expressed as

$$\Psi(Q_1, Q_2, \ldots, Q_n) = 0 \qquad [2.20]$$

where Q_i $(i = 1, 2, \ldots, n)$ denotes the ***generalised force*** applied, and Ψ is a function.

(4) When the applied loads reach a point on the limit surface, the structure or the cross-section concerned will reach its limit state, and the associated plastic deformation at this state should obey the normality flow role; that is, the generalised plastic strain vector must be perpendicular to the limit surface at that point.

2.2.2 Example: a beam under bending and tension

As an example of the geometric significance of the limit surface and associated flow role, consider a rigid, perfectly plastic beam of rectangular cross-section, which is simultaneously subjected to bending moment M and axial force N, as shown in Fig. 2.8(a). Based on a generic stress profile sketched in Fig. 2.8(b), where c denotes the deviation of the neutral axis of the lon-

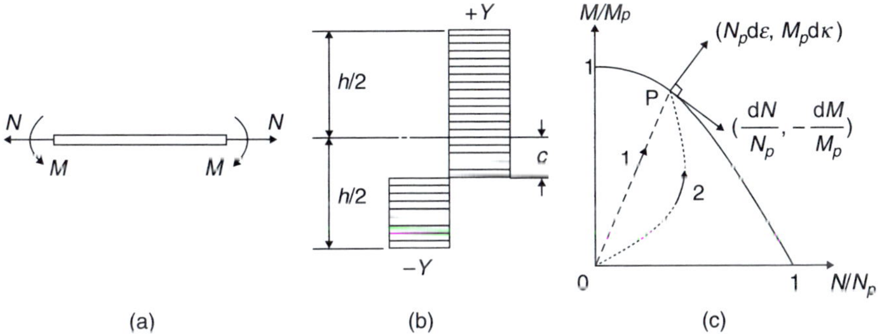

2.8 (a) A beam subjected to bending moment and axial force; (b) stress profile across the thickness in a plastic state; (c) interactive limit curve in the (M, N) plane.

gitudinal stress from the central axis of the beam due to the application of the axial force, the bending moment and the axial force are related to c by

$$M = Yb\left\{\left(\frac{h}{2}\right)^2 - c^2\right\}$$

$$N = 2Ybc \qquad [2.21]$$

Note that in the present case, the fully plastic bending moment and the fully plastic tensile force can be expressed as $M_p = Ybh^2/4$ and $N_p = Ybh$, respectively. Thus, eliminating c from the two expressions of Eq. [2.21] leads to (e.g. refer to Hodge, 1959; Chen and Atsuta, 1976; or Yu and Zhang, 1996)

$$\frac{M}{M_p} + \left(\frac{N}{N_p}\right)^2 = 1 \qquad [2.22]$$

and the limit curve is a parabola in the (M, N) plane, as shown in Fig. 2.8(c).

Notwithstanding any particular loading path (e.g. path 1 and path 2 sketched in Fig. 2.8(c)), when the combination of generalised forces M and N reaches point P on the limit curve, the beam achieves its limit state. The subsequent plastic deformation of the beam will follow the associated flow rule; that is, the general strain vector $(N_p\mathrm{d}\varepsilon, M_p\mathrm{d}\kappa)$ is along the normal direction of the limit surface

$$\frac{N_p\mathrm{d}\varepsilon}{M_p\mathrm{d}\kappa} = -\frac{\mathrm{d}(M/M_p)}{\mathrm{d}(N/N_p)} \qquad [2.23]$$

where $\mathrm{d}\varepsilon$ and $\mathrm{d}\kappa$ represent incremental axial strain and incremental curvature, respectively. The right hand side of Eq. [2.23] can be found from a differentiation of Eq. [2.22]

$$\frac{\mathrm{d}(M/M_p)}{\mathrm{d}(N/N_p)} = -2\left(\frac{N}{N_p}\right) \qquad [2.24]$$

Combining Eq. [2.23] with Eq. [2.24] leads to

$$\frac{\mathrm{d}\varepsilon}{\mathrm{d}\kappa} = 2N\frac{M_p}{N_p^{\,2}} \qquad [2.25]$$

When the limit state (i.e. Eq. [2.22]) is satisfied at a plastic hinge, the incremental dissipation by combined bending/stretching is given by $\mathrm{d}D = M\mathrm{d}\theta + N\lambda\mathrm{d}\varepsilon$ at the plastic hinge of finite length λ, where λ denotes the ***effective length*** of the plastic hinge. As will be discussed later, λ is usually in the same order as the beam's thickness h, e.g. $\lambda \approx (2 \sim 5)h$ (refer to Nonaka, 1967; Jones, 1989b; or Stronge and Yu, 1993). Thus, Eq. [2.25] can be rewritten as

$$\frac{\mathrm{d}\varepsilon}{\mathrm{d}\theta/\lambda} = 2N\frac{M_p}{N_p^{\,2}} \qquad [2.26]$$

It is seen from the above derivation that at a plastic hinge, the partitioning of the energy dissipation, i.e. the ratio between the dissipation rate in stretching and that in bending, is found to be proportional to the axial force applied at the hinge

$$\frac{\dot{D}_s}{\dot{D}_b} \equiv \frac{N_p\dot{\varepsilon}}{M_p\dot{\kappa}} = \frac{N_p\dot{\varepsilon}}{M_p\dot{\theta}/\lambda} = 2\left(\frac{N}{N_p}\right) \qquad [2.27]$$

Here the dot above a quantity denotes the rate with respect to the physical time, or with respect to a process parameter, defined in a particular analysis.

To quantify the influence of axial force on the energy dissipation in a structure, examine a simple example shown in Fig. 2.9, in which force F applied at the tip of a cantilever of length L inclines at 45° with the axis of the cantilever. Hence, the bending moment and axial force at root A are $M_A = FL/\sqrt{2}$ and $N_A = F/\sqrt{2}$, respectively. If the interactive yield criterion [2.22] is satisfied at cross-section A, that is

$$\frac{M_A}{M_p} + \left(\frac{N_A}{N_p}\right)^2 = \frac{FL}{\sqrt{2}M_p} + \frac{1}{2}\left(\frac{F}{N_p}\right)^2 = 1 \qquad [2.28]$$

then cross-section A is fully plasticised under the combined bending and tension and so it can be called a ***generalised plastic hinge***.

If the influence of axial force is neglected, then Eq. [2.28] results in a limit force $F = \sqrt{2}M_p/L$. Accordingly, the second term on the left hand side of Eq. [2.28] is estimated as

$$\left(\frac{N_A}{N_p}\right)^2 = \frac{1}{2}\left(\frac{F}{N_p}\right)^2 \approx \frac{1}{2}\left(\frac{\sqrt{2}M_p/L}{N_p}\right)^2$$
$$= \left(\frac{M_p}{N_pL}\right)^2 = \left(\frac{h}{4L}\right)^2 << 1 \qquad [2.29]$$

With the help of Eq. [2.26], the incremental energy dissipation at generalised plastic hinge A is found to be

$$\mathrm{d}D_A = M_A\mathrm{d}\theta + N_A\mathrm{d}\varepsilon = M_A\mathrm{d}\theta\left[1 + 2\left(\frac{N_A}{N_p}\right)^2\right]$$
$$\approx M_A\mathrm{d}\theta\left[1 + 2\left(\frac{h}{4L}\right)^2\right] \qquad [2.30]$$

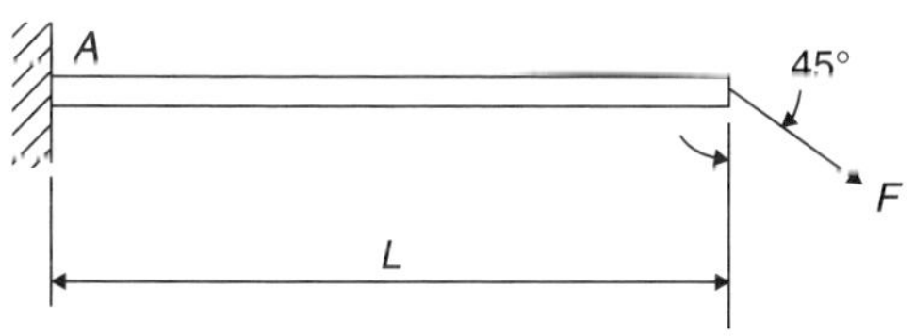

2.9 A cantilever loaded by an inclined force at its tip.

It is seen that the influence of the axial force on the energy dissipation at the generalised hinge is very small, provided the beam is slender ($h << L$).

2.2.3 Statically admissible stress field and lower bound theorem

Suppose a rigid, perfectly plastic solid/structure is subjected to a set of loads (tractions) at specified positions on its surface, and suppose the effect of the body force is negligible, then a stress field (i.e. a set of stress distribution in the solid/structure) is termed ***statically admissible*** if

(1) it satisfies the equations of equilibrium in the entire solid/structure;
(2) it does not violate the yield condition in the entire solid/structure;
(3) it satisfies the stress boundary condition.

Obviously, the statical admissibility of a stress field only concerns the satisfaction of the requirements from static equilibrium and yield condition for the magnitude and distribution of the stress itself. The statical admissibility of stress fields cares for neither the material's constitutive relationship nor any requirement related to the associated strains and displacements. Hence, in general, a statically admissible stress field may not result in a strain/displacement field for the solid/structure under its limit state.

If a statically admissible stress field is constructed for a particular solid/structure problem, and this stress field is in equilibrium with an external load whose location and direction are pre-specified, then this load must be a lower bound of the actual limit loads of that solid/structure, that is

$$P^o \le P_s \qquad [2.31]$$

where the superscript o pertains to a value resulting from a statically admissible field, and P_s denotes the limit load of the solid/structure. This statement is called the ***lower bound theorem***. Readers may refer to Prager and Hodge (1951) or Martin (1975) for the proof of the theorem.

To assist in understanding the above concept and theorem, take a beam as an illustrative example. Suppose a beam of length $2L$ is clamped at one end, simply supported at the other and subjected to a concentrated transverse force P at its mid-span, as shown in Fig. 2.10(a).

Based on the elementary theory of elastic beam, solving this statically undetermined beam results in

$$M(x) = \begin{cases} P\left[\frac{5}{16}(2L-x)-(L-x)\right] & \text{for} \quad 0 \le x \le L \\ P \times \frac{5}{16}(2L-x) & \text{for} \quad L \le x \le 2L \end{cases} \qquad [2.32]$$

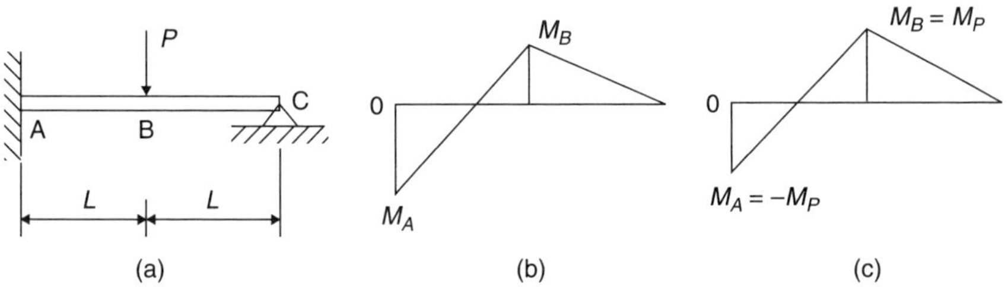

2.10 (a) A beam with one end clamped and the other simply supported; (b) elastic bending moment diagram; (c) bending moment diagram when moment at both A and B reaches the fully plastic bending moment.

so that the bending moment diagram contains straight segments in AB and in BC, while the bending moments at both point B and point A are local maximum (by magnitude), as shown in Fig. 2.10(b), where

$$M_A = M(0) = -\frac{3}{8}PL \qquad \text{and} \qquad M_B = M(L) = \frac{5}{16}PL \tag{2.33}$$

By simply assuming the magnitude of bending moment, M_A to be equal to the fully plastic bending moment of the beam M_p, the bending moment diagram shown in Fig. 2.10(b) provides a statically admissible moment field because the three conditions required in the definition are all satisfied. That $|M_A| = M_p$ leads to

$$P^o = \frac{8M_p}{3L} \approx 2.67\frac{M_p}{L} \leq P_s \tag{2.34}$$

Here the load calculated from a statically admissible field, P^o, must be smaller than, or at most equal to, the actual limit force P_s according to the lower bound theorem.

However, it is worthwhile noting that the statically admissible moment diagram for the problem is non-unique. For instance, another statically admissible moment diagram can be easily constructed by assuming both M_A and M_B (by magnitude) to be equal to M_p, as shown in Fig. 2.10(c). Note that the vertical reaction force at the clamped end A in the above elastic solution is equal to $Q_A = (M_B - M_A)/L = (11/16)P$ (see Eq. [2.33]) and since the variation of bending moment from A to B is $2M_p$, the following is obtained

$$Q_A L = \frac{11}{16}PL = 2M_p \tag{2.35}$$

Consequently, the external force, which is in equilibrium with the proposed statically admissible moment diagram (Fig. 2.10(c)), is found to be

$$P^o = \frac{32M_p}{11L} \approx 2.91\frac{M_p}{L} \leq P_s \tag{2.36}$$

Obviously, the lower bound P^o given by Eq. [2.36] is better than that given by Eq. [2.34], because the former is closer to the actual limit force P_s.

In constructing a bending moment diagram for a lower bound solution, however, it is also important to check that the bending moment does not violate the yield condition at any cross-section of the beam concerned, which comes from condition (2) in the definition of a statically admissible field. That is, $|M(x)| \leq M_p$ must be satisfied at any cross-section along the entire span of the beam.

For instance, if we take the elastic bending moment diagram (Fig. 2.10(b)) and let $M_B = M_p$, then the corresponding external force would be equal to $P = (16/5)M_p/L = 3.2M_p/L$, which is higher than that given in Eq. [2.36]. However, it is not a lower bound of the limit force because, whilst $M_B = M_p$ is assumed, $|M_A| = (6/5)M_B = 1.2M_p > M_p$ results, implying a violation of the yield condition at cross-section A.

2.2.4 Kinematically admissible velocity/displacement field and upper bound theorem

To approach a limit state of a solid/structure, alternatively we may consider a velocity or displacement field which possesses the required properties. Suppose a rigid, perfectly plastic solid/structure is subjected to a set of loads (tractions) at specified positions on some portions of its surface, whilst the displacements are specified on other portions of the surface, then a velocity/displacement field (i.e. a velocity or displacement distribution in the solid/structure) is termed ***kinematically admissible***, if

(1) it is continuous in the solid/structure, except that the tangential components may have discontinuity along finite-numbered, discrete, plastic hinge-lines in 2-D cases;
(2) it satisfies the velocity/displacement boundary condition;
(3) the external loads do positive work on this velocity/displacement field.

Obviously, the kinematical admissibility of a velocity/displacement field concerns only the satisfaction of the requirements of geometric compatibility for the distribution of the velocity/displacement itself. The kinematical admissibility cares for neither the material's constitutive relationship nor any requirement to the related stress field. Hence, generally speaking, a kinematically admissible velocity/displacement field may not result in a stress field for the solid/structure under its limit state.

If a kinematically admissible velocity/displacement field is constructed for a particular solid/structure problem, then associated with this field, the energy balance (in terms of velocity and displacement, respectively) requires

$$\dot{E}_{in} \equiv P^* v^* = \dot{D} \equiv \sum_{i=1}^{n} M_p |\dot{\theta}_i| \quad \text{or} \quad \equiv \sum_{i=1}^{n} M_o |\dot{\theta}_i| L_i \qquad [2.37]$$

and

$$E_{in} \equiv P^* \Delta^* = D \equiv \sum_{i=1}^{n} M_p |\theta_i| \quad \text{or} \quad \equiv \sum_{i=1}^{n} M_o |\theta_i| L_i \qquad [2.38]$$

where Eq. [2.10] or [2.11] is employed to demonstrate the calculation of the energy dissipation (or its rate) along plastic hinges in 1-D cases or along plastic hinge-lines in 2-D cases. Superscript * pertains to the kinematically admissible field; v^* in Eq. [2.37] denotes the kinematically admissible velocity at the point where load P applies, while Δ^* in Eq. [2.38] denotes the kinematically admissible displacement at the point where load P applies. E_{in} and D denote the external work done and the internal energy dissipation, respectively.

Calculated from the energy balance (using either Eq. [2.37] or Eq. [2.38]), the magnitude of the external load can be determined as P^*; then this load must be an upper bound of the actual limit loads of that solid/structure; that is

$$P^* \geq P_s \qquad [2.39]$$

This statement is called the ***upper bound theorem***. Readers may refer to Prager and Hodge (1951) or Martin (1975) for the proof of the theorem.

When only a beam (or another 1-D structural member, such as a ring or an arch) is concerned, a collapse mechanism can be constructed by simply assigning one or more plastic hinges in the beam with the supporting conditions being satisfied. The number of plastic hinges in the collapse mechanism should be equal to the degree of the static indeterminacy of the beam plus one. For instance, a cantilever beam subjected to a tip force is statically determinate, so only one plastic hinge needs to be assigned in forming a collapse mechanism; a beam clamped at both ends has two degrees of static indeterminacy, so three plastic hinges should be assigned in a collapse mechanism.

To elaborate upon the procedure of applying the upper bound theorem, we still take the beam shown in Fig. 2.10(a) as an illustrative example. Evidently, this beam has one degree of static indeterminacy, so two plastic hinges are required in a collapse mechanism. Figure 2.11 depicts such a mechanism, in which the plastic hinges are located at both cross-sections A and B.

Based on this collapse mechanism, a kinematically admissible displacement field can be constructed by assuming point B to have a transverse displacement Δ^*, which could be a virtual one but should be along the same

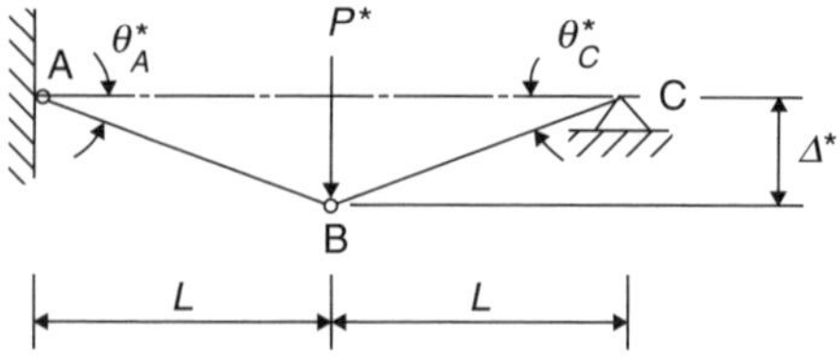

2.11 A collapse mechanism for the beam shown in Fig. 2.10(a).

direction as the applied force P so as to satisfy condition (3) of the definition, i.e. $P^* \Delta^* > 0$. As the displacement is assumed to be small, the relevant rotation angles at plastic hinge A and at support C are both equal to $\theta_A^* = \theta_C^* = \Delta^*/L$. Consequently, the rotation angle at plastic hinge B is $\theta_B^* = \theta_A^* + \theta_C^* = 2\Delta^*/L$.

Now we can use the argument of energy balance, i.e. Eq. [2.38], and obtain

$$P^* \Delta^* = M_p \theta_A + M_p \theta_B = 3M_p \Delta^*/L \qquad [2.40]$$

Note that the simply-supporting end C is not a plastic hinge, so it dissipates no energy although there is a rotation. Equation [2.40] leads to an upper bound of the limit force as

$$P^* = 3M_p/L \geq P_s \qquad [2.41]$$

By combining Eqs [2.41] and [2.36] we have an estimate for the limit force

$$2.91M_p/L = P^o \leq P_s \leq P^* = 3M_p/L \qquad [2.42]$$

Note that the difference between the upper and lower bounds is only 3 %, so that the limit force P_s has been appropriately estimated. It is also confirmed that the estimate of $P = 3.2M_p/L$ obtained from a statically *inadmissible* moment diagram given at the end of subsection 2.2.3 does not provide a lower bound of the limit force.

2.3 Effects of large deformation

2.3.1 Background

In Sections 2.2.3 and 2.2.4 the statically admissible stress field and the kinematically admissible velocity/displacement field were described. It must be noted that these fields are constructed based on the *original* (i.e. *undeformed*) configuration of the structure. For instance, the equilibrium in the beam (reflected in its bending moment diagram) shown in Fig. 2.10 was based on the straight (undeformed) configuration of the beam; and the geometric relation between deflection Δ^* and rotation angle θ^* (see Fig. 2.11) was formulated by assuming that both of them are small. In these two examples, the elongation of segments AB and AC was neglected because the formulation was based on the straight configuration. As a matter of fact, the classical limit analysis and bound theorems can only be used to determine the ***incipient collapse mechanism***, the ***initial limit load*** or its bounds.

However, when a structure or a material is designed or used for the purpose of energy absorption, it is usually expected to experience large plastic deformation under external loading. Therefore, the effects of large deformation should be taken into account when the theory of limit analysis is applied to analysing energy-absorbing structures.

2.3.2 Analysis of an illustrative example

To demonstrate how large deformation will affect the classical limit analysis, in this section we will take the following problem as an example. Suppose initially a round-headed rigid indenter of radius R is in contact with the middle point of a straight rigid-plastic beam of length $2L$; the beam is clamped at its ends (A and A′), but allowed to have axial motions along the clamps, as shown in Fig. 2.12(a). Assume that all the contact surfaces are frictionless.

When the indenter is pressed down transversely to the beam by force P, it is easily seen that the incipient collapse mechanism of this beam contains two plastic hinges at the clamped ends A and A′, as well as a plastic hinge at the middle section C of the beam. Following the procedure given in Section 2.2.4 and calculated from this incipient collapse mechanism (that is a kinematically admissible displacement field), the initial collapse force is found to be

$$P_s = 4M_p/L \tag{2.43}$$

with M_p being the fully plastic bending moment of the beam.

With the increase of the deflection, the deformed configuration of the beam is as sketched in Fig. 2.12(b), where a central portion of the beam, BB′, wraps the round head of the indenter. Since the beam is supposed to be rigid, perfectly plastic, the circular segment BB′ must be in its plastic pure bending state. Hence, in addition to the plastic hinges at clamped ends A and A′, at which $M_A = -M_p$, plastic hinges must form at cross-sections B and B′, at which $M_B = M_p$ holds.

The free body diagram of the large deformation mechanism is sketched in Fig. 2.12(c), from which it is evident that with the increase of the indenter's displacement, both the location and the direction of the indenter's force will change. Note that the variation in bending moment from plastic hinge A to plastic hinge B is $2M_p$, so that

$$\frac{P}{2}(L - R\sin\beta) = 2M_p \tag{2.44}$$

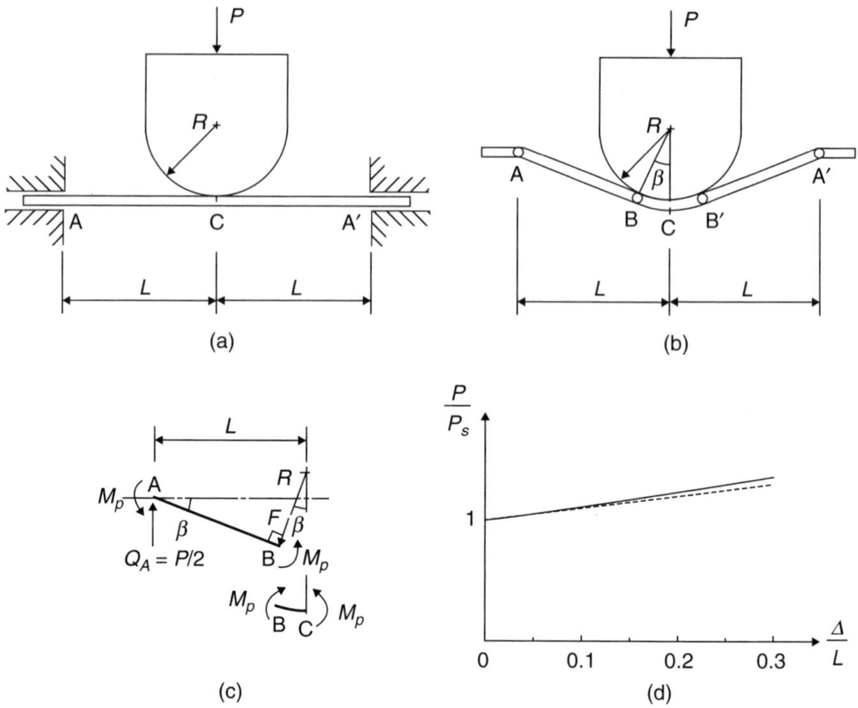

2.12 A beam clamped at both ends without axial constraint: (a) the initial configuration when it is pressed by a round-headed indenter; (b) deformed configuration; (c) free-body diagrams of segment AB and arc BC; (d) the variation of the load-carrying capacity with central deflection.

where β denotes the angle formed by arc BC. Thus, the load-carrying capacity of the beam in large deformation is

$$P = \frac{4M_p}{L - R\sin\beta} = \frac{P_s}{1-(R/L)\sin\beta} \qquad [2.45]$$

where P_s is the initial limit force given by Eq. [2.43].

On the other hand, the maximum deflection of the beam at mid-point C, Δ, can be expressed by angle β as

$$\Delta = (L - R\sin\beta)\tan\beta + R(1-\cos\beta) = L\tan\beta + R - R/\cos\beta \qquad [2.46]$$

or in non-dimensional form

$$\frac{\Delta}{L} = \tan\beta + \frac{R}{L}\left(1 - \frac{1}{\cos\beta}\right) \qquad [2.47]$$

Equations [2.45] and [2.47] provide a relation between force P and maximum deflection Δ through geometric parameter β. By taking R/L = 1/2, this relation is shown in Fig. 2.12(d) by the solid line.

If parameter β is small and R is in the same order as L, $(1 - 1/\cos \beta) \approx -\beta^2/2$, so that Eq. [2.47] indicates that $\beta \approx \Delta/L$. Substituting it into Eq. [2.45] results in

$$\frac{P}{P_s} \approx 1 + \frac{R}{L^2}\Delta \qquad [2.48]$$

which provides an approximate linear relation between P and Δ, as shown in Fig. 2.12(d) by the broken line in case of $R/L = 1/2$.

The above analysis is based on the assumption that axial motion is allowed at the clamped ends of the beam. If axial constraints are applied to the clamped ends, as shown in Fig. 2.13(a), then the deflection of the beam will be accompanied by the elongation of the beam's axis. Using the geometry shown in Fig. 2.12(c), the axial strain of the beam's axis is found to be

$$\varepsilon = \left\{\left(\frac{L - R\sin\beta}{\cos\beta} + R\beta\right) - L\right\}\Big/L = \left(\frac{1}{\cos\beta} - 1\right) - \frac{R}{L}(\tan\beta - \beta) \qquad [2.49]$$

If parameter β is small and R is in the same order as L, it can be approximated by $\varepsilon \approx \beta^2/2$, so that the incremental strain is $d\varepsilon \approx \beta d\beta$. Noting that the incremental rotation angle at plastic hinge A is $d\theta = d\beta$ and using Eq. [2.26], we have

$$2\frac{N}{\lambda}\frac{M_p}{{N_p}^2} = \frac{d\varepsilon}{d\theta} = \frac{\beta d\beta}{d\beta} = \beta \approx \frac{\Delta}{L} \qquad [2.50]$$

If the beam has a rectangular cross-section and the effective length of the plastic hinge is taken as $\lambda = 2h$, then Eq. [2.50] leads to an estimate of the axial force at hinge A, N_A, induced by the large deflection of the beam

$$\frac{N_A}{N_p} = 4\beta \approx 4\frac{\Delta}{L} \qquad [2.51]$$

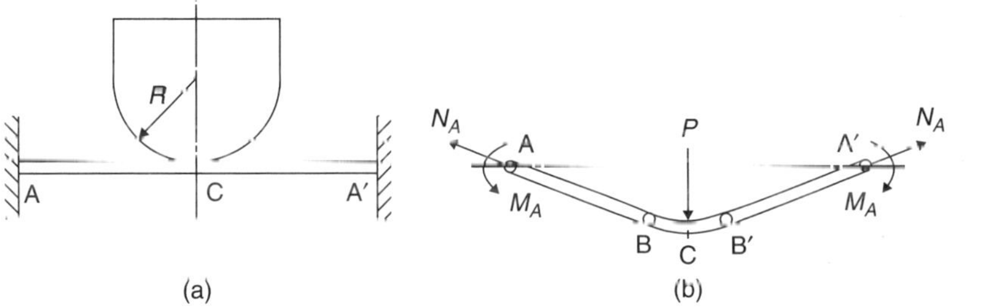

2.13 A beam clamped at both ends with axial constraint: (a) the initial configuration when loaded by a round-headed indenter; (b) the axial forces developed in the beam during its large deflection.

It is seen from Fig. 2.13(b) that the axial force N_A will enhance the load-carrying capacity of the beam by adding a term $2N_A \sin\beta \approx 2N_A\beta \approx 2N_A\Delta/L$. Therefore, using Eq. [2.48] and $\beta \approx \Delta/L$, we have

$$P \approx \frac{4M_p}{L}\left(1+\frac{R}{L^2}\Delta\right)+8N_p\left(\frac{\Delta}{L}\right)^2 = P_s\left(1+\frac{R\Delta}{L^2}+\frac{8\Delta^2}{hL}\right) \qquad [2.52]$$

It is clear that the contribution of the axial force to the collapse load depends on the slenderness of the beam, $2L/h$. For example, for a short beam of $2L/h = 16$, the axial force will result in the collapse load being doubled when the central deflection reaches the beam's thickness ($\Delta = h$), which is really significant.

2.3.3 Various aspects of effects of large deformation

From the analysis of this illustrative example it is observed that large deformation of a structure may affect its collapse mechanism and collapse load in many ways, such as:

- all the geometric relationships and the equations of equilibrium have to be formulated according to the current (instantaneous) configuration rather than the original (undeformed) configuration;
- the loading positions and directions of the external loads may vary with the deformation process – in the case of external loads applied by tools (e.g. punch, die, indenter, etc.), the friction between the structure and the tool's surface may also alter the equilibrium and contribute to the energy dissipation;
- the instantaneous collapse mechanism of the structure may contain moving plastic hinges (or moving plastic regions), which evolute from their stationary counterparts in the incipient collapse mechanism;
- the axial forces (in 1-D structural components) or membrane forces (in 2-D structural components) may be induced by large deformation and greatly enhance the load-carrying capacity of the structure.

Whether all or some of these effects appear in an energy-absorbing structure depends on its configuration, slenderness, supporting conditions and the way of loading. For instance, as shown in the above example, the large deformation behaviour of a beam with axial constraint is severely affected by the axial forces induced by the large deflection, whilst a similar beam without axial constraint is not affected. In the former case, the small deformation formulation may result in big errors when the maximum deflection of the beam reaches the thickness, whilst in the latter case, the small deformation formulation may be applicable until the deflection reaches the same order of the beam's length.

For plates and shells under double-curvature bending, e.g. a circular plate under axisymmetric bending, previous studies have revealed that membrane stress will become important and even dominant as long as the maximum deflection reaches the order of the thickness of the plate or shell, regardless of the boundary condition. For instance, Timoshenko and Woinowsky-Krieger (1959) showed that as long as the central deflection of a simply-supported circular plate is equal to the thickness of the plate, the membrane strain in the middle plane of the plate will reach the same order as the bending strains. Regarding rigid-plastic circular plates, Calladine (1968) demonstrated that membrane stresses induced by large deflection will greatly enhance the load-carrying capacity of the plate even it is simply-supported.

2.3.4 Concluding remarks

(1) For 1-D structural components without axial constraint or for plates under cylindrical bending, the effects of large deformation are mainly reflected by the geometric changes, which become important when the maximum deflection of the component is comparable to the component's characteristic length (e.g. the length of a beam or the radius of a circular ring).
(2) For 1-D structural components with axial constraint or for 2-D components (plates and shells) under double-curvature deformation, the effects of large deformation are mainly associated with the axial/membrane forces induced by the large deflection, which become dominant in the component's load-carrying capacity when the deflection exceeds the thickness of the component.
(3) Usually, slender structural components suffer more in the geometric effect relevant to their large deformation, whilst stubby ones suffer more in the effect of axial/membrane forces – which is also caused by their large deformation.
(4) Depending on the configuration of the structure and the way of loading, sometimes a large deformation may also cause changes in the location and direction of the external loads, as well as changes in the collapse mechanisms (e.g. those containing moving plastic hinges).
(5) In all these cases, the various effects of large deformation need to be carefully incorporated into theoretical modelling of energy-absorbing structures.

2.4 Effects of dynamic loading

The limit analysis of structures illustrated in Section 2.2 is entirely based on quasi-static loading. However, in view of the numerous engineering applications of energy-absorbing structures as impact protectors, their behaviour has to be examined under dynamic loading. Jones (1989a) has comprehensively analysed the dynamic effects for structures under impact loading. In this section, various effects of dynamic loading will be briefly examined.

2.4.1 Stress wave propagation and its effects on energy absorption

Elastic stress wave

Upon the application of dynamic loading, the suddenly gained stress or the suddenly gained particle velocity at the points on the loaded surface will be propagated away from the surface in the form of stress waves.

If the normal stress applied is smaller than the yield stress of the material, Y, then the stress wave is elastic and propagates at a speed of

$$c_L = \sqrt{\frac{E}{\rho}} \qquad [2.53]$$

where c_L denotes the speed of the ***longitudinal elastic stress wave***, E and ρ are the Young's modulus and the density of the material, respectively.

For longitudinal elastic stress wave, there exists a relation between the stress σ and particle velocity v

$$\sigma = \sqrt{E\rho}v = \rho c_L v = \frac{E}{c_L}v \qquad [2.54]$$

where the quantity σ/v is termed stress wave ***impedance***. Therefore, a ***yield velocity*** can be defined as a material's property as follows

$$v_y = \frac{Y}{\sqrt{E\rho}} = \frac{Y}{\rho c_L} = \frac{Y}{E}c_L \qquad [2.55]$$

Only when the particle velocity gained from dynamic loading (e.g. from impact) is below this yield velocity will the stress wave be purely elastic. For mild steels, $c_L \approx 5100$ m/s and typically if $E/Y = 500$, then the yield velocity v_y is about 10 m/s.

Plastic stress wave

If the stress created by the dynamic loading is beyond the yield stress of the material, or the particle velocity gained from impact is beyond the yield velocity, then in addition to the elastic stress wave propagating with velocity c_L, plastic stress waves will be initiated and propagate away from the loaded region. Readers are advised to consult Johnson (1972) or Johnson and Yu (1989) to acquire more information on plastic wave propagation. Only the most fundamental results are summarised herein to facilitate the discussion on energy-absorbing structures in the following chapters.

(1) If the material displays linear strain-hardening as shown in Fig. 2.2(b) or Fig. 2.3(b), then the longitudinal plastic wave will propagate with a speed of

$$c_p = \sqrt{\frac{E_p}{\rho}} \quad [2.56]$$

where E_p denotes the strain-hardening modulus of the material. For structural metals, typically E_p is usually 2 or 3 orders smaller than E, so that propagation of the plastic wave is at least an order slower than that of the elastic precursor.

(2) If the material displays non-linear strain-hardening and obeys a stress–strain relation $\sigma = \sigma(\varepsilon)$ in its plastic range, then the longitudinal plastic wave will propagate with a speed of

$$c_p = \sqrt{\frac{\mathrm{d}\sigma/\mathrm{d}\varepsilon}{\rho}} \quad [2.57]$$

where $(\mathrm{d}\sigma/\mathrm{d}\varepsilon)$ represents the tangential modulus of the material, which in general varies with stress/strain which is brought about by the stress wave.

(3) For decreasingly strain-hardening material, whose tangential modulus $(\mathrm{d}\sigma/\mathrm{d}\varepsilon)$ reduces with increasing strain as shown in Fig. 2.14(a), the longitudinal plastic wave which brings about a higher stress (or a higher strain) will propagate with a lower speed. Consequently, the plastic wave is scattered, as sketched in Fig. 2.14(b).

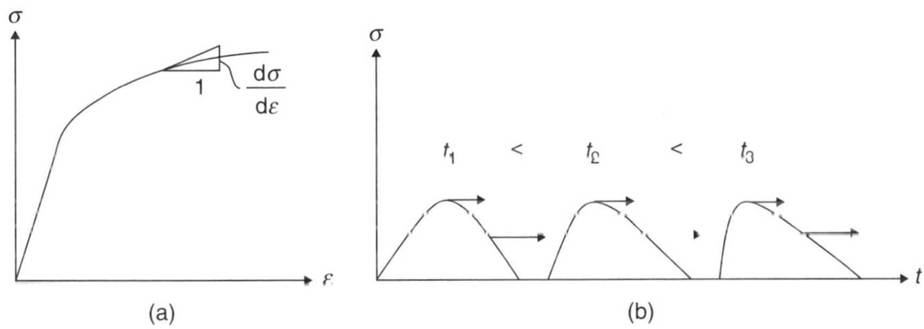

2.14 Stress waves in decreasingly strain-hardening materials: (a) the tangential modulus reduces with increasing strain; (b) change in wave shape during propagation.

(4) For increasingly strain-hardening material, whose tangential modulus ($d\sigma/d\varepsilon$) increases with increasing strain as shown in Fig. 2.15(a), the longitudinal plastic wave which brings about a higher stress (or a higher strain) will propagate with a higher speed. Consequently, the plastic wave is convergent and will eventually form a ***shock wave***, as sketched in Fig. 2.15(b), which is characterised by strong discontinuity in stress and particle velocity at the shock front.
If the applied dynamic load brings about a stress marked by point C in Fig. 2.15(a), then the shock wave will propagate at a speed determined by the slope of the straight line (shown by the broken line in Fig. 2.15(a)) between the initial yield point A and point C, and can be expressed as

$$c_{shock} = \sqrt{\frac{[\sigma]/[\varepsilon]}{\rho}} \qquad [2.58]$$

where $[\sigma]$ and $[\varepsilon]$ denote the jump (discontinuity) in stress and strain, respectively. Examples of applying shock wave theory in structures will be found in Sections 4.6 and 10.4.

Effects of stress waves on deformation mechanisms and energy absorption

Elastic and plastic waves may affect the energy absorption of materials and structures in various complex ways, depending on the dynamic loads, the structure's configuration and the material's properties. Herein we illustrate only a few of those effects that are frequently encountered in the analyses of energy-absorbing structures.

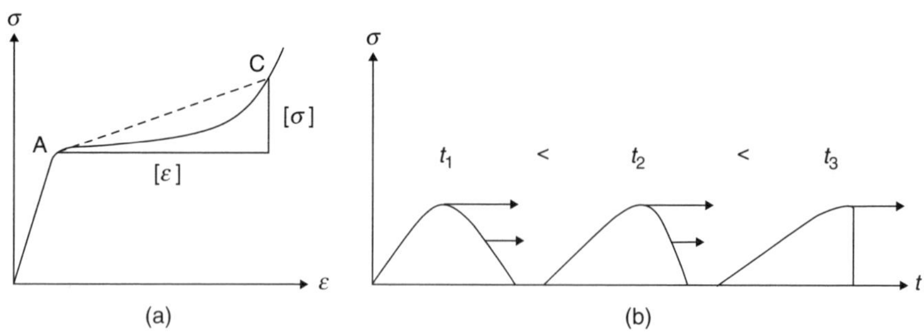

2.15 Stress waves in increasingly strain-hardening materials: (a) the stress–strain curve; (b) change in wave shape during propagation and formation of a shock wave.

(1) At the dynamically loaded region (e.g. the region where a structure is impinged by a rigid projectile), the high stress brought about by the strong plastic compression waves may cause a local plastic collapse – for example (see Section 10.4 for details), a layer of honeycomb cells may be plastically collapsed at the loading end. After some energy has been dissipated in this localised zone, the rest of honeycomb may experience only elastic deformation.

(2) When a compressive elastic wave produced by a dynamic load or impact reaches the distal surface (e.g. the distal end of a long bar), it will reflect from that surface. If the surface is free (i.e. with no constraint to its displacement), the reflected wave will be a tensile one which propagates back from the distal surface. For brittle materials such as concrete and rocks, whose tensile strength is low, this reflected tensile wave may cause fracture of material some distance away from the free surface; consequently a portion of material will be separated and fly away. This kind of failure is termed ***spalling*** and the flying layer will bring a notable portion of the input energy away by its kinetic energy.

(3) When a compressive elastic wave produced by a dynamic load or impact reaches the distal surface where no displacement is allowed (e.g. a clamped distal end of a long bar), the reflected wave will also be compressive and it will result in the magnitude of the compressive stress being doubled. The increase in the stress magnitude may create a plastic compressive wave, so plastic deformation and energy dissipation would first occur in the region close to the distal fixed surface rather than in the loading region. In the case of cellular materials (e.g. honeycombs), cells may first be plastically collapsed at this clamped distal end.

(4) All the above cases usually occur when the dynamic loading is along the longitudinal direction of the structural components, such as a long bar subjected to a compressive force pulse or impact along its axial direction. If a slender structural component (e.g. a beam or a thin plate) is subjected to a dynamic loading in its transverse direction, then the stress waves along that direction will disappear in a brief time period as a result of the frequent wave reflections between two close surfaces of the beam (or the plate) along its thickness direction. However, elastic flexural waves (i.e. bending waves) will propagate away from the loading region.

As is well known, the elastic flexural wave has a frequency-dependent speed when it propagates along a beam, so it is dispersive although there is no plastic dissipation. A few numerical studies on the dynamic response of a cantilever beam to a tip impact (Symonds and Fleming,

1984; Reid and Gui, 1987; Hou *et al.*, 1995) have found that the dynamic behaviour of an elastic-plastic cantilever under impact is distinctly different from that predicted by a rigid-plastic analysis (Parkes, 1955). A thorough investigation made by Yu *et al.* (1997) has confirmed that these differences can be attributed to the interaction between reflected elastic flexural waves from the distal clamped end and the travelling plastic 'hinge'. Thus, it is evident that, although the elastic flexural wave does not directly dissipate energy, it does affect the energy dissipation in the cantilever beam by altering the deformation pattern and influencing the evolution of plastic regions.

2.4.2 Inertia and its effects on energy absorption

The roles of inertia: fundamental consideration

As pointed out above, when a structural component is subjected to a dynamic load in its transverse direction, the stress waves along this direction will usually die out within a brief time period (typically in the order of microseconds). After that, since the structural component starts to deform dynamically whilst each cross-section moves as an entirety, the inertia of the component becomes a dominant factor in its dynamic performance. This has also been pointed out by Jones (1989a).

To demonstrate the significant difference between the dynamic performance of structural components and their quasi-static performance due to the role played by inertia, examine the rigid-plastic model shown in Fig. 2.6(b) while assuming that the movable block in the model has concentrated mass m, as re-drawn in Fig. 2.16(a). Obviously, this rigid-plastic model cannot sustain any quasi-static force greater than F_y, which is the resistant force (limit force) of the structural model.

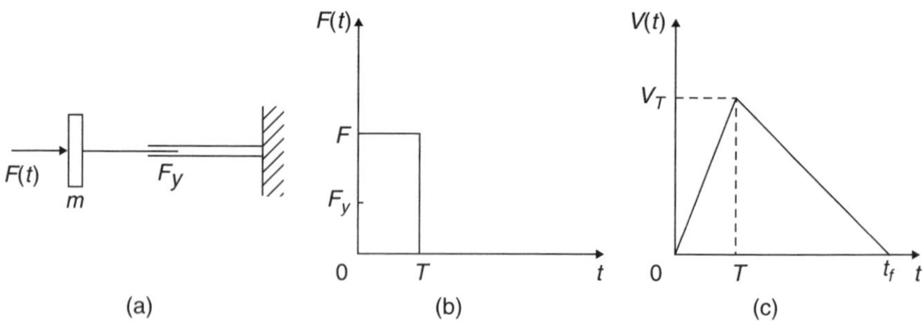

2.16 (a) A rigid-plastic model with mass block m subjected to a force pulse $F(t)$; (b) a rectangular force pulse; (c) the variation of velocity of the block with time.

Now consider a case of dynamic loading and the applied force $F(t)$ is supposed to be a rectangular pulse (Fig. 2.16(b)), that is

$$F(t) = \begin{cases} 0 & t \le 0 \\ F & 0 \le t \le T \\ 0 & t > T \end{cases} \qquad [2.59]$$

where T denotes the period of application of force F.

In $0 \le t \le T$, the mass block undergoes an acceleration phase in the dynamic response of the model and its equation of motion is

$$m\ddot{u} = F - F_y \qquad [2.60]$$

where u denotes the displacement along the direction of the applied force. Hence, at $t = T$, the velocity of the block is $v_T = [(F - F_y)/m]T$.

In the subsequent deceleration phase ($t > T$), the equation of motion of the block is

$$m\ddot{u} = -F_y \qquad [2.61]$$

so that the velocity of the block is

$$\dot{u} \equiv v = v_T - \frac{F_y}{m}(t - T) = \frac{F}{m}T - \frac{F_y}{m}t \qquad [2.62]$$

Thus, at $t_f = FT/F_y$, $v = 0$ and the motion of the block ceases. The variation of the velocity of the block with time is depicted in Fig. 2.16(c).

The total displacement of the block is found from the area of the triangle shown in Fig. 2.16(c), which gives

$$\Delta_f = v_T t_f / 2 = \frac{F(F - F_y)}{2mF_y}T^2 \qquad [2.63]$$

so that the energy dissipation of the system during the dynamic response to the applied force pulse is

$$D = F_y \Delta_f = \frac{F(F - F_y)}{2m}T^2 = \frac{p \times p_o}{2m} \qquad [2.64]$$

where $p = FT$ and $p_o = (F - F_y)T$ represent the total impulse and the 'overloading' impulse, respectively.

It is evident that here D is inversely proportional to the mass m, clearly indicating the effect of inertia on the energy dissipation in the system. When applying this analysis to a structural component, of course, the important but difficult issue is to determine the 'equivalent mass' m for the component concerned.

The role of inertia in dynamic performance of structural components

As a simple example for a structural component, examine a free-free beam of length $2L$, as shown in Fig. 2.17(a). Since the beam has no support at all, it cannot sustain any quasi-static load in its transverse direction.

However, if a transverse step force, F, as shown in Fig. 2.17(b), is applied to the mid-point of the free-free beam, the force will make the beam move transversely with an acceleration $a_o = F/2\rho L$, where ρ denotes the mass per unit length of the beam. According to the d'Alambert principle, the inertia force $\rho a_o = F/2L$ is uniformly distributed along the beam, as shown in Fig. 2.17(c). Together with the applied step force F, this inertia force produces a bending moment diagram as shown in Fig. 2.17(d), in which the maximum bending moment appears at the mid-span of the beam and is equal to $M_{\max} = (F/2)(L/2) = FL/4$. Therefore, a plastic hinge will appear at the mid-span of the beam, if the magnitude of force F reaches a 'dynamic' collapse force

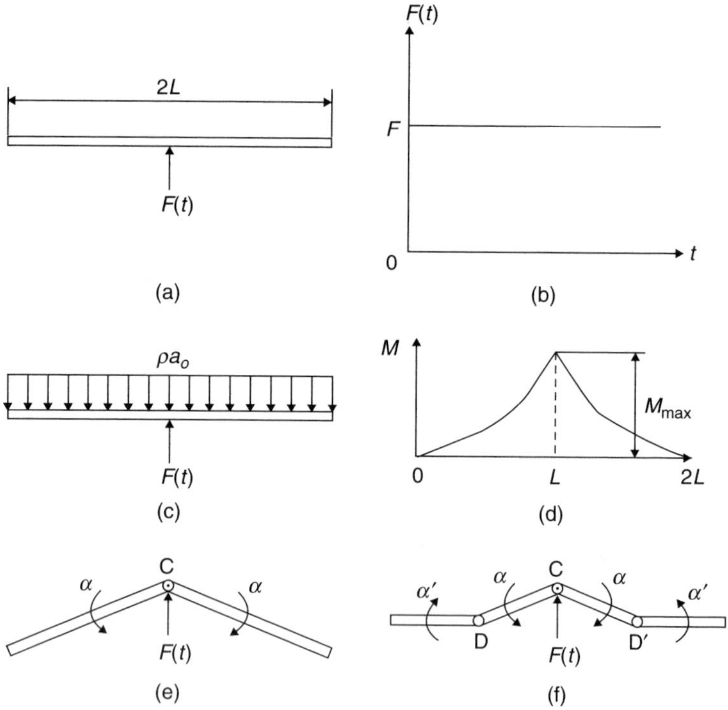

2.17 (a) A free-free beam loaded by a dynamic force at its mid-span; (b) a step force; (c) the inertia forces developed in the beam; (d) bending moment diagram; (e) deformation mechanism with one plastic hinge; (f) deformation mechanism with three plastic hinges.

$$F_d = 4M_p/L \tag{2.65}$$

Here it is worthwhile noting that there is no 'static' collapse force for the free-free beam concerned.

The dynamic deformation mechanism is as shown in Fig. 2.17(e), from which the angular acceleration of a half of the beam is found to be

$$\alpha = \mathrm{d}^2\theta/\mathrm{d}t^2 = (FL/4 - M_p)/J = 3(FL - 4M_p)/(\rho L^3) \tag{2.66}$$

where $J = \rho L^3/12$ is the moment of inertia of a half of the beam about its own centre. Consequently, the energy dissipation rate is

$$\mathrm{d}D/\mathrm{d}t = 2M_p(\mathrm{d}\theta/\mathrm{d}t) = 2M_p\alpha t = 6M_p(FL - 4M_p)t/(\rho L^3) \tag{2.67}$$

whilst the input energy rate is

$$\mathrm{d}E_{in}/\mathrm{d}t = Fv_C = F(a_o + \alpha L/2)t = 2Ft(FL - 3M_p)/(\rho L^2) \tag{2.68}$$

where a_o denotes the transverse acceleration of the mass centre of the beam and v_C denotes the velocity of the mid-span C of the beam. By combining Eq. [2.67] with Eq. [2.68], the ratio of the plastic dissipation to the total input energy is found to be

$$\begin{aligned} R_p \equiv D/E_{in} &= 3(M_p/FL)(FL - 4M_p)/(FL - 3M_p) \\ &= 3(f-4)/f(f-3) \end{aligned} \tag{2.69}$$

where $f \equiv FL/M_p$ denotes non-dimensional force applied, and this equation is valid only when $f \geq 4$ (i.e. $F \geq F_d$). It is known from Eq. [2.69] that the maximum value of ratio R_p occurs when $f \equiv FL/M_p = 6$, at which $R_p = 1/3$. When f increases further, this ratio reduces, e.g. $R_p = 2/9$ when $f = 12$.

A further analysis (refer to Lee and Symonds, 1952) indicates that when the applied force further increases to $f \equiv FL/M_p \geq 22.9$, then two more plastic hinges will appear on the sides of the mid-span and the deformation mechanism will contain three hinges, as shown in Fig. 2.17(f).

Figure 2.18 depicts the variation of the energy dissipation ratio $R_p \equiv D/E_{in}$ with the non-dimensional force $f \equiv FL/M_p$. Clearly, the dependence is rather complex because of the alteration of the deformation mechanisms. More details of the analysis can be found in Yang *et al.* (1998) and Yu (2002).

It is evident from this illustrative example that when a structural component is accelerated by dynamic loads, its inertia will create shear forces and bending moments on top of those produced by the applied loads; consequently, its dynamic deformation mechanism and energy-absorption behaviour may be significantly altered, for instance:

- the dynamic load-carrying capacity of a structural component could be significantly different from its static counterpart (in fact, the free-free

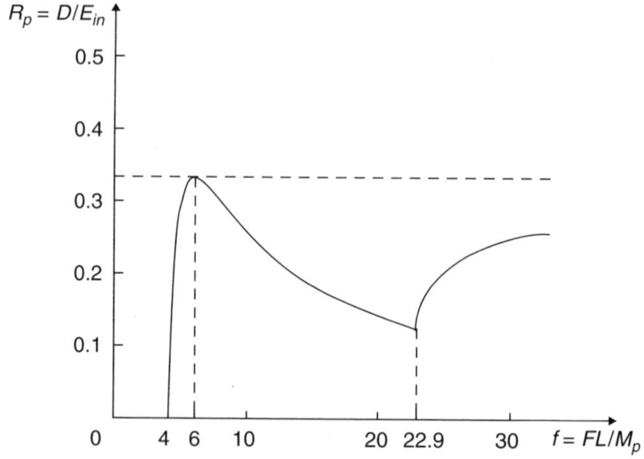

2.18 The ratio of the plastic dissipation D to the input energy varies with the magnitude of the applied step force F. (Yang *et al.*, 1998)

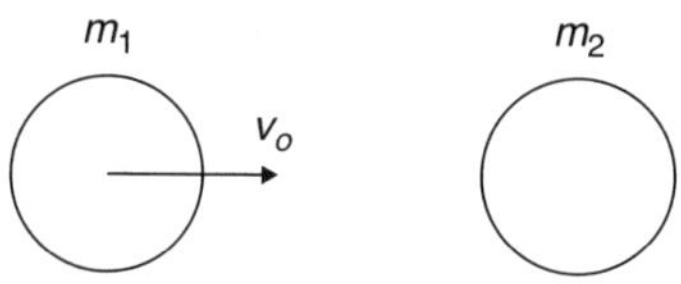

2.19 Central collision of two spheres.

beam analysed has no static load-carrying capacity, but it can sustain dynamic loading);

- the ratio of plastic energy dissipation to the input energy may non-monotonically vary with the magnitude of the applied force;
- the dynamic deformation mechanism may be different from the quasi-static collapse mechanism and it may vary with the magnitude of the applied force.

Energy loss during collision

If a projectile collides with a structural component, no matter whether the projectile is rigid or elastic or elastic-plastic, a portion of the initial kinetic energy of the projectile will be lost during the collision, whilst the momentum of the system remains conservative.

In the classical analysis of central collision between two rigid spheres, as shown in Fig. 2.19, it is assumed that before collision the sphere of mass m_1 was moving with velocity v_o, and the sphere of mass m_2 was stationary. If the central collision is assumed to be completely inelastic, then the conservation of linear momentum requires

$$m_1 v_o = (m_1 + m_2) v' \tag{2.70}$$

where v' denotes the common velocity of two spheres after the inelastic collision. Hence

$$v' = \frac{m_1}{(m_1 + m_2)} v_o \tag{2.71}$$

Obviously, when the two spheres are regarded as a system, there is a loss in the kinetic energy of the system during the collision

$$K_{loss} = K_o - K' = \frac{1}{2} m_1 v_o^2 - \frac{1}{2}(m_1 + m_2) v'^2$$
$$= \frac{m_1 m_2}{2(m_1 + m_2)} v_o^2 = \frac{m_2}{(m_1 + m_2)} K_o \tag{2.72}$$

Clearly, $0 < K_{loss} < K_o$. It is also seen that the larger the mass ratio m_2/m_1, the larger the relative energy loss K_{loss}/K_o.

In the case where a structural component is struck by a projectile, m_2 would represent an equivalent mass of that component. Obviously, a loss in the kinetic energy will take place during impact, especially if the structure's mass is much greater than that of the projectile.

Thus, it is concluded that the structure's inertia will notably alter the input energy during a collision/impact. This important issue will be further addressed in Chapter 7, in which a typical inertia-sensitive energy-absorbing structure will be comprehensively discussed.

2.4.3 Strain-rate and its effects on energy absorption

Strain-rate sensitivity of materials

The application of a rapidly rising dynamic load to a structure will make it deform rapidly, resulting in high strain-rates. On the other hand, numerous previous studies have shown that the mechanical properties of many engineering materials, such as mild steel and some polymers, are dependent on the strain-rate. For instance, at a strain-rate of $10^3 \mathrm{s}^{-1}$, the dynamic yield stress of low carbon steel could be more than double the quasi-static yield stress, while its ductility is notably reduced.

From the point of view of material science, the material rate-dependence can be of two kinds. Materials such as polymers at temperatures above the glass transition temperature exhibit viscoplastic behaviour that manifests itself as creep; this behaviour is related to viscous flow as polymer chains slip past one another. In polycrystalline materials such as metals, viscous drag due to dislocation motion around barriers exhibits similar behaviour which has been termed rate sensitivity of strain-hardening. In many metals, rate-dependence at temperatures substantially below the melting point is related to an additional stress required to generate and accelerate dislocations; i.e. to initiate changes in the rate of plastic flow.

A number of micromechanics models have correlated this strain-rate sensitivity with the thermal activation of dislocation motion in metals or with the rapid alignment of polymer chains. However, for engineering applications, it would be more useful to have some explicit phenomenological rate-dependent constitutive equations in accounting for the effects of strain-rate on the materials' yield stress and flow stress.

Among various phenomenological rate-dependent constitutive equations for engineering materials, the ***Cowper–Symonds relation*** (Cowper and Symonds, 1957) has been most popularly employed in structural impact problems. This relation represents a rigid, perfectly plastic material with dynamic yield or flow stress that depends on strain-rate. Thus the ratio of dynamic yield stress Y^d to static yield stress Y is

$$\frac{Y^d}{Y} = 1 + \left(\frac{\dot{\varepsilon}}{B}\right)^{1/q} \qquad \dot{\varepsilon} > 0 \qquad [2.73]$$

where B and q are material constants. Actually, B represents a characteristic strain-rate, at which $Y^d = 2Y$, while the material constant q is a measure of the rate sensitivity of the material. Some representative values of B and q for use in the Cowper–Symonds relation are listed in Table 2.1. These values were obtained for strain $\varepsilon = 0.05$; they may not be accurate for strains that are either very large or very small in comparison with $\varepsilon = 0.05$.

Rate effects in bending of beams

For pure bending of beams or other slender components the relationship between the dynamic fully plastic bending moment, $M_p{}^d$, and the rate of curvature can be obtained from Eq. [2.67] and the condition that plane sections remain plane. The mathematical derivation mainly involves an integration over the cross-section of the beam (refer to Stronge and Yu 1993), which leads to a relationship as follows

Table 2.1 Parameters for representative rate-sensitive materials

Material	B (s^{-1})	q	B_q (s^{-1})	Reference
Mild steel	40	5	65	Forrestal and Wesenberg, 1977
Stainless steel	100	10	160	Forrestal and Sagartz, 1978
Titanium (Ti 50A)	120	9	195	Stronge and Yu, 1993
Aluminium 6061-T6	1.70×10^6	4	2.72×10^6	Symonds, 1965
Aluminium 3003-H14	0.27×10^6	8	0.44×10^6	Bodner and Speirs, 1963

$$\frac{M_p^d}{M_p} = 1 + \frac{2q}{2q+1}\left(\frac{h\dot{\kappa}}{2B}\right)^{1/q} = 1 + \left(\frac{h\dot{\kappa}}{2B_q}\right)^{1/q} \qquad \dot{\kappa} > 0 \qquad [2.74]$$

where material constant $B_q \equiv B(1 + 1/2q)^q$ is a combination of constants B and q, which is also listed in Table 2.1. Clearly, Eq. [2.74] is similar to Eq. [2.73] in its form.

Influence of rate dependence on deformation mechanism

The inclusion of a rate-dependent yield moment (i.e. a rate-dependent fully plastic bending moment, such as that given by Eq. [2.68]) in the dynamic models of beams in response to dynamic loading would significantly alter the deformation mode and deformation history.

For example, for a cantilever beam struck by a rigid mass at its tip, the complete rigid-plastic solution (Parkes, 1955) consists of two response phases: Phase I contains a travelling plastic hinge which moves from the tip towards the root of the cantilever and Phase II is a rigid-body rotation of the cantilever about a stationary plastic hinge located at the root. However, if the cantilever is made of rate-dependent material then, as shown by Bodner and Symonds (1960, 1962) and Ting (1964), instead of two separate phases, the cantilever response becomes a single continuous motion. In order to satisfy the equations of motion, the yield condition and flow rule for a rate-dependent cantilever, the plastically deforming region must initially extend over the full length of the cantilever. This is in contrast to the rigid-plastic beam model where plastic deformation is concentrated at a plastic hinge. During the response of a viscoplastic cantilever, the plastically deforming region shrinks and an unloading region extends from the tip where the impact took place. This deformation mode is sketched in Fig. 2.20.

It is observed from this typical example that the inclusion of a rate-dependent yield moment completely changes the kinematics of the system. Therefore, the deformation mechanism and the partitioning of energy

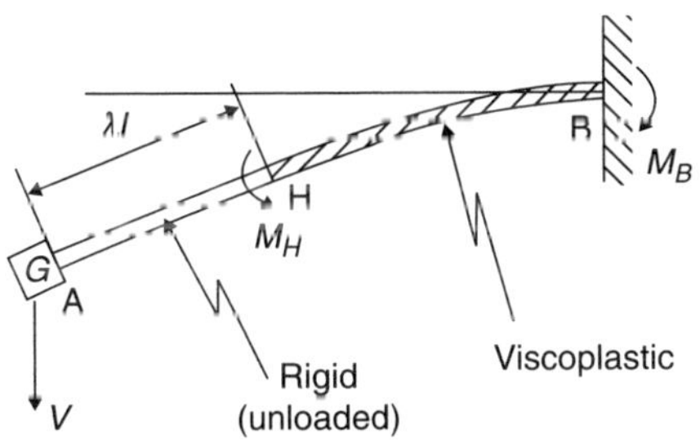

2.20 Dynamic deformation mechanism of a cantilever made of rate-dependent material.

dissipation in the structural component are all notably affected by the rate-dependence of the material.

2.5 Energy method

In analysing the deformation mechanisms and energy-absorption capacity of various structures, the energy method is a very powerful tool and is consequently widely employed. However, a number of issues need to be clarified to enable us to make appropriate and efficient use of the energy method in our target problems.

2.5.1 Energy method used in determining incipient collapse load and mechanism

Determination of incipient collapse load

For an elastic-plastic structure, the energy balance takes the general form

$$E_{in} = W^e + D \tag{2.75}$$

where E_{in}, W^e and D denote the input energy (i.e. the external work done), elastic strain energy stored in the structure and the plastic energy dissipation, respectively. As stated in Section 2.1.6, in the case of $E_{in} >> W^e$, we may neglect W^e in the equation and employ rigid-plastic models for the structural analysis.

As seen in Section 2.2.4 on the kinematically admissible velocity/displacement field and upper bound theorem, upper bounds of collapse load of a structure can be determined by considering the energy balance equation [2.38] or its rate form Eq. [2.37], that is

$$E_{in} = D \tag{2.76}$$

or

$$\dot{E}_{in} = \dot{D} \tag{2.77}$$

Since the external work E_{in} or its rate must be proportional to the applied load P, calculating the energy dissipation D related to a kinematically admissible displacement field will result in an upper bound of the collapse load of the structure, i.e. $P^* \geq P_s$.

As an example, examine a beam of length L subjected to uniformly distributed load q, shown in Fig. 2.21(a). The beam has one end clamped and the other end simply supported. Obviously, two plastic hinges are required to form a collapse mechanism. As shown in Fig. 2.21(b), it is assumed that one of the hinges is located at the clamped end A, while the other is located at a cross-section C, that is ξL $(0 < \xi < 1)$ away from the clamped end.

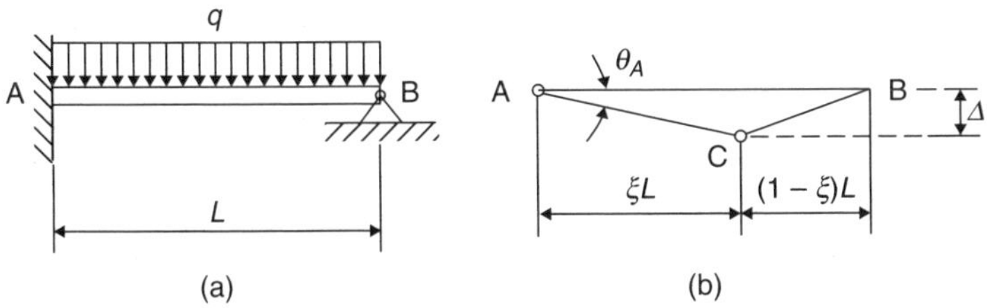

2.21 A beam clamped at one end and simply supported at the other: (a) uniformly distributed load; (b) the incipient collapse mechanism.

Suppose the interior hinge section C has a virtual displacement Δ in the transverse direction, then the energy balance equation [2.76] results in

$$q^* \times \frac{1}{2}\Delta L = M_p(\theta_A + \theta_C) = M_p\left(2\frac{\Delta}{\xi L} + \frac{\Delta}{(1-\xi)L}\right)$$
$$= M_p\frac{\Delta}{L}\times\frac{2-\xi}{\xi(1-\xi)} \qquad [2.78]$$

so that a upper bound of the collapse load is given by

$$q^* = \frac{2-\xi}{\xi(1-\xi)}\times\frac{2M_p}{L^2} \qquad [2.79]$$

To find the lowest upper bound, take $\partial q^*/\partial\xi = 0$, leading to $\xi^2 - 4\xi + 2 = 0$, which has a real root in the interval $0 \le \xi \le 1$ as $\xi = 2 - \sqrt{2} = 0.586$. Accordingly, the lowest upper bound is found as

$$q^* = 2(3+2\sqrt{2})\times\frac{M_p}{L^2} = 11.656\frac{M_p}{L^2} \qquad [2.80]$$

Since the above calculation process only involves infinitesimal velocity/displacement or virtual velocity/displacement, the so-called ***upper bound method*** in limit analysis is indeed an energy method used in determining the incipient collapse load.

Determination of incipient collapse mechanism

It is worth noting that, based on the upper bound theorem, among a number of the kinematically admissible velocity/displacement fields for the structure concerned, the lowest upper bound of the collapse load can help us to identify one of these fields as either the accurate incipient collapse mechanism (if the lowest upper bound is exactly equal to the collapse load) or the best approximation of it. Therefore, although the energy dissipation in a structure is a scalar value only (same as the incipient collapse load), the energy method employed here is capable of accurately or approximately determining the collapse mechanism which would be a complex velocity/displacement (vector) field. Thus, the energy method (i.e. the upper bound method) can be used to determine both the incipient collapse load and incipient collapse mechanism for many structural problems.

For example, examine a frame shown in Fig. 2.22(a), which is subjected to a force F at height H. Figures 2.22(b) and (c) sketch two possible collapse mechanisms. By using the upper bound method, it can be shown that if $H < L/2$, then the 'local' collapse mechanism shown in Fig. 2.22(b) gives a lower upper bound of the collapse load; if $H > L/2$, then the 'global' collapse mechanism shown in Fig. 2.22(c) provides a lower upper bound. Therefore, by comparing the values resulting from different collapse mechanisms, we are able to select one of them as the 'real' or 'appropriate' collapse mechanism for the structure under particular loading.

However, attention also needs to be paid to the fact that sometimes two or even more collapse mechanisms of a solid/structure may be associated with the same collapse load. In the above example (Fig. 2.22), when force F exactly applies at $H = L/2$, two mechanisms shown in Figs 2.22(b) and (c) result in the same collapse load $F_s = 8M_p/L$. Hence, in this particular case, the upper bound theorem does not help in identifying an appropriate collapse mechanism.

This example evidently indicates that, although the incipient collapse load for a solid/structure is unique, the associated collapse mechanism may not be, and consequently, a post-collapse analysis of the structure may be required. If this situation arises with an energy absorber then we may have to trace the respective processes of large deformation following two or more incipient collapse mechanisms.

2.5.2 Energy method used in case of large deformation

Load-carrying capacity of a structure in its large deformation

In the case of large deformation, the analysis is aimed at examining the variation of the load-carrying capacity of a structure during its deformation

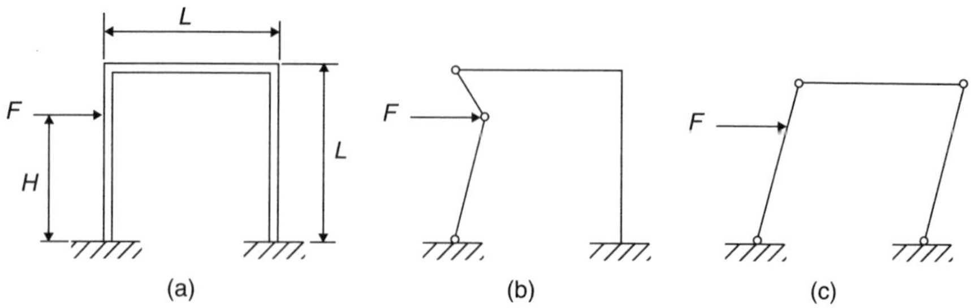

2.22 (a) A rigid-plastic frame subjected to a force F; (b) a 'local' collapse mechanism; (c) a 'global' collapse mechanism.

process, instead of merely determining its incipient collapse load. In other words, now the objective is to find the limit load as a function of the deformation

$$P = P(\Delta) \quad [2.81]$$

where Δ is a representative displacement in the structure during its large deformation process. When load P is a concentrated force, it would be straightforward and convenient to take the displacement at the loading point as Δ, with the same direction as P.

By assuming a large deformation mechanism of the structure under load P, the energy balance directly leads to

$$E_{in} = \int_0^{\Delta_f} P(\Delta)\mathrm{d}\Delta = D = \int_0^{\Delta_f} \mathrm{d}D \quad [2.82]$$

where Δ_f is the final displacement in the process, and the internal energy dissipation D is calculated as an integration of the incremental plastic dissipation dD during the deformation process.

Differentiation of Eq. [2.82] gives

$$P(\Delta) = \mathrm{d}D/\mathrm{d}\Delta \quad [2.83]$$

If there is no unloading during the assumed large deformation process, by neglecting the influence of the plastic deformation history of the structure, the internal energy dissipation D can be calculated from the final configuration of the structure.

To show an example of using the energy method in the case of large deformation of structure, re-examine the beam pressed by a round-headed indenter, as shown in Fig. 2.12(a). With the increase in the deflection, the deformed configuration of the beam is as sketched in Fig. 2.12(b), where angle β serves as a process parameter. The internal energy dissipation is

$$D = 2D_A + D_{BB'} = 2M_p\beta + M_p\frac{1}{R}2R\beta = 4M_p\beta \quad [2.84]$$

where D_A is the dissipation at plastic hinge A, and $D_{BB'}$ is that along the bent arc BB′.

The force P is found from Eq. [2.83] as

$$P(\Delta) = \mathrm{d}D/\mathrm{d}\Delta = 4M_p/(\mathrm{d}\Delta/\mathrm{d}\beta) \quad [2.85]$$

Using the geometric relation between Δ and β given in Eq. [2.46], the following is obtained

$$\frac{P}{P_s} = \frac{\cos^2\beta}{1-(R/L)\sin\beta} \quad [2.86]$$

where P_s = $4M_p/L$ is the initial collapse force as defined in Eq. [2.43]. Together with Eq. [2.47], Eq. [2.86] gives the variation of the load-carrying capacity of the beam during its large deflection.

By comparing Eq. [2.86] with Eq. [2.45] it is found that the load-carrying capacity obtained from the energy method (Eq. [2.86]) is slightly smaller than that obtained from the equilibrium method (Eq. [2.45]). It is observed, therefore, that although the energy method leads to upper bounds for the initial collapse load of the structure whilst the equilibrium leads to lower bounds, this does not promise that the former will provide a higher estimate than the latter for the load-carrying capacity of the same structure during its large deformation.

Other forms of energy dissipation

For transversely loaded beams and plates, the incipient collapse mechanisms involve only bending deformation. However, during their large deformation, other forms of energy dissipation may become important or even dominant, so that the internal energy dissipation may be typically written as

$$D = D_b + D_m + D_{fri} + D_{frac} + \ldots \qquad [2.87]$$

where D_b, D_m, D_{fri} and D_{frac} denote the energy dissipation by bending, membrane deformation, friction and fracture, respectively.

Still taking the beam shown in Fig. 2.12(a) as an example, if a blank-holding force F is applied vertically at the clamped end, then when the beam slides along the clamp the friction force is μF with μ being the coefficient of friction between the beam and the clamp. Since the sliding distance at each end of the beam is (refer to see Eq. [2.49] for the geometric relation)

$$\delta L = (1/\cos\beta - 1)L - (\tan\beta - \beta)R \qquad [2.88]$$

the work done by the friction (i.e. the energy dissipation due to friction) is

$$D_{fri} = 2\mu F \times \delta L \qquad [2.89]$$

If the beam's ends are axially constrained, as shown in Fig. 2.13, then the axial strain and axial force are induced by the large deflection of the beam, as given by Eqs [2.49] and [2.51], respectively. Accordingly, the right-hand side of the expression of energy dissipations (Eq. [2.87]) should have one more non-zero term

$$D_m = 2N_A \times \varepsilon L = 8N_p \varepsilon \Delta \qquad [2.90]$$

where the axial strain is calculated from Eq. [2.49] as a function of parameter β. Obviously, this term proportionally increases with deflection Δ, so when the deflection is large enough, the axial (membrane) deformation will dominate the energy dissipation in the beam.

Selecting deformation character by minimising the external load

When a cylindrical tube subjected to an axial load P collapses into an axisymmetric folding pattern (see Section 6.1), its large deformation load-carrying capacity varies with the displacement periodically, but the energy dissipation in a load cycle can be written in the following form

$$D = D_b + D_m = A + B\lambda^2 \quad [2.91]$$

where λ is the half length of the fold and A and B are coefficients depending on material and geometry. On the other hand, the work done by the external load is

$$E_{in} = CP\lambda \quad [2.92]$$

with C being another coefficient. Therefore, equating Eq. [2.91] and Eq. [2.92] gives

$$P = \frac{A'}{\lambda} + \beta'\lambda \quad [2.93]$$

with $A' = A/C$ and $B' = B/C$. Since P given in Eq. [2.93] is an upper bound, minimising it in terms of λ will result in

$$\lambda = \sqrt{A'B'} \quad [2.94]$$

This example shows how the character length in a large deformation mechanism can be determined by minimising the required load.

To take another example, when a cylindrical tube with longitudinal pre-cracks is compressed in the axial direction, the large deformation mechanism involves both bending energy and fracture energy. The latter is proportional to the number of fractures which occur, whilst the former is inversely proportional to this number. Again, an optimum number for the fractures occurring in the tube can be obtained from the energy balance argument with the external load minimised.

2.5.3 Energy method used in case of dynamic deformation

Kinetic energy in energy balance

In the case of dynamic deformation of structures in response to a pulse loading (e.g. an external force as a prescribed function of time), the kinetic energy of the structure will be involved in the energy balance, so that the equation of energy balance in an elastic-plastic system, Eq. [2.75], should include the kinetic energy of the structure, K; that is

$$E_{in} = D + W^e + K \quad [2.95]$$

where E_{in} denotes the work done by the force pulse $F(t)$. It should be noted that, although the pulse $F(t)$ is prescribed, the relevant displacement is not prescribed, so that the value of E_{in} is known only when the dynamic response of the structure is solved. Therefore, unlike that in the quasi-static deformation, the energy equation [2.95] in the dynamic case will not lead directly to an explicit expression on the relationship between force F and displacement Δ.

In fact, after plastic dissipation has been completed, the kinetic energy K still varies with time, because the remaining energy $(E_{in} - D)$ still periodically transfers its form between the kinetic energy K and the elastic strain energy W^e during the final elastic vibration phase of the system. Only when $(K + W^e) << E_{in}$, can the rigid-plastic idealisation be adopted and Eq. [2.76] used to assess the final deformation of the system after a dynamic (pulse) loading.

Upper bound of final displacement of structures under impulsive loading

Impulsive loading refers to an intense dynamic load, which only applies to a structure in an infinitesimal time period (at $t = 0$) but results in a finite impulse. This kind of dynamic load gives the whole structure or a part of it an initial velocity distribution at $t = 0$, but no force pulse applies afterwards.

Martin (1964) proved that an upper bound to the final displacement of an impulsively loaded structure can be given by (see Stronge and Yu, 1993, Section 2.4.5)

$$\Delta_f \le \Delta_f^+ = \frac{K_o}{P_s} \qquad [2.96]$$

where K_o is the initial kinetic energy of the structure due to the impulsive loading, and P_s denotes the quasi-static collapse force of the structure, while P_s and Δ_f should be at the same point and along the same direction.

For example, for the previous example shown in Fig. 2.12(a), suppose that instead of a quasi-static loading by the round-headed indenter, the beam is impinged by a round-headed projectile of mass G and initial velocity V_o, then by applying Eq. [2.96] the final deflection of the beam can be estimated by

$$\Delta_f \le \Delta_f^+ = \frac{K_o}{P_s} = \frac{GV_o^2}{2P_s} = \frac{GV_o^2 L}{8M_p} \qquad [2.97]$$

where the incipient collapse load of the beam, $P_s = 4M_p/L$ given in Eq. [2.43], is employed.

It should be noted that the above upper bound on the final displacement is based on a small deformation assumption. It may not provide an upper bound for the final displacement if a structure experiences a large deformation, when the geometric change and membrane forces become important. In fact, it has been demonstrated in Section 2.3.2 that the load-carrying capacity of the beam varies significantly with the deflection of the beam especially when the ends of the beam are axially constrained; thus the final deflection estimated by Eq. [2.97] would have a big error.

Initial loss in kinetic energy during collision

It has been pointed out in Section 2.4.2 that when two rigid-plastic bodies/structures collide with each other, there is an initial loss in the kinetic energy, denoted as K_{loss}. In this case, Eq. [2.95] will take the form

$$E_{in} = K_o - K_{loss} = D \qquad [2.98]$$

where the elastic strain energy is assumed to be negligible in the system, because both bodies/structures involved are taken as rigid-plastic. When applying the upper bound of the final displacement, the initial energy of the system, K_o in Eq. [2.95], should be replaced by $(K_o - K_{loss})$.

The above initial loss in the kinetic energy can be avoided, if elastic-plastic structural models are used, or if an elastic-plastic contact spring is introduced between two rigid-plastic bodies/structures. A further discussion is given in Section 7.1.

3 Dimensional analysis and experimental techniques

Dimensional analysis occupies an important place in engineering analysis. It forms the basis of small-scale model tests. This chapter introduces the concept and method of dimensional analysis with reference to plastic collapse of structures, discusses the similitude requirements for model testing and presents several experimental techniques for studying the energy absorption of structures.

3.1 Dimensional analysis

3.1.1 Physical variables, fundamental dimensions and dimensional homogeneity

Any physical measurement or variable must have two characteristics: a numerical or quantitative value and a qualitative unit. For example, when we say the yield stress of a steel is 250×10^6 Pa, the number 250×10^6 specifies the quantity while the unit 'Pa' (Pascals) is qualitative. These qualitative units are expressed in terms of several ***fundamental dimensions***. Here 'Pa' or N/m^2, which is force over length squared, consists of two fundamental dimensions: force F and length L. Similarly, the unit of 'energy' has two fundamental dimensions: also force F and length L. Generally in mechanical problems, there are three fundamental dimensions: length L, force F (or mass M) and time T. Other fundamental dimensions such as temperature or electric charge will be present in problems involving thermal or electrical effects.

Dimensional homogeneity (or consistency) means that any properly established relationship must be true regardless of the choice of units for the ***physical variables***. Any constant should be purely numerical without a unit. For example, the collapse load of a fully clamped beam with a concentrated load acting at mid-span is given by Eq. [2.43] as $P_s = 4M_p/L$. Here the constant 4 does not have any units and the equation is always true whether the force and length are measured in Newtons and metres, kilo-

Newtons and kilometres, or even a set of consistent imperial units. On the other hand, Eqs [8.5] and [8.6] are not dimensionally consistent: they are true only for the given set of units for thickness, stress and energy – involving the two fundamental dimensions of force and length.

The concepts of physical variables, fundamental dimensions and dimensional homogeneity form the basis of dimensional analysis, which we discuss in the next section.

3.1.2 Dimensional analysis

First, we will consider an example to illustrate a typical dimensional analysis process. Suppose a thin ring of radius R is subjected to two equal, opposite inward point loads P as shown in Fig. 3.1. The cross-section of the ring is rectangular with thickness h and width B. We want to find the initial collapse load P_o for a given ring material. This problem can be easily studied using the plasticity theory presented in Chapter 2 and indeed we will do this in Chapter 4. However, for our present purpose, we will pretend that the analytical solutions are not known; we only have experimental results for tubes of various dimensions and materials.

From our insight into the physics of this problem, we recognise that yield stress Y may be chosen as the only material parameter; Young's modulus is irrelevant as we are dealing with a plastic collapse problem. The geometrical parameters are radius R, thickness h and width B. Hence we may argue that the collapse load P_o is a function of the above mentioned physical variables which describe geometric and material properties of the structures. We can write the following equation

$$P_o = F_1(Y, R, B, h) \qquad [3.1]$$

or

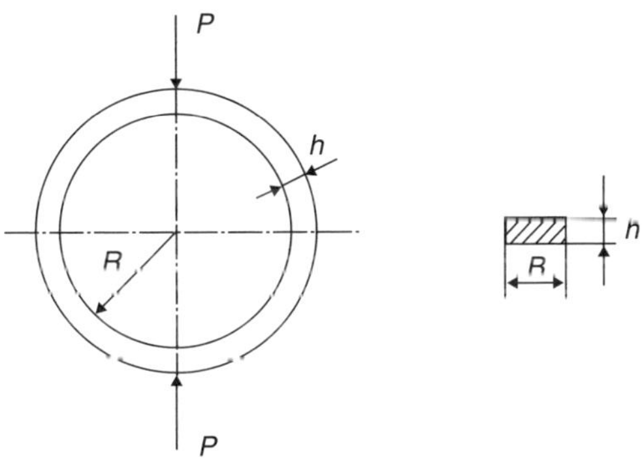

3.1 A circular ring under two opposite point loads. The cross-section of the ring is rectangular ($h \times B$).

$$F_2(P_o, Y, R, B, h) = 0 \quad [3.2]$$

where F_1 and F_2 are functions. There are a total of five physical variables in Eq. [3.2]. From the requirement for dimensional homogeneity, we deduce that Eq. [3.2] can be expressed in dimensionless form. To choose suitable ***dimensionless groups*** (usually called ***π groups***), we follow the simple procedure below.

The two fundamental dimensions applicable here are force F and length L. We obtain a particular dimensionless group by involving each physical variable raised to its own specific power. Hence, for this example a dimensionless group DG is of the form

$$DG = Y^a R^b B^c h^d P_o^e \quad [3.3]$$

The values of powers a, b, c, d and e are obtained from the fact that DG is dimensionless. Substituting the units for each physical variable in terms of the two fundamental dimensions, e.g. P_o as F and Y as F/L^2, we have

$$DG = \left(\frac{F}{L^2}\right)^a (L)^b (L)^c (L)^d (F)^e \quad [3.4]$$

or

$$DG = (F)^{a+e} (L)^{-2a+b+c+d} \quad [3.5]$$

Because DG is dimensionless, we have

$$\begin{aligned} a + e &= 0 \\ -2a + b + c + d &= 0 \end{aligned} \quad [3.6]$$

The above two equations have five unknowns whose values cannot be determined uniquely. If, however, we assume the values for three unknowns, we can then solve the other two. Let $a = -1$, $b = c = 0$, then $d = -2$ and $e = 1$. Hence, we could choose as our first dimensionless group

$$\pi_1 = \frac{P_o}{Yh^2} \quad [3.7]$$

Similarly, if we let $a = 0$, $b = 1$, $c = -1$, then $d = 0$, $e = 0$, i.e.

$$\pi_2 = \frac{R}{B} \quad [3.8]$$

Finally, if $a = 0$, $b = 0$, $c = 1$, then $e = 0$, $d = -1$, i.e.

$$\pi_3 = \frac{B}{h} \quad [3.9]$$

In principle, we may choose a number of other dimensionless groups, but we argue for the moment that there are only three independent dimensionless groups. Other groups can be obtained by various combinations of the above three groups. For example, if we choose $a = 0$, $b = 1$, $c = 0$, then $d = -1$, and $e = 0$. The corresponding dimensionless group is (R/h). But this is in effect $(\pi_2 \cdot \pi_3)$, and is therefore dependent upon two of the three previous groups, Eqs [3.7–3.9]. Equation [3.2] can now be written in dimensionless form

$$f_1(\pi_1, \pi_2, \pi_3) = 0 \qquad [3.10]$$

or

$$f_1\left(\frac{P_o}{Yh^2}, \frac{R}{B}, \frac{B}{h}\right) = 0 \qquad [3.11]$$

The dimensional analysis is now complete; we have successfully reduced the number of variables from five to three. The functional form of f_1 will need to be determined from experiments. We could, for example, plot π_1 against π_2 for a given value of π_3. For this particular problem, it turns out from experiments (which will be presented later) that π_3 has little effect on π_1 and may be discarded. A simple relationship between the remaining two dimensionless groups can be established as follows

$$P_o \propto Yh^2 f_2\left(\frac{R}{B}\right) \qquad [3.12]$$

Buckingham Pi theorem

The above example demonstrates two fundamental points. First, a dimensionally homogeneous equation of five physical variables, Eq. [3.2], can be written as an equation with a set of three dimensionless groups, Eq. [3.11]. Second, the number of independent dimensionless groups is only three, when five physical variables are involved with two fundamental dimensions. In general, the ***Buckingham theorem*** (Buckingham, 1914) states that a dimensionally homogeneous equation with a number of physical variables can be reduced to an equation with several dimensionless groups. The number of independent dimensionless groups is equal to the difference between the number of physical variables and the number of the fundamental dimensions. This is reflected in our example above where the number of independent dimensionless groups is three (= 5 – 2).

Remarks on the dimensional analysis

At this stage, we may highlight several points in relation to the choice of dimensionless groups and the use of dimensional analysis.

(1) One needs to have a sound insight into the nature of the engineering problem under study. All the physical variables in question must be included; but irrelevant ones should be discarded. An intelligent choice of the physical variables will simplify the problem. The following example illustrates this point.
(2) The dimensionless groups must be independent.
(3) In our example above, we have chosen force and length as the two fundamental dimensions. This is convenient in static loadings. Alternatively, mass, length and time could be chosen as the three fundamental dimensions, especially in dynamic cases. The final result will be the same for both the methods, because force can be related to mass through Newton's second law (which involves acceleration, or length/time^2).
(4) The exact form for each dimensionless group is not unique. It is more a matter of personal choice. Dimensionless groups can be obtained following the procedure explained in the above example, or indeed by inspection. Other, more rigorous approaches are available (Bridgman, 1922; Gibbings, 1982; Sedov, 1993; Harris and Sabnis, 1999).
(5) There are, nevertheless, some common practices used with regard to the dimensionless groups: (a) quite often, several different forms may be tried before those producing the simplest results are finally selected; (b) it is preferable to choose a dimensionless group that retains a certain physical meaning. Examples are the aspect ratio of a rectangular cross-section and the relative thickness of a circular tube section (h/R). (c) Mathematical formulation of a much simplified, though probably crude, model may give us a lot of clues to the way each physical variable should come into a dimensionless group.
(6) Dimensional analysis is particularly useful when an engineering problem is so complicated that it is not possible to obtain exact solutions (which could be due to a lack of information on the exact mechanics involved). Dimensional analysis enables us to reduce the total number of variables, and to design an experimental programme so as to cover a wide range of dimensionless parameters instead of merely certain physical variables. For example, to study the behaviour of a tube, we should vary, as widely as possible, the value of h/R, rather than just h or R. Two tests with the same value of h/R, though with different values of h and R, would not reveal much extra information.

Importance of physical insight – an example

In our earlier example for the collapse load of a circular ring under two point loads (Fig. 3.1), we argued that five physical variables are involved. In the beginning, the only knowledge we had was that the collapse load was

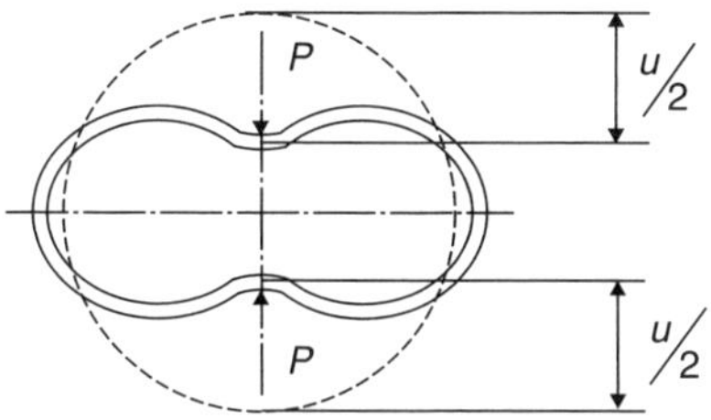

3.2 Collapse of an initially circular ring.

independent of the elastic behaviour of the material and hence Young's modulus was not relevant. Now suppose we have a better understanding of this problem: the collapse of this tube is largely governed by bending of the tube wall without any stretching. Therefore, the effect of thickness h and the material property Y should be present in the form of bending resistance, which in this case is the fully plastic bending moment per unit width, M_o (= $Yh^2/4$). Consequently, we can now argue that the initial collapse load P_o is only a function of M_o, the tube width B and radius R, i.e.

$$F_3(P_o, M_o, R, B) = 0 \quad [3.13]$$

Now there are only four variables and we may choose the following two dimensionless groups

$$f_3\left(\frac{P_o}{M_o}, \frac{R}{B}\right) = 0 \quad [3.14]$$

It is straightforward to show that this equation is in effect the same as Eq. [3.12]. An elegant choice of physical variables based on physical insight into the problem greatly simplifies the analysis of experimental results.

A further example: energy absorption in laterally loaded circular rings

The previous example deals with the initial collapse load and hence displacement is not involved. Now we are interested in energy absorption during the whole crushing process – the subject of this book. Take the same ring and loading as in the previous example Fig. 3.2. We need to select a parameter to describe the crushing. Let the total displacement between the two loading points be u. The work done by the two loads, W, is equal to the energy dissipated by plastic bending of the tube. Hence, considering all physical variables

$$F_4(W, M_o, R, B, u) = 0 \quad [3.15]$$

The number of independent dimensionless groups involved is again three (= 5 – 2) and we could choose the following dimensionless groups

$$f_4\left(\frac{W}{M_oB}, \frac{B}{R}, \frac{u}{R}\right) = 0 \quad [3.16]$$

The above equation suggests that we should plot $W/(M_oB)$ against u/R for various values of B/R. Figure 3.3(b) is such a plot for a number of experiments of which some 'raw' test data are shown in Fig. 3.3(a). Again, the curves indicate that B/R has little effect and therefore this parameter can

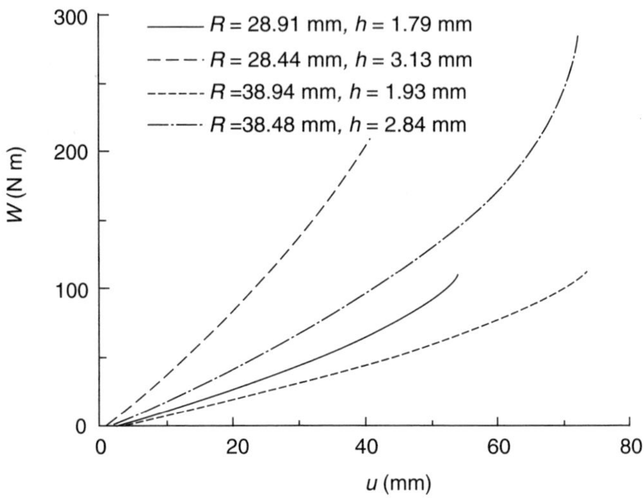

(a)

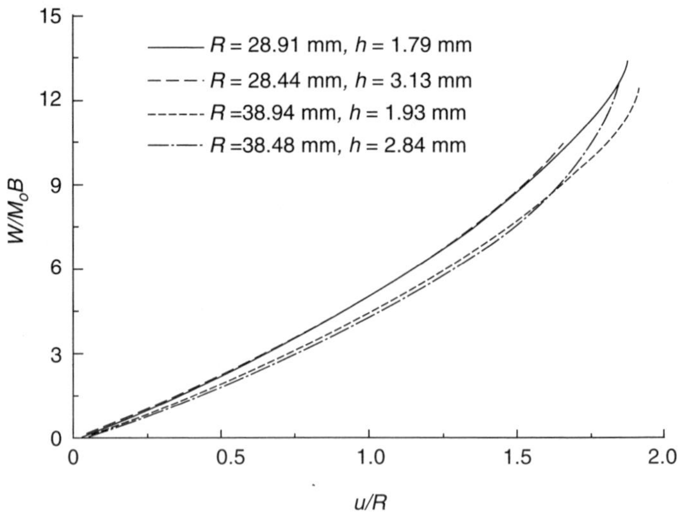

(b)

3.3 (a) Energy–deflection curves for tubes of different values of thickness; (b) non-dimensionalised energy–deflection curves.

be discarded as irrelevant. An analytical study of this problem to be presented in Chapter 4 also demonstrates this.

3.2 Small-scale structural models

3.2.1 Similarity requirements

The above dimensional analysis demonstrates that a relationship involving physical variables, such as Eq. [3.2], can be conveniently simplified to another relationship in terms of a reduced number of dimensionless groups, e.g. Eq. [3.10]. Now consider two physical structures which have the same geometry, but with different dimensions and subjected to loadings of different magnitudes. If the value of each dimensionless group is exactly the same for both structures, we then say that the two structures and loadings are similar. The smaller sized structure may be conventionally called a ***small-scale model*** of the other larger structure.

Considering the circular ring in our example, we could obtain a small-scale model by applying the same ***scaling factor***, S_l, for all the linear dimensions of the original structure (called ***prototype***). Thus

$$R_m = S_l R_p, \quad h_m = S_l h_p, \quad B_m = S_l B_p \qquad [3.17]$$

These two structures are sketched in Fig. 3.4 with a scale factor $S_l = 0.5$. Immediately, we can see that the values of π_2 (Eq. [3.8]) and π_3 (Eq. [3.9]) are the same for both the prototype and the model, i.e.

$$\pi_{2m} = \pi_{2p}, \quad \pi_{3m} = \pi_{3p} \qquad [3.18]$$

From Eq. [3.10], we conclude that

$$\pi_{1m} = \pi_{1p} \qquad [3.19]$$

Expressing this in actual physical variables

$$\left(\frac{P_o}{Yh^2}\right)_m = \left(\frac{P_o}{Yh^2}\right)_p \qquad [3.20]$$

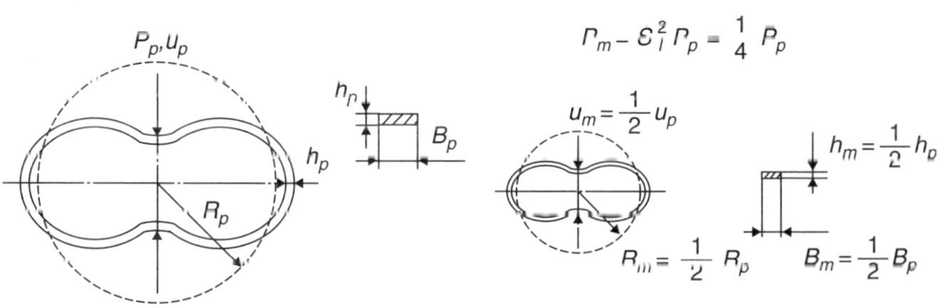

3.4 Sketch of a prototype and a half sized model ($S_l = 1/2$).

Therefore

$$P_{om} = \left(\frac{Y_m h_m^{\;2}}{Y_p h_p^{\;2}} \right) P_{op} = S_m S_l^{\;2} P_{op} \qquad [3.21]$$

Here $S_m = Y_m/Y_p$ is the scale factor for material properties of the model and prototype. If the model is made of the same material as the prototype, then $S_m = 1$.

The above argument can also be applied to the example of energy absorption corresponding to a given deflection. Referring to Eq. [3.16], geometric similarity requires that the value of the third dimensionless group, u/R, be the same for both the model and prototype (in addition to those specified in Eq. [3.17]). Thus

$$u_m = S_l u_p \qquad [3.22]$$

Because the value of B/R is the same for both the model and prototype, we conclude from Eq. [3.16] that this must also be true for the last remaining dimensionless group, $W/(M_oB)$. Hence

$$\left(\frac{W}{M_o B} \right)_m = \left(\frac{W}{M_o B} \right)_p$$

or

$$W_m = \left(\frac{M_{om} B_m}{M_{op} B_p} \right) W_p = S_m S_l^{\;3} W_p \qquad [3.23]$$

Equations [3.21] and [3.23] indicate that when the model and prototype are made of the same material ($S_m = 1$), the two loads are related by S_l^2 while the energy absorbed is proportional to S_l^3. In plain words, if the structures are geometrically similar, the strain ε is the same. The characteristic area of a structure is proportional to S_l^2. When the material is the same (with the same yield stress Y), the load (which is equal to stress multiplied by area) is then proportional to area only, or S_l^2. Similarly, energy absorbed is proportional to the deforming volume multiplied by the energy density ($Y\varepsilon$), which is the same for the model and prototype. Therefore, energy is proportional to S_l^3.

The above process can be applied to various other physical variables of interest. In Table 3.1 we summarise the quantities likely to be encountered in the study of energy absorption. In principle, we should ideally keep the values of all dimensionless groups the same for both model and prototype. However, in practice, it is sometimes very difficult or even impossible to scale certain quantities according to Table 3.1. We next discuss several such cases.

Table 3.1 Summary of scale factors between model and prototype

Physical variable	Dimension	Scale factor in general cases	Scale factor (same material; gravity insignificant)
Linear dimension	L	S_l	S_l
Area	L^2	S_l^2	S_l^2
Volume	L^3	S_l^3	S_l^3
Material stress–strain parameters (*E*, *Y*, . . .)	FL^{-2}	S_m	1
Material density	FT^2L^{-4}	S_ρ	1
Mass	FT^2L^{-1}	$S_\rho S_l^3$	S_l^3
Load	F	$S_m S_l^2$	S_l^2
Pressure	FL^{-2}	S_m	1
Stress	FL^{-2}	S_m	1
Strain	–	1	1
Displacement	L	S_l	S_l
Elastic and plastic energy	FL	$S_m S_l^3$	S_l^3
Elastic wave speed	LT^{-1}	$S_m^{1/2} S_\rho^{-1/2}$	1
Velocity	LT^{-1}	1	1
Angular velocity	T^{-1}	S_l^{-1}	S_l^{-1}
Time (impact duration or time elapsed)	T	S_l	S_l
Acceleration	LT^{-2}	S_l^{1}	S_l^{-1}
Acceleration due to gravity (*g*)	LT^{-2}	1	Neglected
Inertia force	F	$S_\rho S_l^2$	S_l^2
Momentum	FT	$S_m S_l^3$	S_l^3
Kinetic energy	FL	$S_\rho S_l^3$	S_l^3
Strain-rate	T^{-1}	S_l^{-1}	S_l^{-1}
Elastic fracture surface energy	FL^{-1}	$S_f S_l^2$	S_l^2
Ductile tearing energy	FL^{-1}	unclear	unclear
Material's microstructural dimension	L	1	1

3.2.2 Quantities difficult to scale exactly

Gravity load

Table 3.1 specifies that the load applied to a model should be proportional to S_l^2 in order to maintain complete similarity. Gravity load is given by mass multiplied by gravitational acceleration, *g*. The mass scales as S_l^3, which means that gravity load will in fact also scale as S_l^3 (because *g* is a constant), rather than the required S_l^2. We cannot therefore strictly satisfy the similitude requirements. Fortunately, in practice, gravity loads are usually very small in comparison with other loads and they can be neglected. If, however, in a particular case gravity loads are significant and need to be considered in scaling, adopting a different material for the model may overcome this difficulty; however, the model density must ensure that the overall scaling factor $S_\rho S_l^3$ is equal to $S_m S_l^2$. This may again present some practical problems. From this argument, if two materials have similar mechanical properties ($S_m \approx 1$), then a small-scale model will need to have a higher value of density than its prototype.

Strain-rate effect

A characteristic strain-rate may be expressed as the ratio between the impact velocity and a representative length of a structure. It is therefore easy to see that when the impact velocity is the same, a small-scale model (which has a smaller linear dimension) will experience a higher strain-rate than a prototype; this is also indicated in Table 3.1. If the material is rate sensitive, the strain-rate will in turn affect the material's properties, such as the yield stress Y and ultimate stress as discussed in Section 2.4. As a result, even if the materials are the same for the model and prototype, their mechanical properties will be different. A small-scale model will have an apparently higher yield stress, leading to a smaller deflection than predicted by the similarity law. This can be illustrated by the following example.

Suppose a mild steel bar has an idealised elastic, perfectly plastic stress–strain behaviour, with a static yield stress of 250 MN/m^2. Under high strain-rate, the dynamic yielding stress can be obtained using the Cowper–Symonds relationship (Eq. [2.73]). Let the length of the bar be 400 mm and the cross-section be 50 mm^2. The bar is under uniaxial tension with uniform deformation. Now, if the external work done on the bar is 2 kJ, the energy density is then 2 kJ/(0.4 × 0.05 × 0.05) m^3 = 2 × 10^6 J/m^3. The plastic strain for this prototype is then $\varepsilon_p = 2 \times 10^6$ J/m^3/250 MN/m^2 = 0.008.

Now suppose we take a scale factor of $S_l = 0.1$. The energy input should be $S_l^3 \times 2$ kJ = 2 J. Because the energy density remains the same, the strain for the model is $\varepsilon_m = \varepsilon_p = 0.008$. Apply the load at a rate of 200 mm/s for both the model and prototype according to Table 3.1. The strain-rate for the prototype is then 0.5 s^{-1} and for the model 5 s^{-1}. The corresponding dynamic yielding stresses based on Eq. [2.73] with $B = 40.4$ s^{-1} and $q = 5$, are 354 MN/m^2 and 415 MN/m^2, respectively. These lead to a new plastic strain $\varepsilon_p = 0.00565$ and $\varepsilon_m = 0.00482$. This demonstrates that, due to the strain rate effect, the model experiences a strain of 85 % that of its prototype. There is a potential danger of underestimating the deflection of a prototype based on the small-scale tests, if we assume complete similarity (and hence equal strain among all the models and the prototype).

Deformation with fracture

Structural plastic deformation and failure are often accompanied by fracture and ductile tearing. With fracture of brittle materials, the fracture energy is usually related to the newly created surface area by a material constant, hence $W_f \propto S_l^2$. However, the plastic energy is proportional to volume, hence $W_p \propto S_l^3$. Total energy is then composed of these two parts; there is no simple scaling law to relate this total energy to the scale factor S_l. Atkins elaborated this point with practical examples (1988). There are two extremes: if fracture energy is dominant, then total energy will be approximately proportional to S_l^2. Alternatively, if plastic deformation is the major energy dissipation mechanism, then total energy may be assumed to be proportional to S_l^3.

For ductile tearing there is no simple material constant to characterise the tearing energy (see Chapter 8 for details). As a result, simple scaling law does not appear to be applicable. However, if the tearing process is dominated by plastic deformation at the crack tip zone, then yield stress may be assumed to be the only significant material constant and the total energy is related to S_l^3. The last example in Chapter 8 demonstrates this point with a metal plate cut by a wedge.

Remarks on lack of similarity

In addition to the three major quantities mentioned above, which do not obey elementary scaling, other sources exist that may lead to a departure from complete similarity. These include accidental overlook of certain important variables, a deliberate rejection of a variable incorrectly considered non-critical, normal and frictional forces related to gravity, the material's microstructures and certain highly localised deformation with heat generation.

The departure from elementary scaling law is usually referred to as the '***size effect***'. Fundamentally, such departure means that the values of all the dimensionless groups are not kept the same for both the model and prototype. In principle, we could remedy this problem by studying the effect on the others of changing one or two specific dimensionless groups, i.e. by exploring the functional dependence of dimensionless groups. We can then deduce the significance of non-similarity of certain variables on other parameters of interest. However, in practice this may be difficult to do when the number of dimensionless groups is large.

A small scale model is used in a case study of energy absorption of a roadside guardrail in Section 12.4. Structural scaling has also been employed in other investigations of impact deformation of metallic structures (Duffey, 1971; Booth *et al.*, 1983).

3.3 Experimental techniques

3.3.1 Universal testing machine

For a low velocity impact, plastic deformation modes are usually very similar to those produced with quasi-static loading. Generally, it is convenient to start the investigation by conducting quasi-static tests first, for two reasons. First, the experimental setup for quasi-static loading is simpler than that for impact tests. Second, a quasi-static test enables us to observe, with relative ease, the detailed deformation history.

It is well known that plastic energy is dependent upon loading and deformation history, rather than just final deformation as is the case for elastic deformation. Hence, continuous monitoring of load, displacement and strain at various characteristic positions, as well as observation of the deforming structures, should lead to a much better understanding of the possible mechanisms of plastic deformation. It would be much more difficult to obtain such information for a dynamic test.

Quasi-static tension or compression tests can conveniently be performed using a standard universal testing machine. The crosshead speed is usually set in the range 3–5 mm per minute (0.05–0.08 mm/s). For a structure with a characteristic length of 100 mm, a speed of 3 mm/min produces (approximately) a strain rate of $5 \times 10^{-4}\,s^{-1}$, which can be regarded as static loading. Controlled hydraulic servos can produce higher ram speeds up to, say, 800 mm/s. This would correspond to a strain rate of $16\,s^{-1}$ for a 50 mm long specimen and hence studies of strain-rate effect in the range between 10^{-4} and $10\,s^{-1}$ can be conducted in this way. The MTS universal testing machine at Swinburne University of Technology is such an example. Load-displacement curves and other variables such as strain can be recorded using a personal computer.

3.3.2 Drop hammer, sled and pendulum

Impact tests can be conducted by means of a ***drop hammer***, a ***pendulum*** or an inclined ***sled***. In the case of a drop hammer, a mass is lifted to a certain height and then released to cause an impact upon a structure placed at the base of the rig. The free fall maximum velocity achievable is governed by the total height of the drop hammer. (Friction in the vertical guiderail is usually minimised with roller bearings.) Other means such as compressed air or springs can be employed at the top of the drop hammer in order to accelerate the mass to a higher impact velocity. Typical instrumentation used includes an accelerator attached to the impact mass, a device for measuring the velocity just before impact (usually by measuring the time interval to travel a known distance), a displacement transducer to record the

3.5 Photograph of a drop hammer (Dynatup) at Hong Kong University of Science and Technology.

movement of the impact mass and a dynamic load cell usually placed underneath the structure to be tested. Figure 3.5 shows the drop hammer (Dynatup 8250) at Hong Kong University of Science and Technology. It is about 1.5 m high and has an adjustable impact mass of up to 44.89 kg. When gravity driven, the impact velocity is in the range 0.61–3.66 m/s and it increases to 3.66–13.41 m/s when pneumatically assisted.

Figure 3.6 shows another method for conducting impact tests, a sled testing facility at Cranfield Impact Centre. This sled facility consists of a

3.6 Sled testing facility at Cranfield Impact Centre (Courtesy of Dr M. Sadeghi).

ramp, which is inclined at 11° to the horizontal, upon which a trolley floats on four air pads giving frictionless motion and repeatable impact speeds. Low impact speeds are achieved solely by using gravity, but for higher speeds elastic cords are utilised. The rig has a minimum mass of 780 kg and a maximum mass of 2000 kg, with a maximum speed of 13.5 m/s and maximum energy of 125 kJ.

Pendulums can also be used to apply impact loads. The swinging arms need to be designed so that the impact face only, translates. They should be long enough to minimise the radial movement at the impact face. One such rig shown in Fig. 3.7 is available at Cranfield. The large pendulum facility consists of two side supports with a pendulum hung between them. The two arms of the pendulum ensure a parallelogram action that constrains the impact face to remain vertical at all times. The rig has a minimum mass of 467 kg and a maximum mass of 1000 kg, with a maximum speed of 10 m/s and maximum energy of 50 kJ.

In all the test methods described, the velocity of the striker after impact is not a constant, but varies with the displacement until the striker comes completely to rest. The strain-rate is, therefore, not constant. However, these test methods simulate real impact events reasonably well.

3.3.3 Split Hopkinson pressure bar

For strain rates in the range of 10^2–$10^4\,s^{-1}$, material constitutive relationships can be obtained by using a ***split Hopkinson pressure bar (SHPB)***

3.7 Pendulum testing facility at Cranfield Impact Centre (Courtesy of Dr M. Sadeghi).

(Hopkinson, 1914; Kolsky, 1953). Figure 3.8 shows a sketch of a compressive split Hopkinson, a typical record of signals from the strain gauges and the resulting stress–strain curve.

The specimen is placed in between incident and transmitter bars. An elastic pressure pulse is produced in the incident bar by impact from a striking bar. At the interface between the incident bar and the specimen (interface 1), this elastic stress wave is partially reflected and partially transmitted to the short specimen, thus deforming the specimen plastically. Similarly, at the interface between the specimen and the transmitter bar (interface 2), the stress wave is partially reflected and partially transmitted. The incident stress needs to be of sufficient duration. The material's stress–strain behaviour at high strain rates can then be deduced from the measurement of strain–time history at the incident and transmitter bars, respectively. The corresponding theory is described below.

Stress propagation in this setup is assumed to be one-dimensional. For a specimen of length L, the strain rate within the specimen is

$$\dot{\varepsilon} = \frac{d\varepsilon}{dt} = \frac{v_1(t) - v_2(t)}{L} \qquad [3.24]$$

where $v_1(t)$ and $v_2(t)$ are particle velocities at interfaces 1 and 2, respectively.

From Eq. [2.54] and Hooke's law, particle velocity and strain are related by

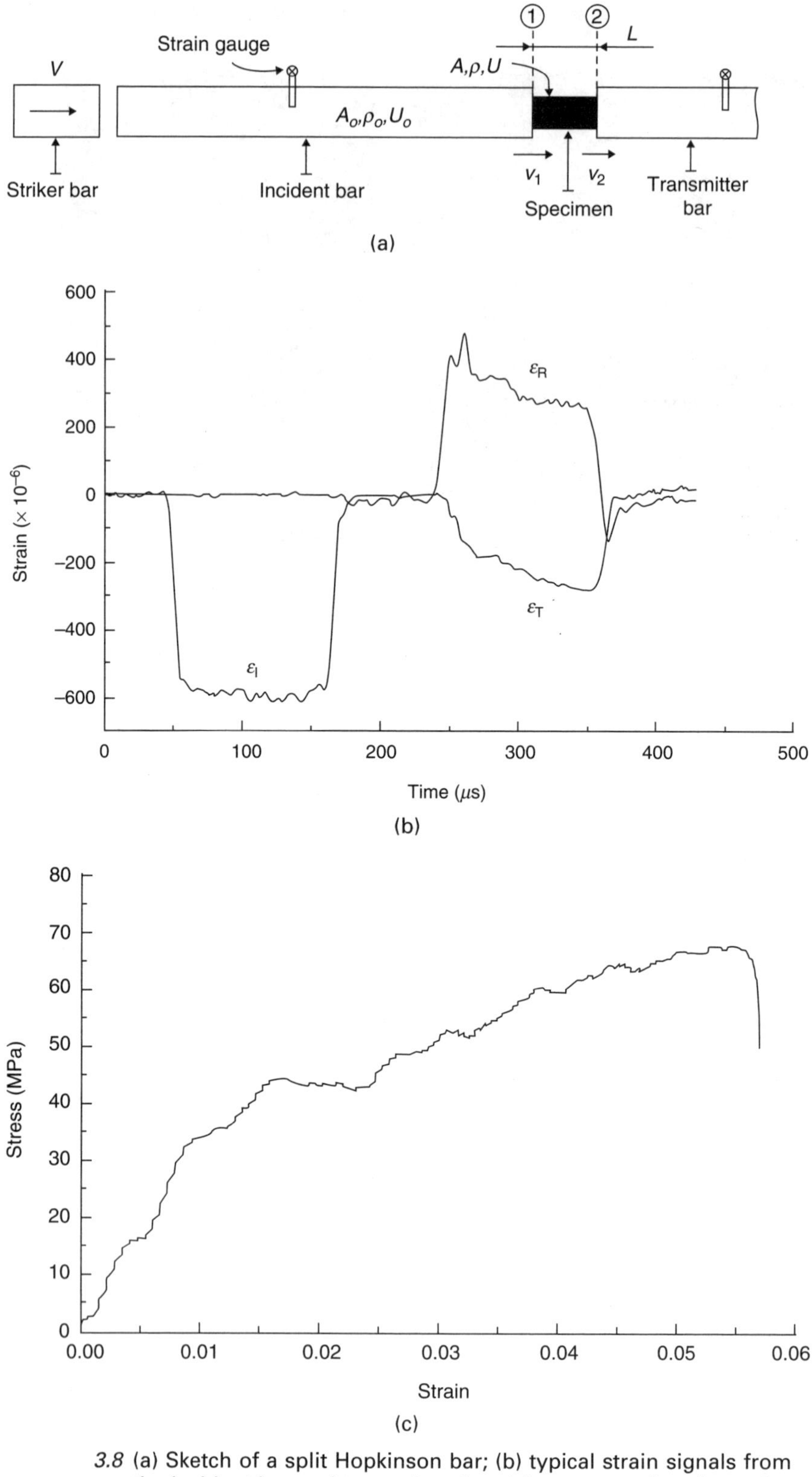

3.8 (a) Sketch of a split Hopkinson bar; (b) typical strain signals from the incident bar and transmitter bar; (c) corresponding dynamic stress–strain curve.

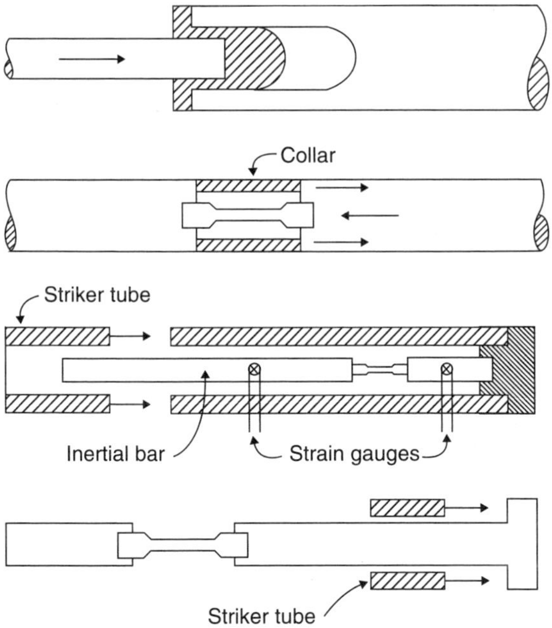

3.9 Sketch of various arrangements for tensile tests (reproduced with kind permission by John Wiley & Sons Inc.).

$$v = c_L \varepsilon \tag{3.25}$$

The strain at interface 1 is equal to the difference between the incident strain, ε_I, and the reflected strain, ε_R. Hence

$$v_1 = c_L(\varepsilon_I - \varepsilon_R) \tag{3.26}$$

Similarly at interface 2

$$v_2 = c_L \varepsilon_T \tag{3.27}$$

where ε_T is the strain at the transmitter bar. Consequently

$$\dot{\varepsilon} = \frac{d\varepsilon}{dt} = \frac{c_L(\varepsilon_I - \varepsilon_R - \varepsilon_T)}{L} \tag{3.28}$$

The strain of the test specimen is

$$\varepsilon(t) = \frac{c_L}{L}\int_0^t (\varepsilon_I - \varepsilon_R - \varepsilon_T)dt \tag{3.29}$$

Forces at the two interfaces are

$$F_1 = A_o E_o(\varepsilon_I + \varepsilon_R)$$
$$F_2 = A_o E_o \varepsilon_T \tag{3.30}$$

Table 3.2 Classification of testing techniques according to strain-rates (Meyers, 1994)

Strain rate (s^{-1})	Common testing methods	Dynamic considerations	
10^7	HIGH-VELOCITY IMPACT – Explosives	SHOCK-WAVE PROPAGATION	Inertial forces important
10^6	– Normal plate impact – Pulsed laser – Exploding foil		
10^5	– Incl. plate impact (pressure-shear)	SHEAR-WAVE PROPAGATION	
	DYNAMIC-HIGH	PLASTIC-WAVE PROPAGATION	
10^4	– Taylor anvil tests – Hopkinson bar – Expanding ring		
10^3	DYNAMIC-LOW	MECHANICAL RESONANCE IN	
10^2	High-velocity hydraulic, or pneumatic machines: cam	SPECIMEN AND MACHINE IS IMPORTANT	
10^1	plastometer		
10^0	QUASI-STATIC Hydraulic, servo-hydraulic	TESTS WITH CONSTANT CROSSHEAD VELOCITY	Inertial forces negligible
10^{-1}	or screw-driven testing machines	STRESS THE SAME THROUGHOUT LENGTH OF SPECIMEN	
10^{-2}			
10^{-3}			
10^{-4}			
10^{-5}	CREEP AND STRESS-	VISCOPLASTIC RESPONSE	
10^{-6}	RELAXATION	OF METALS	
10^{-7}	– Conventional testing machine		
10^{-8}	Creep testers		
10^{-9}			

where A_o and E_o are the cross-sectional area and elastic modulus of the identical incident and transmitter bars. From approximate equilibrium of the specimen, $F_1 \approx F_2$. Therefore

$$\varepsilon_T = \varepsilon_I + \varepsilon_R \qquad [3.31]$$

The average compressive stress in a specimen of cross-sectional area A is then

$$\sigma = \frac{F_1 + F_2}{2A} = E_o \frac{A_o}{A} \varepsilon_T \qquad [3.32]$$

Equations [3.29–3.32] enable us to calculate the stress–strain relationship and the strain-rate for a split Hopkinson bar test. Figure 3.8(b) shows strain signals for testing a sintered bronze sample with a porosity of 37 %. The specimen had a length of 6.25 mm and a diameter of 12 mm. The velocity of the striking bar was 11.63 m/s. The corresponding stress–strain curve is given as Fig. 3.8(c).

The split Hopkinson bar can also be used to study material properties in tension, torsion and shear. Some sketches of tension test setups are shown in Fig. 3.9. Shear stress–shear strain curves can be obtained by using a torsional bar.

3.3.4 Gas guns and other techniques

Gas guns have been extensively employed to generate impact velocities in the range 100–8000 m/s. For studies of energy absorption, a common one-stage gas gun is sufficient. For example, the one at Swinburne has a maximum velocity of approximately 600 m/s. The barrel is 6 m long and has an inside diameter of 12.58 mm. Maximum operating pressure of 15 MPa (about 150 bar) is used. A cylindrical specimen can be accelerated using this device to cause an impact on a rigid flat anvil. Using ***Taylor theory*** (Taylor, 1948), the average flow stress at high strain-rates can be determined by measuring the impact velocity and specimen dimensions after impact. Strain-rates of the order of $10^4\,s^{-1}$ can be obtained using this technique.

This gas gun has also been used to study the dynamic stresses for porous materials (Lu *et al.*, 2001). Alternatively, a circular tube can be placed inside the barrel and then accelerated to cause an impact on a rigid anvil (B. Wang and G. Lu, 2002), exhibiting various plastic buckling and tearing modes. A maximum velocity of 250 m/s was used in the latter case.

Light gases such as hydrogen and helium could be used to achieve higher velocities. A two-stage gas gun produces considerably higher velocities still (Crozier and Hume, 1957). Several other techniques are available to produce dynamic loadings, such as using explosives or electromagnetic acceleration. Typical strain-rates which can be produced from the common ***impact test methods*** are listed in Table 3.2 (Meyers, 1994).

4
Rings and ring systems

Rings and short tubes are common structural components. Their two-dimensional deformation is relatively simple to analyse theoretically. This chapter presents theoretical and experimental studies of rings and short tubes under in-plane loads. The results illustrate important features of plastic deformation and dynamic effects, which are inherent in more complex structures.

4.1 Ring compressed by two point loads

Consider a rigid, plastic ring compressed by two opposite point loads (Fig. 4.1(a)). Four plastic hinges are required for it to form a collapse mechanism. We could obtain the corresponding load–deflection curve by using an energy approach. Alternatively, this problem can be solved from equilibrium consideration of a segment (Fig. 4.1(b)). Thus for an angle of segment rotation θ, the compression is

$$\frac{\delta}{2} = R - \sqrt{2}R\sin\left(\frac{\pi}{4} - \theta\right) = R + R\sin\theta - R\cos\theta \qquad [4.1]$$

and the current length AB is

$$AB = \sqrt{2}R\cos\left(\frac{\pi}{4} - \theta\right) \qquad [4.2]$$

The force and moments acting on the segment are equivalent to two forces of equal magnitude, $P/2$, acting in opposite directions. Equilibrium requires that these two forces must act along the same line, known as the ***line of thrust***. Hence the shift of force $2M_p/P$, must be AB/2, where M_p is the fully plastic bending moment of the ring. Incorporating this into Eq. [4.2]

$$\frac{2M_p}{P} = \frac{\sqrt{2}}{2}R\cos\left(\frac{\pi}{4} - \theta\right)$$

or

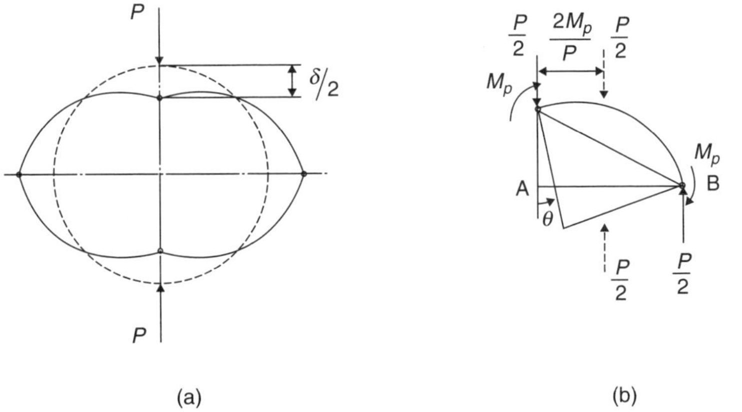

4.1 Collapse mechanism of a ring under two inwards acting point loads (a). Forces acting on a quadrant (b). Four plastic hinges are necessary.

$$P = \frac{2\sqrt{2}M_p}{R\cos\left(\frac{\pi}{4} - \theta\right)} \tag{4.3}$$

When $\theta = 0$, the initial collapse load is therefore

$$P_o = \frac{4M_p}{R} = \frac{8M_p}{D} \tag{4.4}$$

where D is the diameter of the ring. Combining Eqs [4.1], [4.3] and [4.4], the load–deflection curve is given by

$$\frac{P}{P_o} = \frac{1}{\left[1 + 2\frac{\delta}{D} - \left(\frac{\delta}{D}\right)^2\right]^{\frac{1}{2}}} \tag{4.5}$$

This relation indicates that load decreases as deflection increases, which is plotted in Fig. 4.2 as a short dashed line. The above method is referred to as the ***equivalent structure technique*** (Merchant, 1965; Reddy *et al.*, 1987).

4.2 Ring pulled by two point loads

When a ring is subjected to two similar point loads, but these loads act in opposite outward directions, its deformation can be analysed by exactly the same method as above. One key difference is that the position of the two side plastic hinges, corresponding to the maximum bending moment, moves as deflection increases (hinge B in Fig. 4.3(a)). This ***moving hinge*** is the

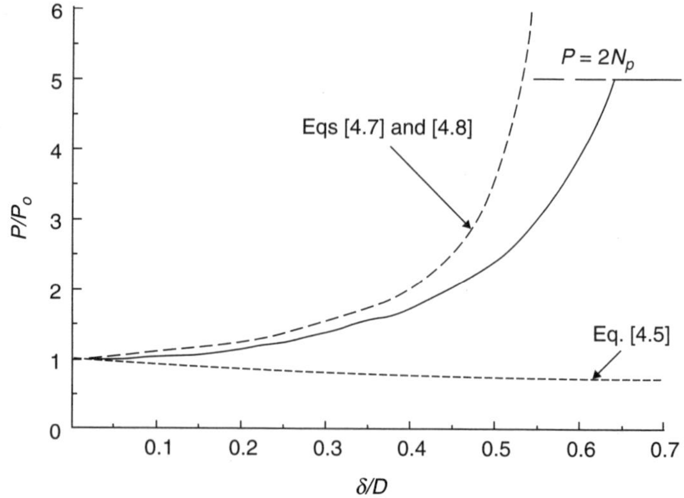

4.2 Non-dimensional load–displacement curves for a ring of rectangular cross-section under two point loads. $D/h = 5$. ----- Eq. [4.5] (compressive loads); — — — Eqs [4.7] and [4.8] (without axial force effect); —— with axial force effect. The force is limited to $P = 2N_p$ (i.e. $P/P_o = D/h = 5$).

same as that encountered in Section 2.3. The undeformed segment is always tangential to the deformed straightened portion at the current plastic hinge B. Furthermore, the total length of the ring remains the same during deformation. These considerations, together with the equivalent structure technique described above, lead to (Yu, 1979)

$$\frac{P}{2}R(1-\sin\theta) = 2M_p \qquad [4.6]$$

or

$$\frac{P}{P_o} = \frac{1}{1-\sin\theta} \qquad [4.7]$$

and

$$\delta = 2R(\cos\theta + \theta - 1) \qquad [4.8]$$

Equations [4.7] and [4.8] give the load–displacement curve shown in Fig. 4.2 by the long dashed line.

The above analysis does not consider the effect of axial force on yielding, which can be important when the ring is thick and when δ/D is close to 1. The effect of axial force on yielding for a rectangular cross-section ($b \times h$) was discussed in Section 2.2.2. One approximate way of incorporating

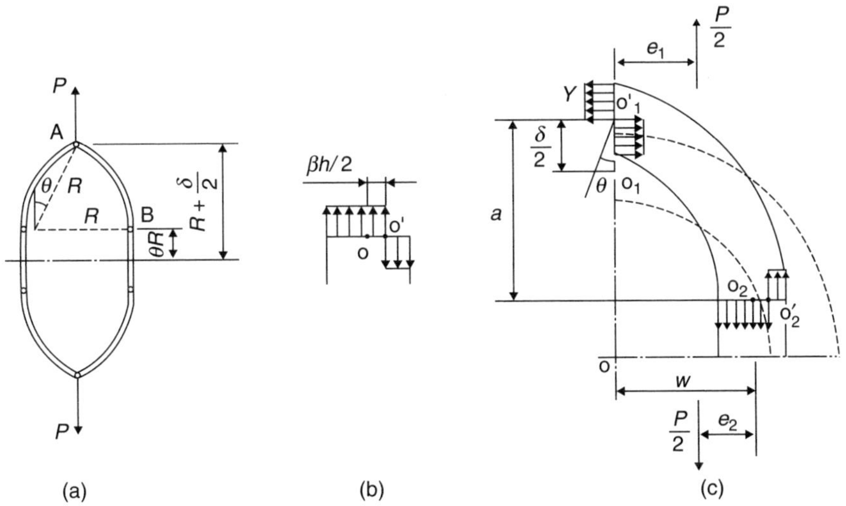

4.3 Collapse mechanism of a ring under two outwards acting point loads (a). Stress state of a plastic hinge with axial force (b). Details of forces acting on a segment (c). During collapse the two middle hinges split into four hinges such as at point B. (Lu, 1993a)

this ***axial force effect*** here is to modify Eq. [4.6] by considering Eq. [2.22]. The axial force at each hinge is $P/2$ and the bending capacity now reduces from M_p to

$$M = M_p\sqrt{1-\left(\frac{P/2}{N_p}\right)^2}$$

Thus, Eq. [4.6] becomes

$$\frac{P}{2}R(1-\sin\theta) = 2M_p\sqrt{1-\left(\frac{P/2}{N_p}\right)^2} \qquad [4.9]$$

As before, $N_p = Ybh$ is the axial yielding force. Combining Eqs [4.9] and [4.8] leads to a load–displacement curve with a maximum load equal to $2N_p$ when $\theta = \pi/2$.

Nevertheless, strictly speaking, this analysis is only approximate because the associated flow (normality) rule has not been taken into account, although the axial force was considered in the yielding criterion. The total length of the ring remains the same in this case. One method of overcoming this shortcoming is presented here (Lu, 1993a).

The ***normality rule*** (associated flow rule) described in Section 2.2.2 (Eq [2.23]) can be understood here, alternatively, from the actual stress distribution over the cross-section, instead of the resultant axial force N and resultant bending moment M. It is clear that corresponding to such a stress state, deformation of the cross-section is a rotation about point O′ rather than the midpoint O, see Fig. 4.3(b). O′ is also called the '***pivot point***'. Thus when the cross-section rotates about point O′ by angle θ, fibres at the mid-point O stretch by $\theta\beta h/2$. Here β is a fraction and $\beta h/2$ is the distance between O and O′. It is easy to see that the resultant axial force is $N = \beta hb$ and the bending moment is $M = (1 - \beta^2)Yh^2b/4$. The axial force at the upper hinge is zero and at the lower hinge it is $P/2$. For a ring under tension (Fig. 4.3(a)), the total external force is

$$P = 2Y\beta hb \tag{4.10}$$

The position of the line of action of forces in the equivalent structure (Fig. 4.3(c)) is given by

$$e_1 = \frac{M_p}{P/2} = \frac{H}{4\beta} \tag{4.11}$$

and

$$e_2 = \frac{M}{P/2} = \frac{H}{4}\left(\frac{1}{\beta} - \beta\right) \tag{4.12}$$

Again from equilibrium considerations, the two opposite forces must act along the same line. From the geometry at the onset of collapse, displacement δ is zero and we have

$$e_1 + e_2 = R = D/2 \tag{4.13}$$

Substituting Eqs [4.11] and [4.12] into [4.13], we obtain the expression for the positive value of β:

$$\beta = -\frac{D}{h} + \sqrt{\left(\frac{D}{h}\right)^2 + 2} \tag{4.14}$$

For example consider the two cases $D/h = 10$ and $D/h = 5$. Equation [4.14] leads to $\beta = 0.0995$ and 0.1962, respectively. The initial collapse load becomes

$$P_o = 2Y\beta hb = 2\frac{D}{h}\left(\sqrt{1 + 2\left(\frac{h}{D}\right)^2} - 1\right)Yhb \tag{4.15}$$

This formula is identical to that given by de Runtz and Hodge (1963) for the initial collapse load of a ring compressed by two rigid flat plates when the effect of axial force on yielding is considered.

As mentioned previously, during the collapse process, the middle hinge will split into two, one moving upwards and the other downwards. This is required by the condition that a plastic hinge always forms at the position of maximum bending moment. The deformation in the upper hinge remains a pure rotation about point O_1', which is the mid-point of the cross-section.

From the stress state for the post-collapse stage as shown in Fig. 4.3(c), Eq. [4.13] becomes

$$e_1 + e_2 = w \tag{4.16}$$

where w is the horizontal distance between O_1' and O_2. Substituting in Eqs [4.11] and [4.12]

$$\frac{h}{4}\left(\frac{2}{\beta} - \beta\right) = w \tag{4.17}$$

Let the total rotation of the rigid segment at this instant be θ. We have

$$w = R(1 - \sin\theta) \tag{4.18}$$

and

$$a = R\cos\theta \tag{4.19}$$

Eliminating θ from Eqs [4.18] and [4.19], we have

$$a = \sqrt{2wR - w^2} = \sqrt{\frac{Dh}{4}\left(\frac{2}{\beta} - \beta\right) - \left(\frac{h}{4}\right)^2\left(\frac{2}{\beta} - \beta\right)^2} \tag{4.20}$$

At this instant, let the centre of the mid-hinge cross-section O_2 move sideways by an increment dw, the corresponding rotational increment of the segment is then dw/a and the corresponding increment of deflection δ is

$$\mathrm{d}\delta = 2\left(w + \frac{\beta h}{2}\right)\frac{\mathrm{d}w}{a} = h\left(\frac{1}{\beta} + \frac{\beta}{2}\right)\frac{\mathrm{d}w}{a} \tag{4.21}$$

But by differentiating Eq. [4.17], we have

$$\mathrm{d}w = -\frac{h}{4}\left(\frac{2}{\beta^2} + 1\right)\mathrm{d}\beta \tag{4.22}$$

The negative sign means that when β increases, w decreases. Substituting Eqs [4.20] and [4.22] into Eq. [4.21], we obtain after some rearrangement

$$\frac{\mathrm{d}\delta}{D} = -\frac{\frac{h}{D}\left(\frac{1}{2\beta^3} + \frac{1}{2\beta} + \frac{\beta}{8}\right)}{\left[\frac{1}{4}\left(\frac{D}{h}\right)\left(\frac{2}{\beta} - \beta\right) - \left(\frac{1}{2\beta} - \frac{\beta}{4}\right)^2\right]^{1/2}}\mathrm{d}\beta \tag{4.23}$$

Therefore

$$\frac{\delta}{D} = \int\frac{\mathrm{d}\delta}{D} = \int_{\beta_o}^{\beta} -\frac{\frac{h}{D}\left(\frac{1}{2\beta^3} + \frac{1}{2\beta} + \frac{\beta}{8}\right)}{\left[\frac{1}{4}\left(\frac{D}{h}\right)\left(\frac{2}{\beta} - \beta\right) - \left(\frac{1}{2\beta} - \frac{\beta}{4}\right)^2\right]^{1/2}}\mathrm{d}\beta \tag{4.24}$$

Thus for a given value of D/h, β_o can be determined from Eq. [4.14] and then Eq. [4.24] gives the value of δ/D corresponding to any value of β. The tension force can then be determined from Eq. [4.10]. This result is shown in Fig. 4.2 as a solid line for the case of $D/h = 5$. This tension force is, of course, lower than the case where the effect of axial force was not considered. Also, maximum non-dimensional displacement, δ/D, before full membrane yielding develops is 0.61, which is higher than the 0.57 given by Eq. [4.8] for $\theta = \pi/2$. This clearly demonstrates stretching of the ring as a result of the axial force. The maximum tensile force is limited to the full axial yielding force ($=2N_p$), and the ring then behaves like a bar under simple tension.

4.3 Built-in semi-circular arch under point loads

We next present an analysis of a built-in semi-circular arch under point loads, acting either inwards or outwards (Gill, 1976). Strictly speaking, this structure is not a ring – the subject of this chapter. Nevertheless, because its analysis resembles that of a constrained tube it should be beneficial to present it here.

4.3.1 Semi-circular arch with an outwards load

Consider a semi-circular arch with an outwards acting load P, as shown in Fig. 4.4(a). The equivalent resultant forces $P'P'$ are shown in opposite directions acting along the directions of AC and CE. Neglecting the axial force effect, we have

$$AA' = BB' = CC'' = EE' = \frac{R}{4+2\sqrt{2}} \qquad [4.25]$$

and

$$P'\frac{R}{4+2\sqrt{2}} = M_p$$

The initial collapse load is therefore

$$P_o = \sqrt{2}P' = \frac{4M_P}{R}(1+\sqrt{2}) \qquad [4.26]$$

Compared with Eq. [4.4], Eq. [4.26] indicates that the initial collapse load of the semi-arch is 2.4 times that of a similar ring. We note that Eq. [4.26] is also true when the two loads act inwards, a case to be discussed later. The post-collapse mechanism (Fig. 4.4(b)) involves two additional hinges: hinge B splits into two moving hinges B_1 and B_2, the line B_1B_2 is straight as shown. From Fig. 4.4(b) and the condition that total length is constant

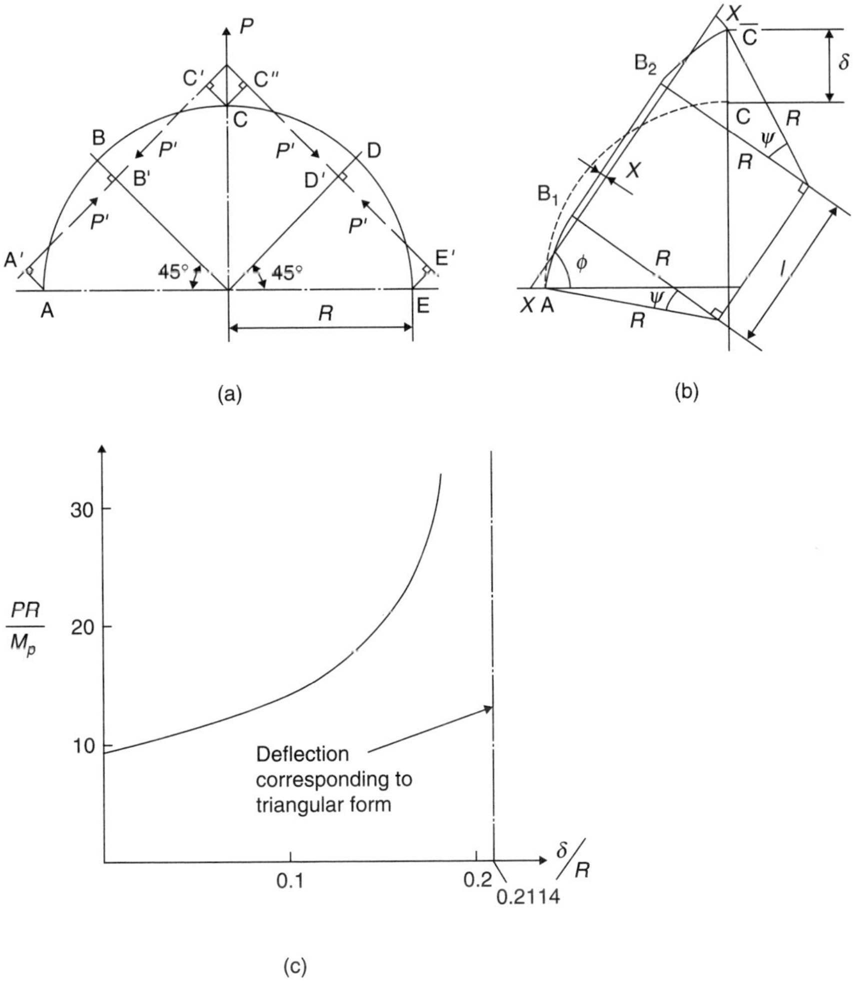

4.4 A built-in semi-circular arch under an outwards acting point load. (a) Initial collapse mode; (b) post-collapse mode; (c) non-dimensional load–deflection curve. (Gill, 1976)

$$\delta = [(2R\sin\psi + l^2) - R^2]^{\frac{1}{2}} - R$$

where $\psi = \pi/4 \quad l/2R$ and l is the straight length B_1B_2. Hence, for convenience l may be used as a variable in calculating the load–displacement curves. Therefore

$$P = \frac{4M_p(\delta + R)}{R(1-\cos\psi)\left[(\delta+R)^2 + R^2\right]^{1/2}} \qquad [4.27]$$

Equation [4.27] is plotted non-dimensionally in Fig. 4.4(c), which indicates that the load P increases with deflection δ. As in the case of a ring subjected to two outwards pointing loads, the actual load would be governed by the plastic membrane forces. This state is reached when $\delta/R = 0.2114$, at which point the arch ACE would have straightened to a triangular shape.

4.3.2 Semi-circular arch with an inward load

When the same arch is subject to an inwards acting load, see Fig. 4.5(a), and point $\overline{C}$ is constrained to move vertically only, the initial collapse load is the same as for outward loading (Eq. [4.26]). In this case, after the initial collapse, four collapse modes exist.

Mode 1 (Fig. 4.5(b)) has five plastic hinges whose positions are fixed. $\overline{BC}$ rotates clockwise, whereas $A\overline{B}$ rotates counter-clockwise. Point $\overline{B}$ always has the maximum bending moment, until $\overline{BC}$ becomes horizontal, invoking mode 2 (Fig. 4.5(c)). In mode 2, another hinge B″ occurs and AB″ does not rotate, but remains fixed. A third mode as shown in Fig. 4.5(d), takes place when B″C″ becomes horizontal. Here two travelling hinges move apart in segment $B_1\overline{\overline{C}}$. The final mode is shown in Fig. 4.5(e), which involves splitting hinges at B into B and D. Non-dimensional load–deflection curves corresponding to the four modes are shown in Fig. 4.5(f). The final shape after bending deformation is triangular; this is shown as a dashed line in Fig. 4.5(a). Interested readers should refer to the original paper by Gill (1976) for a detailed description and analysis.

4.4 Ring compressed by two flat plates

Four plastic hinges are needed for a ring to collapse under compression between two flat plates, and two common modes are shown in Fig. 4.6 (Burton and Craig, 1963; de Runtz and Hodge, 1963). The first mode has four stationary plastic hinges and is more appropriate for mild steel, which has an upper and lower yield point. The second mode involves straightening of the ring at the moving contact point. Both modes have the same force diagram for the undeformed segment and hence lead to the same force–deflection curve. The initial collapse load is the same as for the point loading case (Eq. [4.4]). From equilibrium

$$\frac{1}{2}PR\cos\theta = 2M_p \qquad [4.28]$$

and from geometry

$$\delta = 2R\sin\theta \qquad [4.29]$$

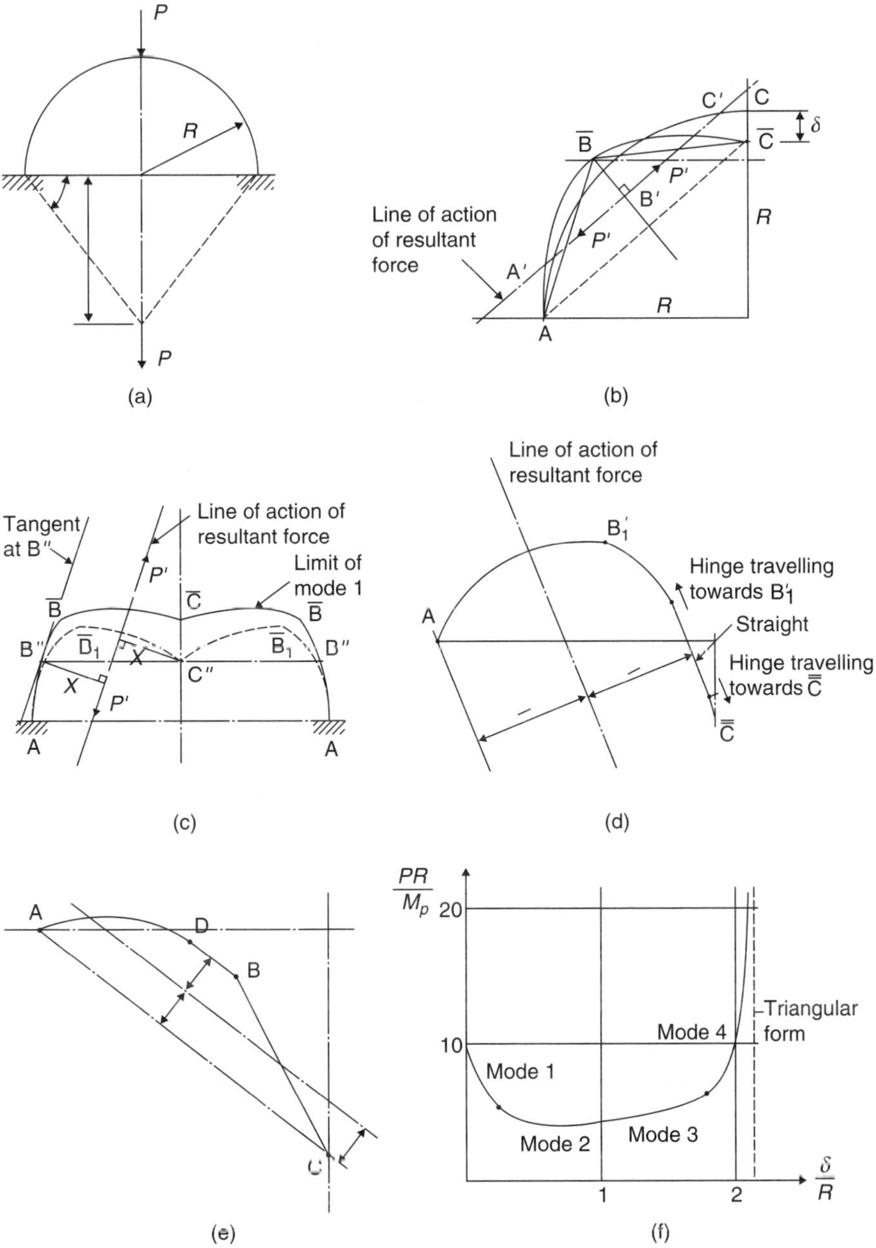

4.5 A built-in semi-circular arch under an inwards acting point load. (a) Initial configuration (solid line) and final configuration (dashed line); (b)–(e) collapse modes 1–4 (see text for details); (f) non-dimensional load–deflection curve. (Gill, 1976)

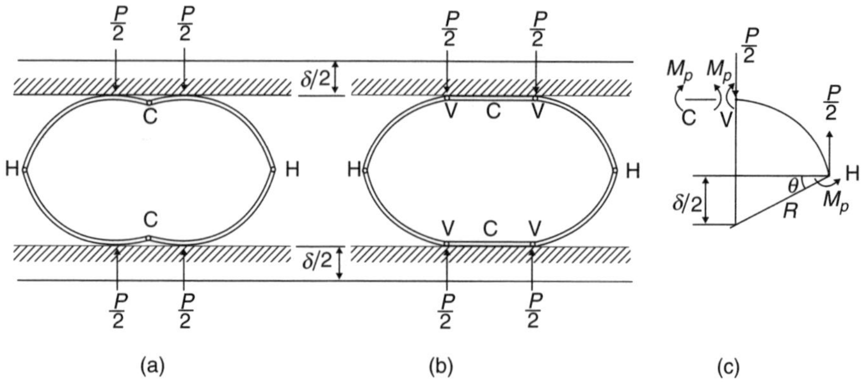

4.6 Collapse mechanisms proposed by: (a) de Runtz and Hodge (1963) and (b) Burton and Craig (1963); (c) also shown are forces on a deforming segment.

Combining Eq. [4.28] with [4.29], and noting Eq. [4.4], we have

$$P = \frac{P_o}{\left[1-(\delta/D)^2\right]^{1/2}} \qquad [4.30]$$

or

$$P = \frac{2Yh^2L}{D\left[1-(\delta/D)^2\right]^{1/2}} \qquad [4.31]$$

where L is the width of the ring or tube. This demonstrates that load increases with deflection (Fig. 4.7). Note that the analysis of rings presented so far is equally applicable to tubes under similar loadings, provided the appropriate value is taken for yield stress. Thus, when the length is not greater than a few thicknesses, a short tube can be taken as a ring; Y in Eq. [4.31] is then equal to the yield stress from a uniaxial tensile test. When the length is larger than the diameter of the tube, Y is taken as $2/\sqrt{3}$ multiplied by the yield stress in simple tension in order to account for the ***plane strain*** condition.

It can be seen from Fig. 4.7 that this force prediction is lower than experimental results. This discrepancy can be accounted for by the strain-hardening effect, which has two implications here. First, the plastic bending moment resistance increases as deformation proceeds (Eqs [2.2] and [2.3]). Second, plastic deformation takes place over a zone instead of being confined within a localised plastic hinge – i.e. the hinge has a certain length. The latter effect leads to a change in load through a small but significant change in geometry and hence moment arm length. A simple way of estimating the strain-hardening effect is to evaluate the average strain involved

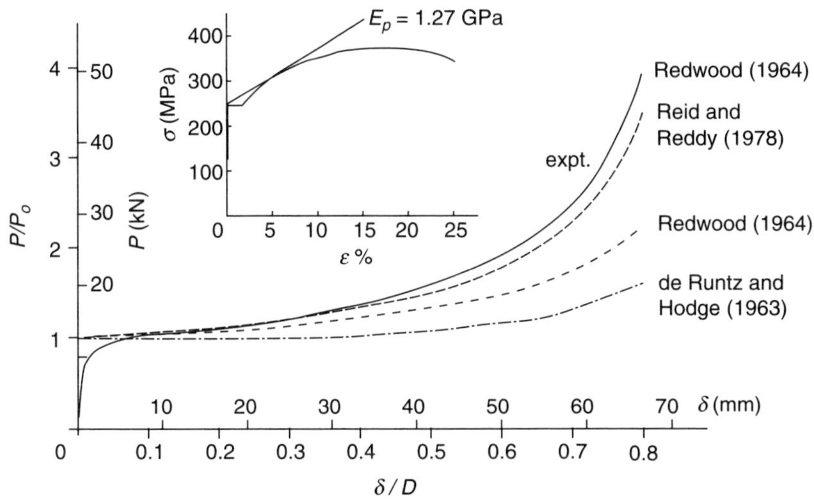

4.7 Non-dimensional load–displacement curves from experiments and theories (h/R = 0.108, R = 42.16 mm, L = 101.6 mm). (Reid, 1983)

in the deformation zone and then incorporate this into an enhancement of bending moment resistance. Assuming that the total ***hinge length*** is λh, which does not change during deformation, the average curvature is then $\kappa = \theta/\lambda h$.

For an assumed ***linear hardening relationship for bending moment***

$$M = M_p + E_p I\kappa = M_p\left(1+\frac{E_p\theta}{3Y\lambda}\right) \qquad [4.32]$$

where I is the second moment of area and E_p is the strain-hardening modulus of the assumed rigid-linear hardening material (see Eq. [2.5]). From Eq. [4.29], $\theta = \sin^{-1}(\delta/D)$. Replacing M_p in Eq. [4.28] with M from Eq. [4.32], we obtain

$$\frac{P}{P_o} = \frac{1}{\left[1-(\delta/D)^2\right]^{1/2}}\left[1+\frac{E_p}{3Y\lambda}\sin^{-1}\left(\frac{\delta}{D}\right)\right] \qquad [4.33]$$

This equation was proposed by Redwood (1964). The value of λ was found to be 5 by measuring the plastic region in the experiment and this value is used in plotting Fig. 4.7.

Clearly Eq. [4.33] gives a better predication than Eq. [4.30], but it is still lower than the experimental results, especially when deflection is large. This is because the plastic hinges were still treated as being very localised and so the geometry is basically the same as previously assumed in Fig. 4.6. Reid and Reddy (1978) investigated this problem and proposed a ***plastica*** theory, which replaced the concentrated hinges with an arc whose length varies with deflection δ. This essentially revealed that the effective moment arm length reduces with deflection, in addition to the enhancement of bending moment.

One quadrant of a tube HV is shown in Fig. 4.8(a), with an enlarged view of the plastic deformation zone HB shown in Fig. 4.8(b). This is similar to Fig. 4.6 except that hinge H is replaced with plastic zone HB. A linear strain-hardening relationship is assumed for the bending moment. The top moving hinge V is assumed to be a concentrated hinge as before. Plastic deformation occurs within zone HB and the moment at B is the initial plastic bending moment M_p. Segment BV remains rigid and rotates during deformation. The governing equation for HB is (Frisch-Fay, 1962)

$$E_p I \frac{d^2\theta}{ds^2} = -\frac{P}{2}\sin\theta \qquad [4.34]$$

At H, $M_H = M_p + Pb/2$. The two governing equations for the system (Fig. 4.8(a)) are

$$P = \frac{4M_p}{R\cos\gamma} \qquad [4.35]$$

and

$$\frac{\delta}{2} = R\sin\gamma - c \qquad [4.36]$$

where γ, b and c are defined as shown in Fig. 4.8(b). The solution procedure for these three equations was given by Reid and Reddy (1978). In particu-

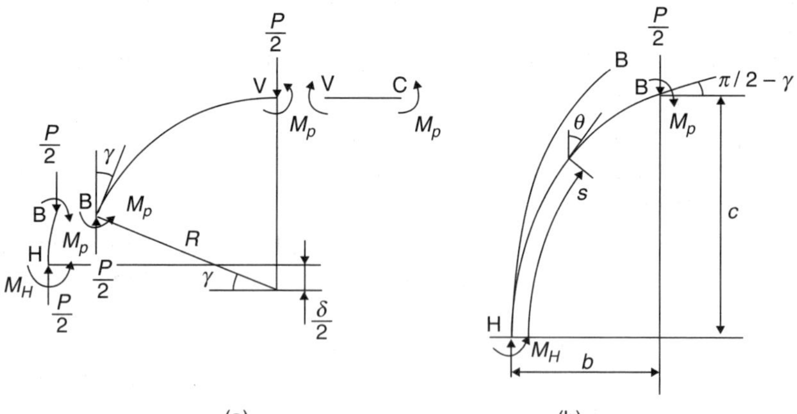

4.8 Tube with strain-hardening material analysed using *plastica* theory – Reid and Reddy (1978): (a) forces on a quadrant of a tube *HV*; (b) deformation of plastic region *HB*.

lar, they identified the following dimensionless parameter, which governs the shape of the load–deflection curve

$$mR = (6Y/E_p hR)^{1/2} \qquad [4.37]$$

High values of mR correspond to a relatively flat load–deflection curve, as predicated by de Runtz and Hodge (1963) for a rigid-perfectly-plastic material. Small values of mR cause the curve to rise up significantly. Their theoretical prediction is in better agreement with experimental results than previous theories, as shown in Fig. 4.7. It may also be noted that, by choosing materials with different strain-hardening effects, the force–deflection curve may be adjusted to be close to the ideal rectangular force–deflection relationship of an energy absorber as described in Chapter 1.

4.5 Laterally constrained tubes

In order to maximise the energy absorption, structures should be arranged to deform with a large volume of material reaching plasticity. Thus, tubes can be constrained laterally so that more plastic hinges form during collapse than for unconstrained cases. Reddy and Reid (1979) investigated one mode, where the horizontal diameter of a tube is prevented from changing. Several more plastic hinges are necessary than in a free tube, in order to form a collapse mechanism. Two types of constraint were used: a tube placed in a grooved block and a tube with two side plates bolted together (Fig. 4.9(a)). The first arrangement introduces friction between the tube and the constraint block, which leads to unsymmetrical deformation between the upper and lower half (Fig. 4.9(b)). Typical load–displacement curves (Fig. 4.10) show an increase in force in the post-collapse stage. Tubes under general lateral constraint of various degrees compressed with cylindrical indenters have been investigated by Shim and Stronge (1986a).

A tube may be placed in a 'V' block and the load-deflection characteristics can be adjusted by varying the block angle (Figs 4.11 and 4.12) (Reid, 1983). For a tube crushed by a point load (V shaped indenter) or a flat plate, the force increases when the block angle α decreases. The sharp drop in force for point loading is a result of geometry change: the moment arm increases with deflection and the strain-hardening effect compensates in the later stage. In all cases, plastic hinges occur at the loading point, contact points and half way between the loading and contact points.

Tubes can be braced with wires for enhanced energy-absorption capacity (Reid *et al.*, 1983b) (Figs 4.13 and 4.14). Annealed mild steel tubes with outside diameter 88.9 mm, wall thickness 1.6 mm and length 50.8 mm were tested. High tensile strength steel wire of 0.3 mm diameter was wound through pairs of holes in the tube to provide tension bracing. For single ***braced tubes*** loaded by a flat plate (Fig. 4.13(a)), load–deflection curves

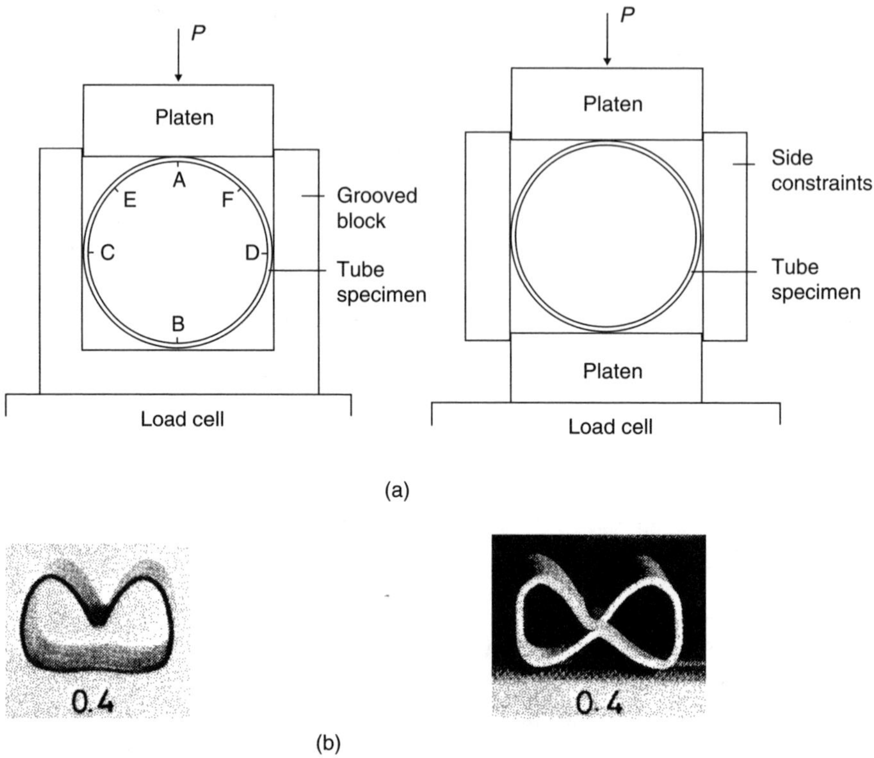

4.9 Constrained tube under flat plate compression: (a) test setup; (b) tubes after testing (Reddy and Reid, 1979).

seem little affected by θ in the range $15° \leq \theta \leq 90°$ (not shown), but the load is much higher when $\theta = 0°$. Double bracing increases the forces for $\theta = 30°$.

For braced tubes, plastic hinges occur at the points of bracing, the loading point, and between the loading and bracing points. This bracing arrangement was further explored recently for ***elliptical tubes*** (Wu and Carney, 1997 and 1998).

The equivalent structure technique can be employed to analyse the collapse behaviour of all the presented cases of laterally constrained tubes. In particular, collapse in Fig. 4.9(a) involves, initially, hinges at A, E, F, C and D, before sliding at C and D occurs. The initial collapse load is the same as a semi-circular arch under point loading (Eq. [4.26]), which is 2.4 times that for a free tube. For a tube supported in a V-block, the initial collapse load is given by

$$P_o = (4M_p/R)\cot[(\pi + \alpha)/8] \qquad [4.38]$$

where α is the V-block angle.

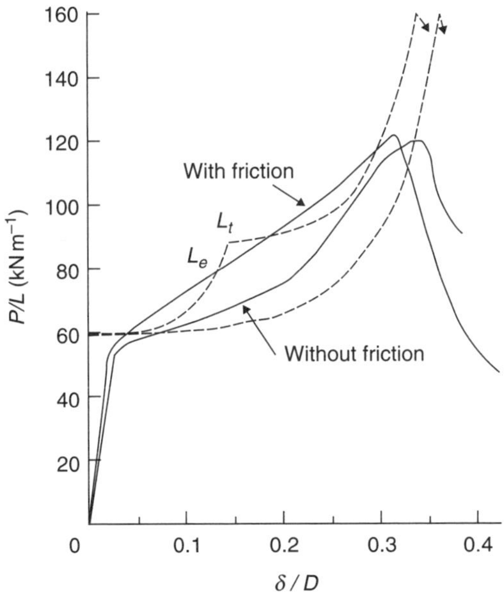

4.10 Comparison between theory and experiment for constrained tubes compressed between flat plates. Aluminium, $D = 25$ mm, $h = 0.9$ mm. The load is higher with friction than without friction. Dotted lines are theoretical results (Reddy and Reid, 1979).

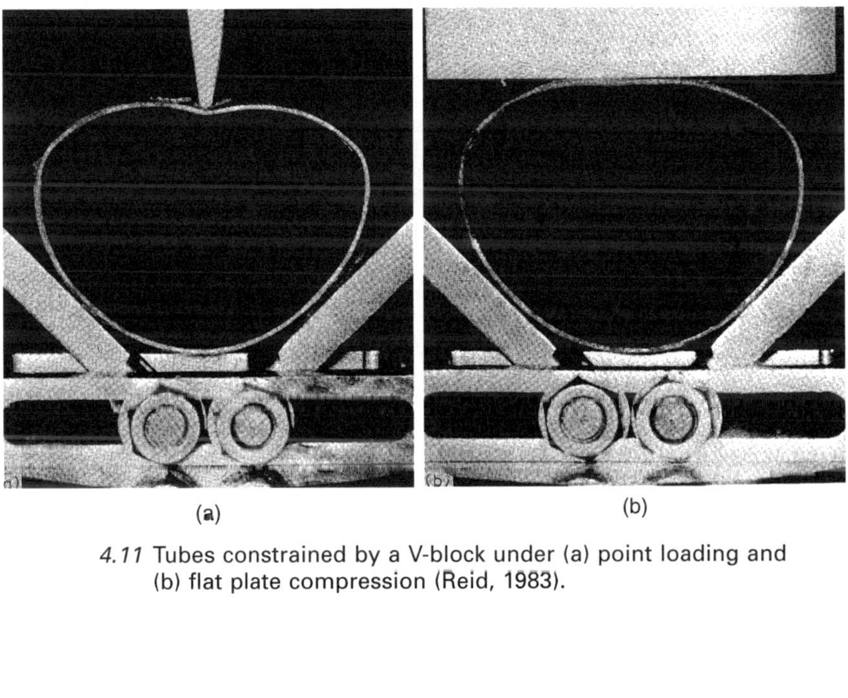

(a) (b)

4.11 Tubes constrained by a V-block under (a) point loading and (b) flat plate compression (Reid, 1983).

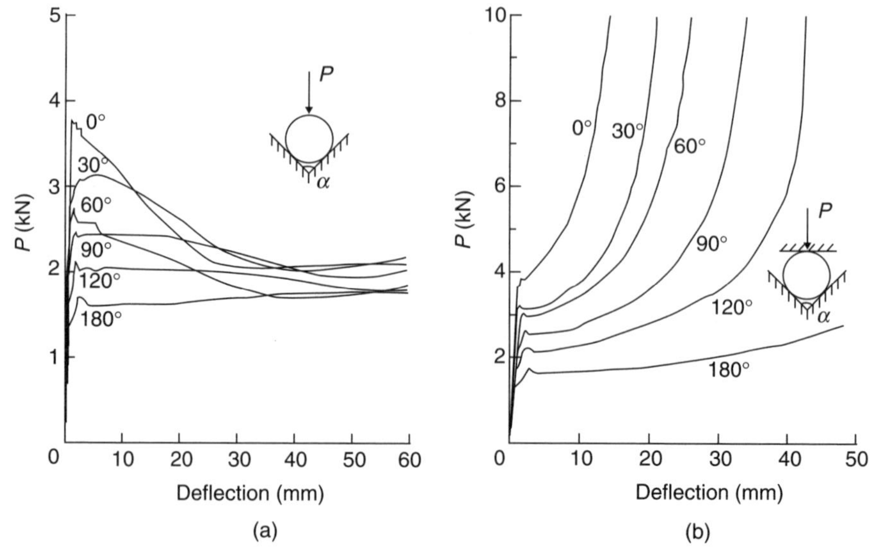

4.12 Load–deflection curves for V-block constrained tubes under: (a) point loading; (b) flat plate loading (mild steel ring, $D = 89$ mm, $h = 3.2$ mm, width 19 mm) (Reid, 1983).

For braced tubes, Reid *et al.* (1983b) obtained the following equations:

For a tube with single bracing whose position is defined by $0 < \theta \leq 30°$ *(Fig. 4.13(a))*

$$P_\theta = \frac{2M_p}{R\sin\left(\frac{\pi}{4}+\frac{\theta}{2}\right)\left[1-\sin\left(\frac{\pi}{4}+\frac{\theta}{2}\right)\right]} \quad [4.39]$$

For a tube with symmetrical double bracing (Fig. 4.13(b)), the collapse load is

$$P_\theta = \frac{4M_p}{R}\begin{cases} \dfrac{\cos\left(\frac{\pi}{4}+\frac{\phi}{2}\right)}{1-\sin\left(\frac{\pi}{4}+\frac{\phi}{2}\right)} & 0 \leq \phi \leq \dfrac{\pi}{6} \\ \dfrac{1-\sin\phi}{1-\cos\phi} & \dfrac{\pi}{6} \leq \phi \leq \dfrac{\pi}{4} \\ 1 & \phi > \dfrac{\pi}{4} \end{cases} \quad [4.40]$$

The post-collapse modes involve moving plastic hinges and contact points between the tube and plates. Nevertheless, the load–deflection behaviour

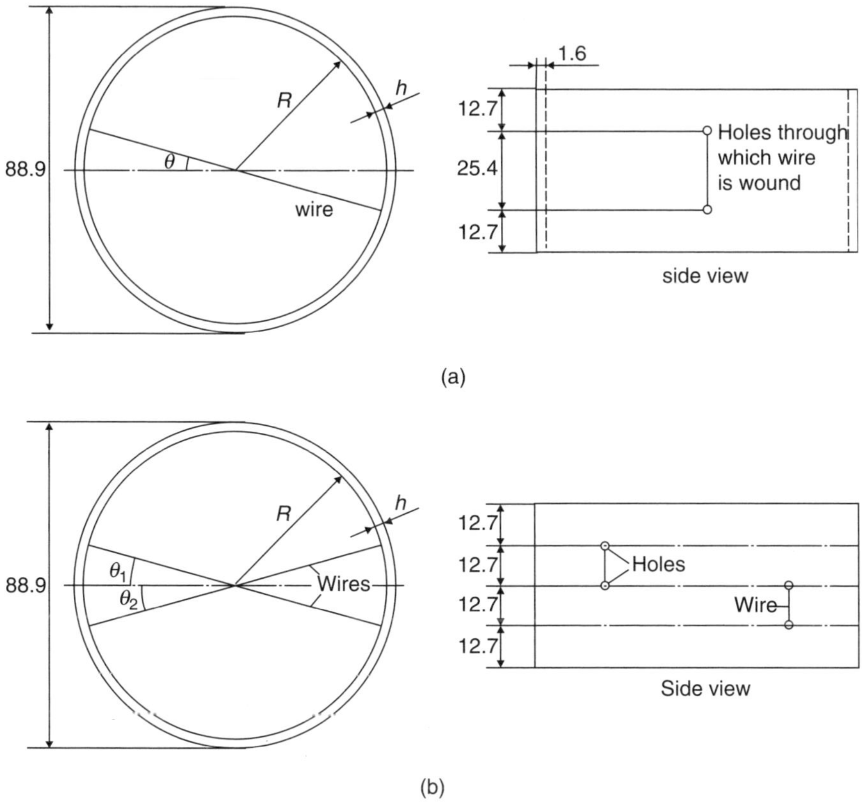

4.13 (a) Single-braced and (b) double-braced tubes, used by Reid *et al.* (1983b) (reproduced with kind permission of Elsevier).

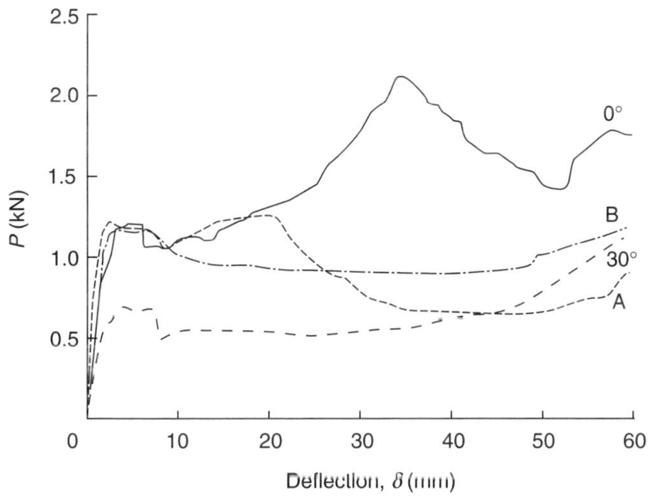

4.14 Load–deflection curves for single braced ($\theta = 0°$ and $\theta = 30°$) and double braced tubes (A: $\theta_1 = 0°$, $\theta_2 = 30°$; B: $\theta_1 = \theta_2 = 30°$) (Reid *et al.*, 1983b) (reproduced with kind permission of Elsevier).

can be obtained using the same method as above, when the strain-hardening effect is ignored.

4.6 One-dimensional ring system under end impact

All the previous discussions have been concerned only with the quasi-static response of tubes. We now present the dynamic response of a ***ring system***. A chain of metal rings subject to end impact was studied by Silva-Gomes *et al.* (1978), with typical mass ratio $m/G \approx 0.01$ and velocity $v_o \approx 4$ m/s, where m is the mass of each ring and G is the mass of the striker. As the velocity was low, deformation occurred uniformly and simultaneously for all rings and therefore the overall response of the system could be determined by analysing each single ring as in the quasi-static case.

Reid, Reddy and their co-workers have conducted further investigations into the impact response of such a ring system, but under higher impact velocities (30~120 m/s) (Reid and Reddy, 1983; Reid *et al.*, 1983a; Reid and Bell, 1984; Reddy *et al.*, 1991). The mass ratio was about 0.2. Two types of ring systems were tested: first a free system where all the rings were aligned freely, without any connection; secondly, a plated system where a mild steel plate of mass $m' = 17.9$ g was inserted between the rings and the system was fastened together by rivets through the rings and plates. Deformed states for thin brass ring systems are shown in Figs 4.15(a) and (b) for free and plated systems, respectively. The ring thickness h is 1.6 mm and the ring is annealed. For the free system, $G = 125$ g ($m/G = 0.22$) in all cases and $v_o \approx 35.5$ m/s, while for the plated system $(m + m')/G = 0.36$ and $v_o \approx 35$ m/s. Both systems have 6~10 rings.

Clearly the deformation (Fig. 4.15 and 4.16), which starts from the proximal end, is not uniform. For the free system, the collapse mode of a ring may start with the four-hinge mode as in Fig. 4.6. However, subsequent deformation leads to a '***wrap-around***' of the ring onto the adjacent one and the mode becomes much more complex. Also, globally, two rings at the distal end experienced substantial deformation. Broadly, this may be explained by adopting a structural wave theory to be presented below, as Reid *et al.* (1983a) argued. Thus, a plastic wave develops at the impact end and then propagates towards the distal end. Early deformation of the two rings at the distal end may be the additive result of the elastic incident and reflected waves at the fixed end, causing a plastic wave travelling backwards.

Deformation of plated systems is more regular. Essentially, deformation starts from the first ring at the proximal end and then propagates towards the distal end. For a large number of rings, this plastic wave stops without reaching the distal end, showing little deformation there.

Compare the stress–strain curve in Fig. 2.15(a) and the load–displacement curve for a ring (Fig. 4.7). They both are convex towards the abscissa,

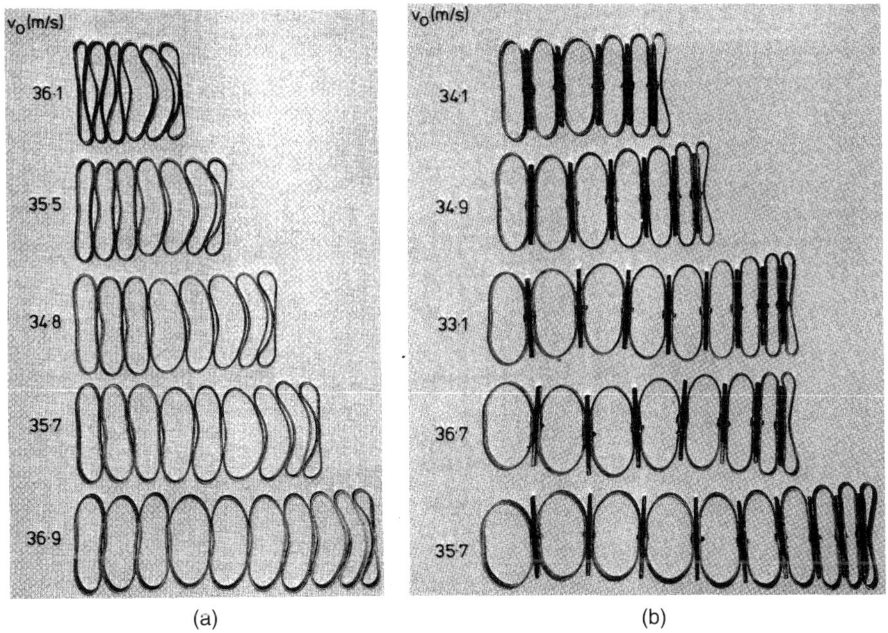

4.15 Final deformed state of brass ring systems after end impact with mass $G = 125$ g at various velocities, v_o; $n = 6$–10: (a) without inserts; (b) with mild steel plate inserts (Reid and Reddy, 1983) (reproduced with kind permission of Elsevier).

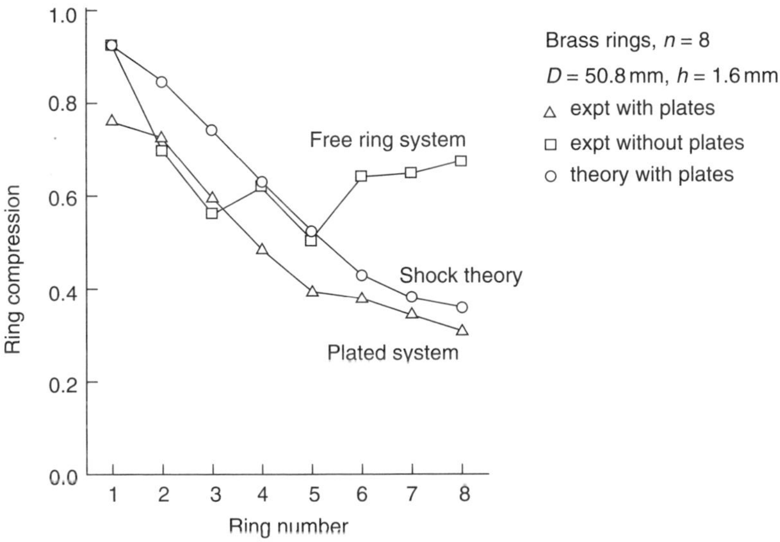

4.16 Experimental and theoretical results for ring compression, δ/D, with $n = 8$ (reproduced with kind permission of Elsevier).

after an initial elastic stage, i.e. for the former case the tangential modulus $d\sigma/d\varepsilon$ increases with strain. By analogy with Fig. 2.15(b), a structural shock wave should also exist, as Reid *et al.* (1983a) proposed, following the ***shock wave theory*** for a bar made of a material with a convex stress–strain curve (Lee and Tupper, 1954).

Consider a ring system subjected to end impact by a mass G (Fig. 4.17). Let the initial collapse load of a ring be P_o and the current load be P. Assume that the quasi-static relationship $P/P_o = f(\delta/D)$ still holds for the dynamic case, where f is a function such as Eq. [4.30] or [4.33]. Ignore the elastic deformation and allow a force discontinuity at the contact point between the deforming ith ring and the $(i+1)$th ring.

For propagation of the initial shock across the ith ring, the governing equations are

$$[G+(i-1)m](u_i - u_{i-1}) = -P_i t_i \qquad [4.41a]$$

$$mu_i = (P_i - P_o)t_i \qquad [4.41b]$$

$$\delta_i = \frac{1}{2}(u_i + u_{i-1})t_i \qquad [4.41c]$$

$$P/P_o = f(\delta/D) \qquad [4.41d]$$

where t_i is the transit time of the shock wave across the ith ring, u_{i-1} and u_i are the initial and final velocities of the deformed part of the system during

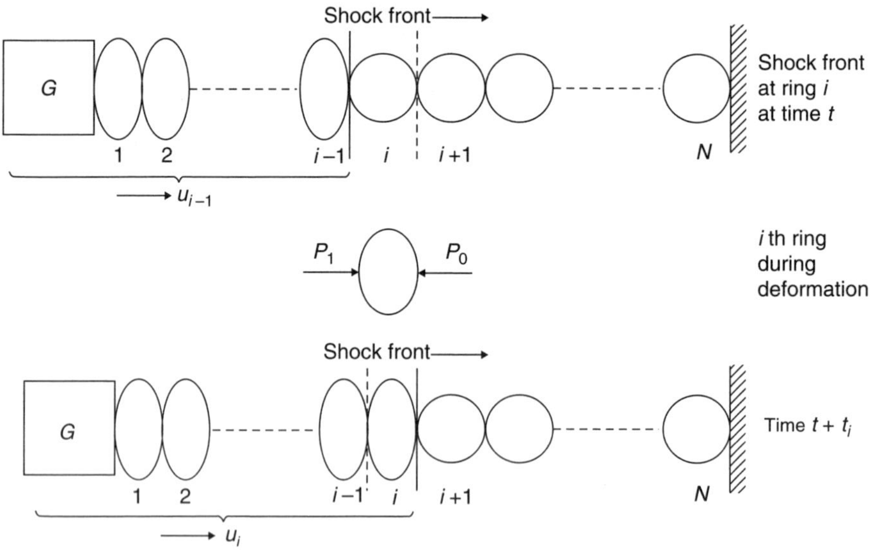

4.17 Model showing propagation of a structural shock wave through *i*th ring.

time interval t_i, and P_i is the force determined from the non-dimensional load–deflection curve of a ring. The first two equations are conservation of momentum for the deformed part of the system and ring i respectively. Equations 4.41(a–d) can be solved numerically for force, velocity and ring compression.

Ring compression predicted by this shock theory is shown in Fig. 4.16 together with the experimental results. In general, agreement between theory and experiment is very good, although the predicted values are about 10% higher than the experimental results. Further modifications incorporating elastic and strain rate effects were made by Reid and Bell (1984). Consideration of elastic waves explains the observed plastic deformation at the distal end of the system. In the context of dynamic crushing of cellular solids, Shim *et al.* (1990) proposed a mass-spring model to account for similar effects, which will be presented in Section 10.4.2.

The more complex deformation of free systems with 'wrap-around' effect may not be well accounted for by the shock theory presented above (see discrepancy between theory and experiment for free system in Fig. 4.16). Detailed finite element analysis carried out by Lim (2001), under the supervision of D. Shu, reproduced the observed non-uniform deformation within the system (Fig. 4.18). ***LS-DYNA3D*** was used in this study.

4.7 Lateral crushing of arrays of circular tubes

Arrays of parallel, thin-walled circular tubes have been compressed between parallel flat plates by Shim and Stronge (1986b). Specimens include steel, brass and aluminium alloy tubes, approximately 0.7 mm thick with both diameter and length about 12.7 mm. Such an arrangement can be an energy-absorbing system itself, but it also represents the micro-structural behaviour of cellular solids (see Chapter 10).

Two packing arrangements of tubes are possible: ***square*** or ***hexagonally packed***. The quasi-static crushing process and corresponding non-dimensional load–deflection curves are shown in Figs 4.19(a) and (b)

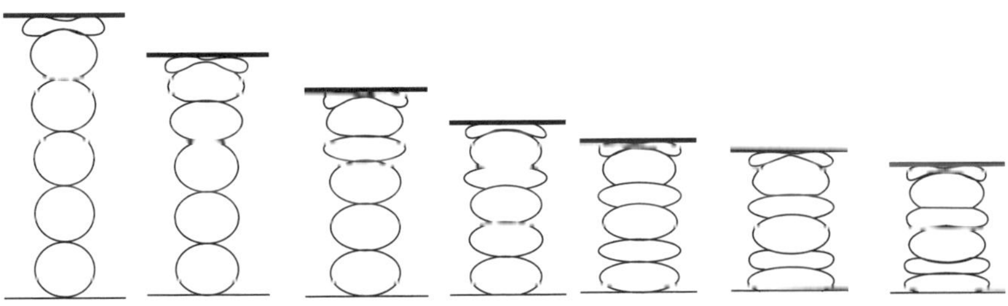

4.18 Deformation stages of a six-ring system from LS-DYNA3D (Lim, 2001).

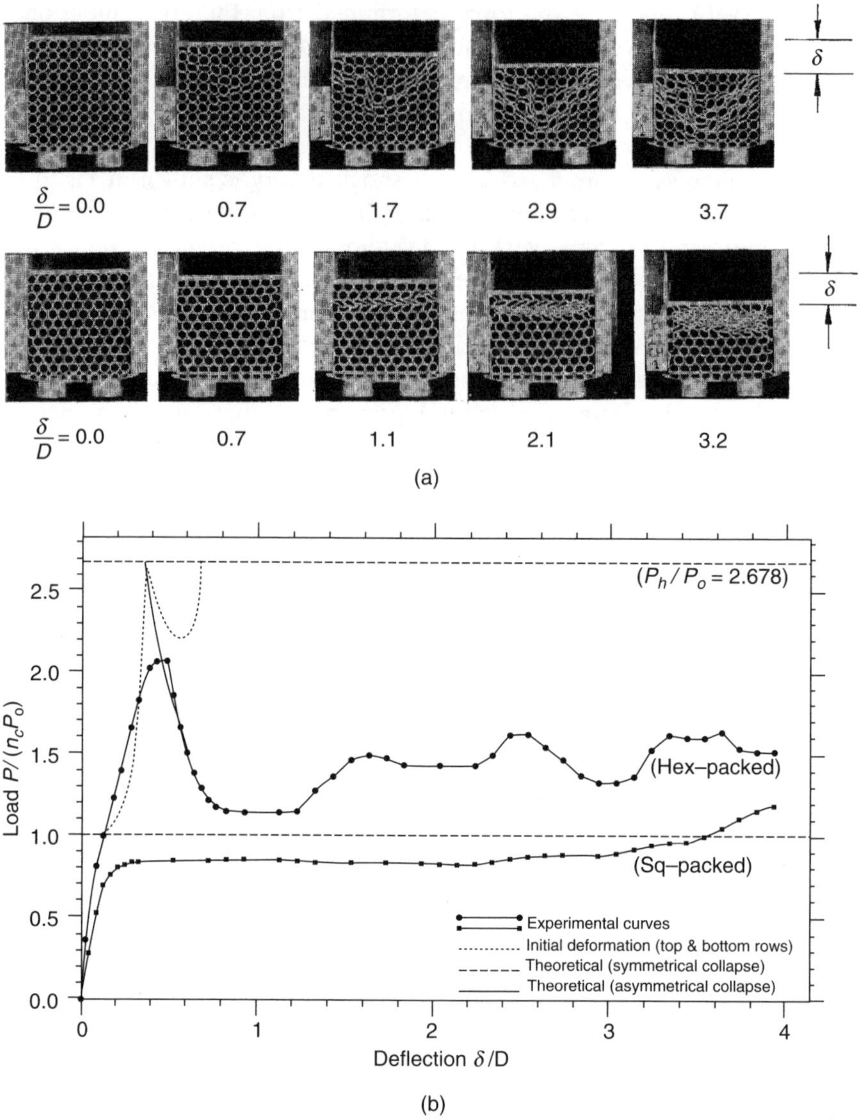

4.19 Compression of brass tube arrays with $n_c = 10$ and $h = 0.71$ mm: (a) top: square-packed; bottom: hexagonally packed; (b) corresponding non-dimensional load–deflection curves (Shim and Stronge, 1986b) (reproduced with kind permission of Elsevier).

respectively, for brass tube arrays. The collapse load for one tube with sides constrained, P_o, is given by Eq. [4.26] (one tube has four evenly spaced equal loads acting on it). The number of tube columns is $n_c = 10$. (Note that for hexagonally packed arrays, this number alternates between 9 and 10 by virtue of the packing configuration.)

For square packed arrays, asymmetrical modes of deformation were observed. A narrow 'V' shaped band forms ahead of the loading platen. This ***localisation*** is associated with a softening behaviour of the ring under asymmetrical collapse mode, as illustrated by Shim and Stronge (1986b). For hexagonally packed arrays, tubes collapse from the top row and then deformation propagates after one row is crushed; the localised band is horizontal.

Theoretical initial collapse load for square packed arrays is simply n_cP_o (P_o being the collapse load for a single tube as given in Eq. [4.26]). For hexagonally packed arrays, each tube is subjected to six equal and evenly spaced forces. Considering the forces acting on a ring (Fig. 4.20) and using the equivalent structure technique, one immediately obtains the initial collapse load as

$$P_{oh} = \frac{4M_p}{R}(3+2\sqrt{3}) \qquad [4.42]$$

Hence, P_{oh} is about $2.678P_o$.

Dynamic tests of the two systems described above were conducted by Stronge and Shim (1987). A dimensionless parameter (***impact energy ratio***) is defined as

$$Q = \frac{Gv_o^2}{2n_cDP_o} \qquad [4.43]$$

where G and v_o are the impact mass and velocity, respectively. The term n_cDP_o represents the idealised energy dissipation capacity of a row of tubes. Different modes of collapse of individual tubes were observed, depending on the packing configuration and the value of Q. The ratio of final crush

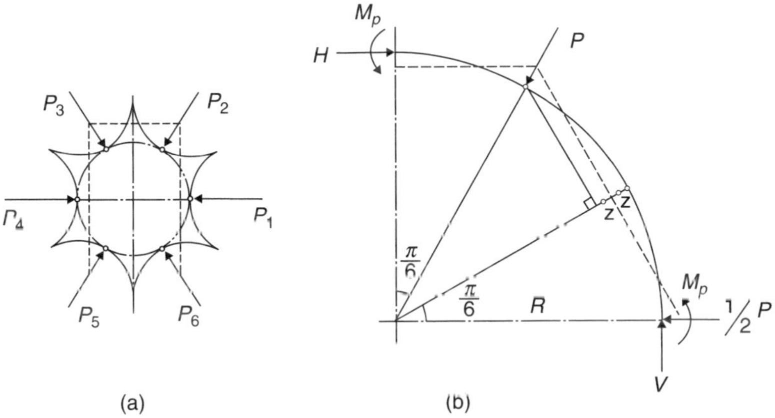

4.20 Forces acting on a tube within a hexagonally packed system (a) and one quadrant (b) (Shim and Stronge, 1986b) (reproduced with kind permission of Elsevier).

height δ/D to Q is almost constant: it is about unity for square packed arrays and 0.6 for hexagonally packed systems.

4.8 Other ring/tube systems

Triangular arrays of metal rings/tubes have been used as energy-absorbing devices (Carney *et al.*, 1982) (Fig. 4.21). The system was subjected to compression along the axis of symmetry and ***oblique compression*** at 15° to the axis. A 30 % reduction in energy absorbed was observed for oblique impact. Carney (1993) discussed practical energy-dissipating devices for highways using tube systems.

A crossed-layer tube system was tested by Johnson *et al.* (1977a). The system consisted of several layers. Within each layer, all circular tubes were parallel and tubes of adjacent layers were perpendicular to each other. When the spacing between tubes within a layer is small, each tube can be considered to be individually in compression with lateral constraints as discussed previously. Moreover, when spacing is large, deformation is no longer uniform along the tube axis direction, but is three-dimensional and the collapse mode is much more complex.

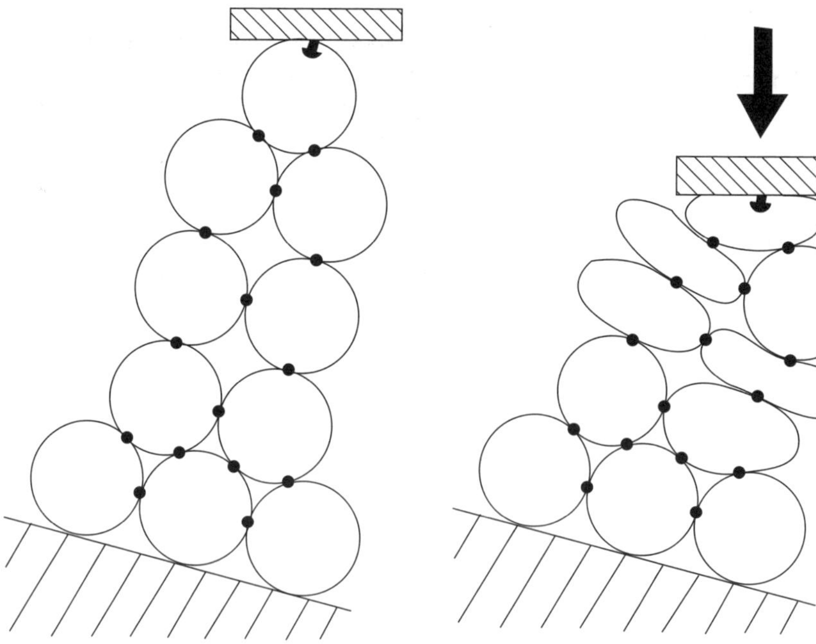

4.21 Triangular ring system used by Carney *et al.* (1982).

4.9 Concluding remarks

It is clear that rings and tubes under in-plane crushing are efficient energy-absorbing components. When crushed between two flat plates, the force–displacement curve is almost flat for a large range of deflections, which is advantageous for an energy-absorber. Further constraint of tubes by means of a V-block or bracing changes the collapse mode and could enhance energy-absorption capacity as more volume would be subject to plastic deformation.

The strain-hardening effect increases loads, not only by enhancing flow stress as a result of high strain, but also by causing plastic deformation to spread over a zone. Thus the geometry of the collapsed structure can be different from the case where plastic deformation is concentrated at localised hinges, which changes the force level. It is therefore possible to adjust the force–displacement curve by selecting materials with different strain-hardening characteristics.

Impact loading can lead to a localised deformation within ring systems, due to the inertia effect. This changes the force level experienced by each tube and hence the total energy absorbed. The main feature observed in this case is localised deformation, which can be explained by the structural plastic shock wave which propagates from the impact end. The strain-rate effect seems less important for such a ring system.

Tubes can be arranged into various tubular systems in order to obtain desirable energy-absorption performance. Examples include closely packed tubes, crossed-layer tubes and triangular formed circular tube systems, which have been used in various practical applications.

5
Thin-walled members under transverse loading

In Chapter 4, rings/tubes underwent two-dimensional plastic deformation; there was no variation of parameters in the third, out-of-plane direction. However, deformation will be three-dimensional when a tube is subjected to local loads. In this chapter, we discuss cases of tubes under local transverse denting by a boss or a wedge and also the bending of rectangular and square tubes, as well as the bending of thin-walled members of channel and angle sections.

5.1 Circular tube under point loading

A cylindrical shell subjected to point loads undergoes local deformation (Fig. 5.1). The plastic deformation zone is of elliptical shape and enlarges with the force magnitude. Maximum deformation occurs at the loading point. Details of contours of constant deflection and the sequence of generator profiles are shown in Fig. 5.2(a) (Morris, 1971) for a tube of free ends, and Fig. 5.2(b) for a tube of fully fixed ends. Two equal and opposite loads were applied by means of two bosses. The tube diameter to thickness ratio is $D/h = 105$, and the length L to diameter ratio is $L/D = 3.4$.

A shorter tube ($L/D = 1.03$) with both ends fully constrained produced much smaller deflections for the same load (Fig. 5.2(b)) than those with free ends (Fig. 5.2(a)). The load–deflection curves for the same end-constrained tube with load applied by different size bosses demonstrates the significance of boss size (Fig. 5.3). $M_o = Yh^2/4$ is fully plastic bending moment per unit length, where Y is the yield stress. Here a ***non-dimensional boss size parameter*** is identified as $\rho = r/\sqrt{Rh}$, where r is the radius of the boss and R is the tube radius. A larger boss leads to a higher load for a given deflection. However, it should be pointed out that the results may not be as sensitive to the indenter size if other indentation arrangements are made, such as using a flat surface to compress the cylindrical shell (Stronge, 1993).

A simple upper-bound calculation was performed by Morris and Calladine (1971), following successful application of their method to a

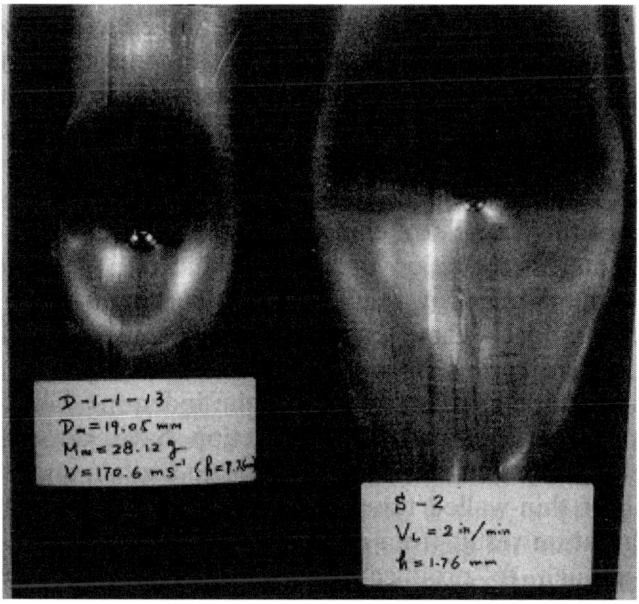

5.1 Photographs of steel tubes after impact (left) and quasi-static indentation (right) (Stronge, 1993).

circular plate under point load (Calladine, 1968). In their calculation for tubes, the elliptical deforming zone is approximated by a series of trapezoidal elements connected by straight plastic hinge lines. By examining an 'area' diagram, the position of the neutral axis was worked out. This calculation took into account the stretching effect of the mid-surface, as well as the large change in geometry. Theoretical results from this scheme broadly agree with those of experiments (Fig. 5.3). Note that in Fig. 5.3, the load is non-dimensionalised with respect to the initial collapse load:

$$P_o = 2\pi M_o \qquad [5.1]$$

which is the same as the collapse load of a centrally loaded circular plate with its edge simply supported (Calladine, 1968).

The tube response is affected by the geometry of the indenter. Stronge (1993) and Corbett *et al.* (1991) used a hemi-spherical indenter to investigate the deflection and perforation of a cylindrical shell, both statically and dynamically. Impact loading leads to a more localised deformation (Fig. 5.1), due to the inertia effect. Empirically, Stronge (1985) proposed that the energy corresponding to the ***ballistic limit***, which is defined as the velocity when the projectile is either stuck in the target or else exits with negligible velocity, for a spherical-nosed missile with diameter 6.35 mm ≤ d ≤ 12.7 mm perforating cold-drawn mild steel tubes (1.2 mm ≤ h ≤ 3.2 mm) is

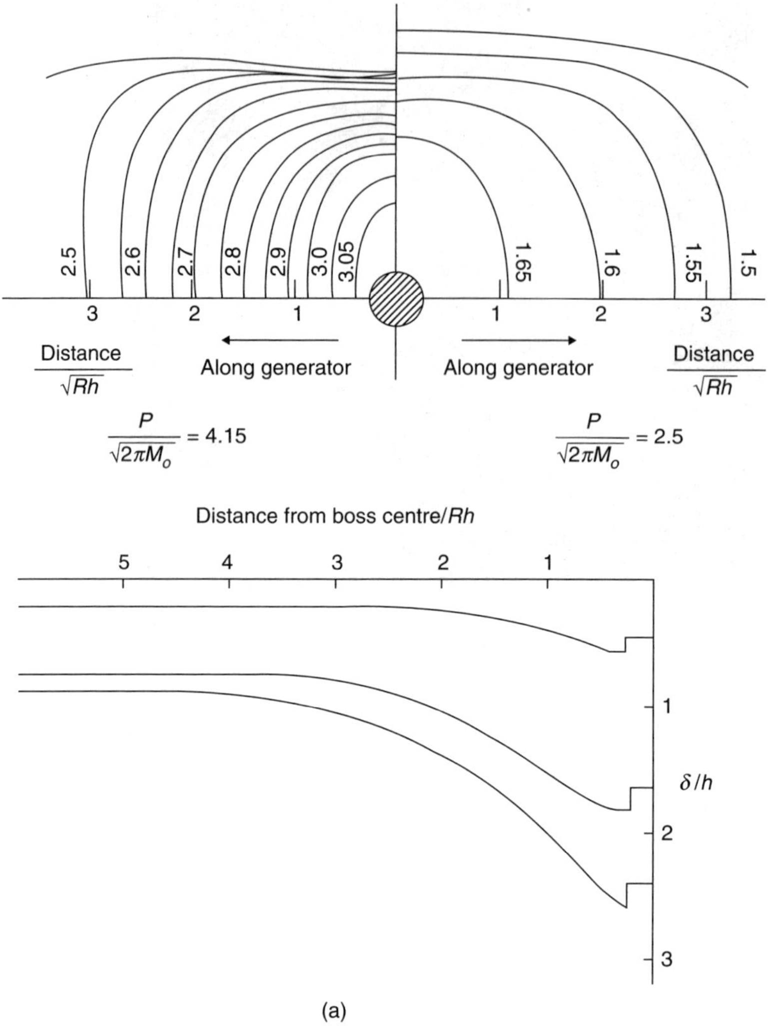

5.2 Top: contours of constant deflection for statically indented tubes at two values of punch force. Bottom: sequence of generator profiles; (a) free ends with $L/D = 3.4$; (b) fully clamped ends with $L/D = 1.03$ (Morris, 1971) (reproduced with kind permission of the Council of the Institution of Mechanical Engineers).

$$W = Ch^{1.63}d^{1.48} \quad [5.2]$$

where C is a constant, but of dimension J mm$^{-3.11}$. Corbett *et al.* (1991) subsequently concluded that the above formula works for $1.66\,\text{mm} \leq h \leq 5.0\,\text{mm}$, but not for rolled and welded tubes.

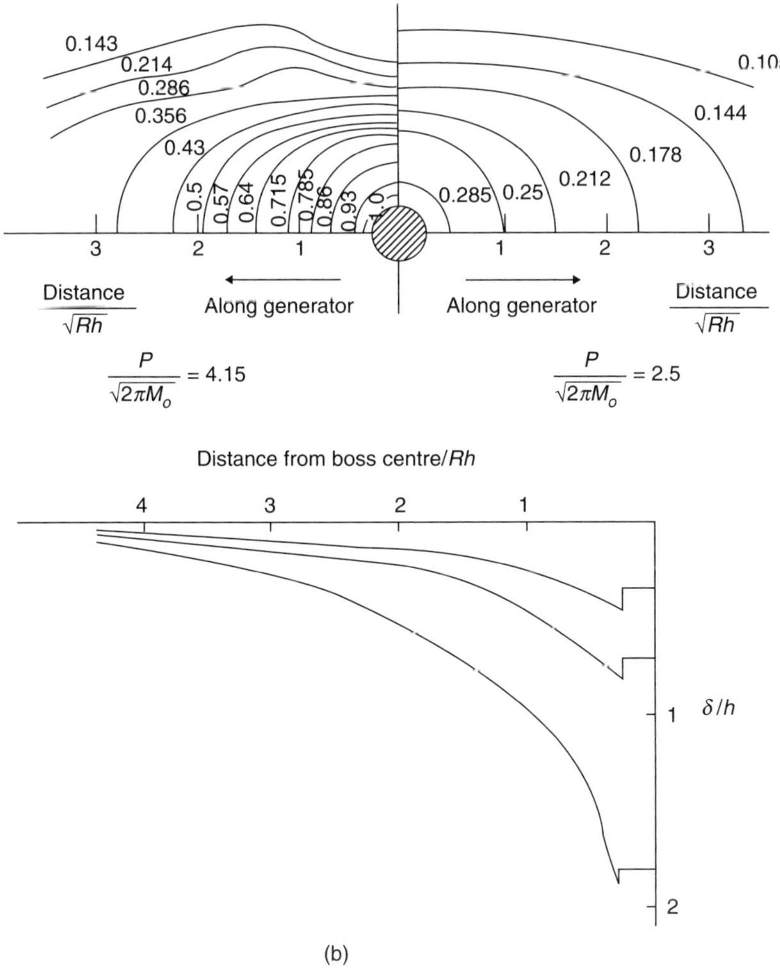

5.2 Continued

5.2 Indentation of a circular tube by a blunt wedge

Another form of tube indentation is achieved by a pressing a wedge with a tip radius of 3~5 mm into the tube wall. The deformed shape is sketched in Fig. 5.4. Plastic deformation is confined to the vicinity of the indenter, ABCD in Fig. 5.4(b) and the size of the plastic zone increases with wedge deflection. Hence, plastic hinges AB, AD, BC and CD travel during deformation and areas ABD and BCD are almost flat. A simple theoretical model due to de Oliveira *et al.* (1982) is described here; experimental results will be presented later.

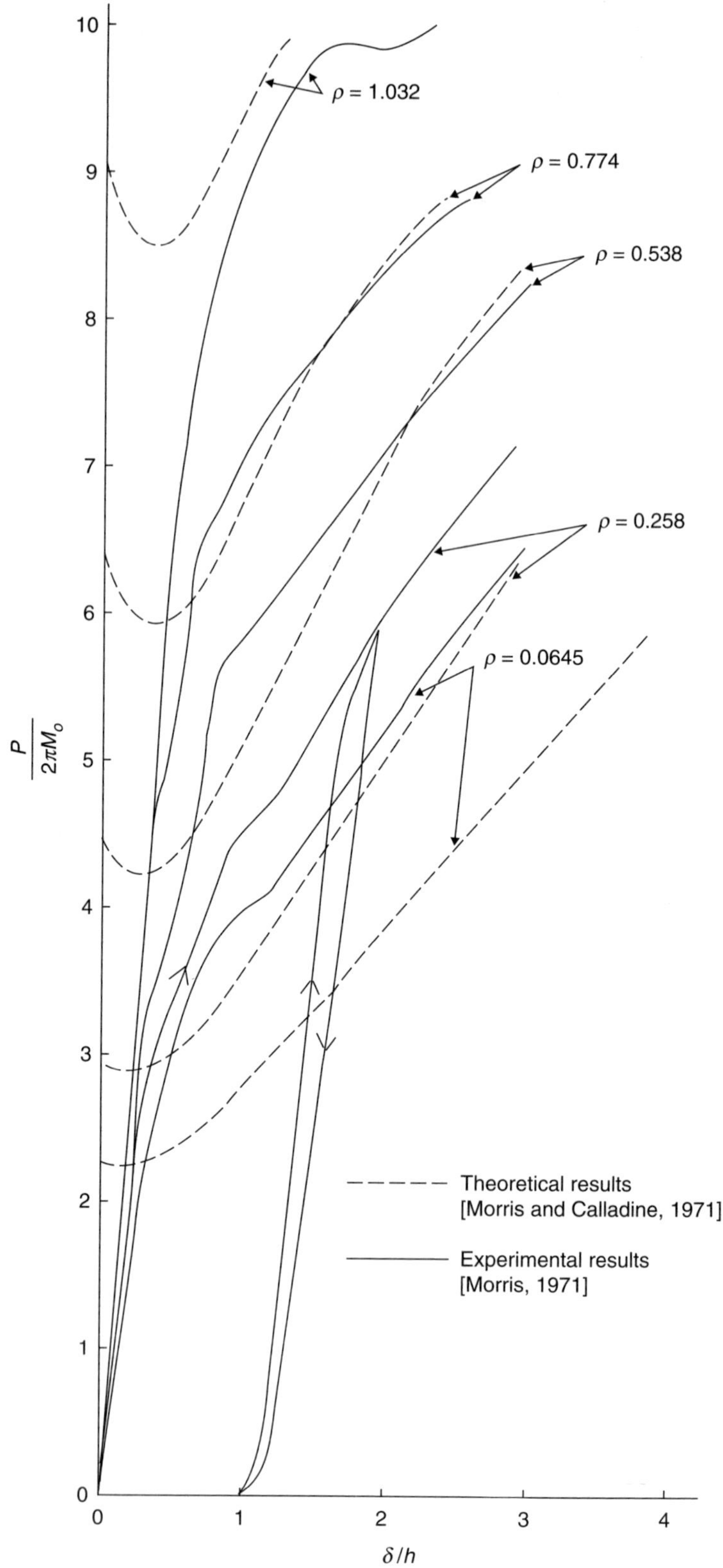

5.3 Experimental and theoretical non-dimensional load–displacement curves for boss-loaded cylindrical shell with fully clamped ends.

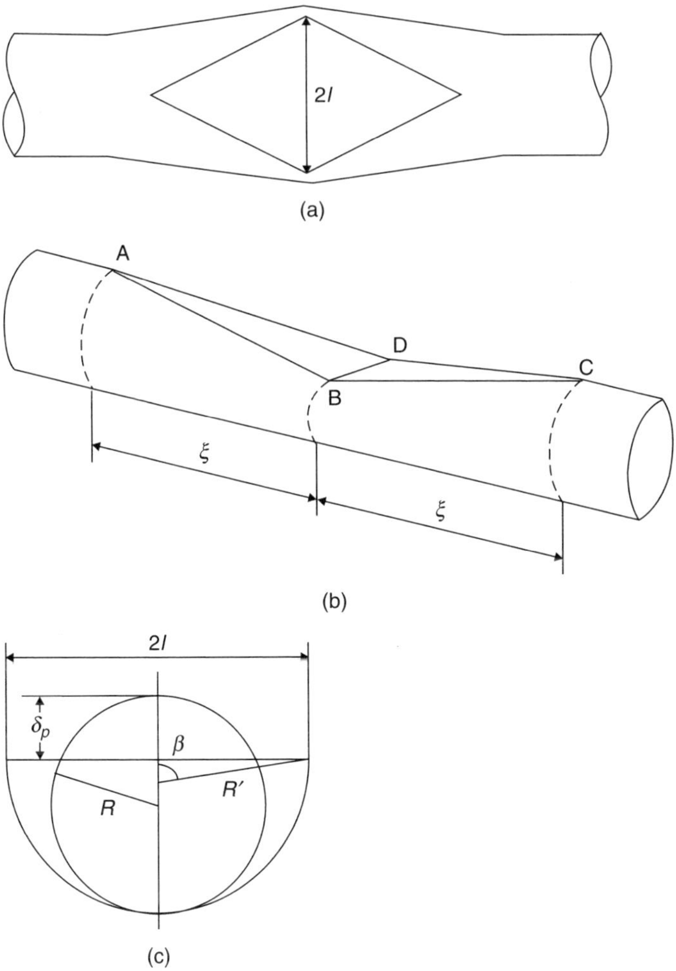

5.4 Indentation of a tube by a wedge indenter. Assumed hinge lines: (a) and (b); and central cross-section (c).

The central cross-section is assumed to be of a circular arc closed with a straight line matching the wedge tip BD, see Fig. 5.4(c). The overall plastic zone size is defined by characteristic length ξ in Fig. 5.4(b) and width BD is $2l$. Strictly speaking, all the travelling plastic hinges must have a (small) radius, instead of being of sharp creases. However, for the present model, this point is not investigated. The ends of the tube can be either free or fully fixed.

For the assumed central cross-section geometry, because the circumferential length remains the same, we have a new radius for the current part:

$$R' = \frac{\pi R}{\pi - \beta + \sin\beta} \quad [5.3]$$

The ***plastic dent depth***

$$\delta_p = \frac{R(2\sin\beta + \pi - 2\beta - \pi\cos\beta)}{\pi - \beta + \sin\beta} \quad [5.4]$$

and the half width of the flat part is

$$l = R'\sin\beta \quad [5.5]$$

The fully plastic bending moment of the central cross-section for the current configuration is straightforward to work out; it is

$$M = \frac{\pi^2(2\sin\alpha - \sin\beta + \sin\beta\cos\beta)}{2(\pi - \beta + \sin\beta)^2} M_t \quad [5.6]$$

where $M_t = YD^2h$, the fully plastic bending moment of the circular tube, and $\alpha = \frac{1}{2}(\pi + \beta - \sin\beta)$. Numerical calculations lead to an approximate equation for the bending moment M

$$M = (1 - \delta_p / D)M_t \quad [5.7]$$

This indicates that the fully plastic bending moment capacity decreases linearly with plastic dent depth. This result is very similar to the behaviour of an equivalent initially square cross-section of side length $a = \pi D/4$, being deformed into a rectangular section with a height reduction of δ_p.

The rate of external work should be equal to rate of the plastic energy dissipation, as discussed in Section 2.2. The unknown parameters ξ and l are obtained by minimising the external work. de Oliveira *et al.* (1982) found that

$$\xi = D\left\{\frac{\pi\delta_p}{4h}\left[1 - \frac{1}{2}\left(\frac{N}{N_p} - 1\right)^2\right]\right\}^{\frac{1}{2}} \quad [5.8]$$

and

$$P = \frac{4M_t}{D}\left\{\frac{\pi h\delta_p}{D^2}\left[1 - \frac{1}{2}\left(\frac{N}{N_p} - 1\right)^2\right]\right\}^{\frac{1}{2}} \quad [5.9]$$

where $N_p = \pi DhY$ is the fully plastic axial force of the tube and N is the axial force generated within the tube. $N/N_p = 0$ and 1 for a tube with free ends and fully fixed ends, respectively.

Wierzbicki and Suh (1988) subsequently improved the above analysis by considering a series of more realistic deformed sectional shapes of the tube, connected by strings in the longitudinal direction. The final equations they obtained are

$$\xi = \frac{D}{2}\left\{\frac{2\pi\delta_p}{3h}\left[1-\frac{1}{4}\left(1-\frac{N}{N_p}\right)^3\right]\right\}^{\frac{1}{2}} \qquad [5.10]$$

and

$$P = 4Yh^2\left\{\frac{\pi D\delta_p}{3hR}\left[1-\frac{1}{4}\left(1-\frac{N}{N_p}\right)^3\right]\right\}^{\frac{1}{2}} \qquad [5.11]$$

The notation used is the same as for Eqs (5.8) and (5.9).

Tube denting experiments were performed by Reid and Goudie (1989), following preliminary tests by Thomas *et al.* (1976). Mild steel seamed tubes ($D = 50.8$ mm, $h = 1.6$ mm) were either simply supported or fully fixed at the ends and then loaded at mid-span by means of a wedge-shaped indenter. In such an experiment, local denting of the tube occurs initially, with little tube global deformation. The total indenter displacement δ_{tl} is always larger than the local plastic dent depth δ_p.

Experimentally, it was found that δ_p is linearly proportional to δ_{tl} (e.g. $\delta_p = 0.835\delta_{tl}$ for a fully fixed tube with length $L = 305$ mm). As deformation proceeds, the fully plastic bending moment at the central section reduces as a result of the change in geometry [Eq. 5.6]. When the indenter force P produces a bending moment equal to this moment capacity (i.e. $PL/4 = M$), structural collapse of the tube occurs, similar to that of a solid beam section under three-point loading (Fig. 2.12). For fully fixed end tubes, axial stretching develops.

From the empirical relation between local plastic dent depth δ_p and ***total indenter displacement*** δ_{tl}, Eq. [5.9] can be re-cast in terms of δ_{tl}. Also, for a given load P the elastic deformation of the tubular beam can be determined using the usual beam theory and hence a 'theoretical' load–deflection curve can be produced. Figure 5.5 compares the results from experiments and theory using this approach (Reid and Goudie, 1989) for 457 mm span tubes with ends free and fully fixed. The curves indicate that elastic-plastic analysis agrees better with the experiments than rigid plastic theory.

Jones and his co workers (Jones *et al.*, 1992; Jones and Shen, 1992) further modified the theoretical analysis and conducted extensive impact tests using a drop hammer on steel tubes with $D/h = 11$~60 and $L/D \approx 10$. Fully clamped tubes were impacted by a rigid wedge indenter at mid-span, quarter span or near a support. Typical results are shown in Fig. 5.6 in terms of plastic deformation and input energy for a mild steel tube: $L = 600$ mm, $D = 60$ mm, $h = 2$ mm. Yield stress is used in the theoretical calculations. The

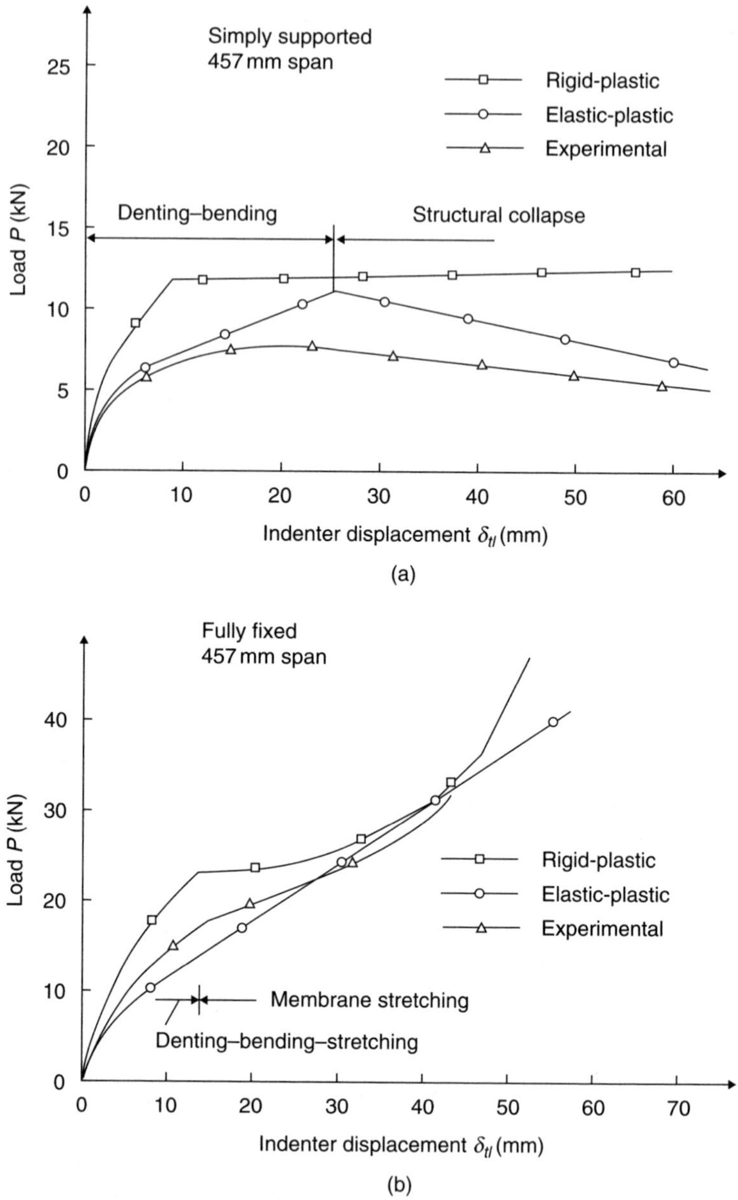

5.5 Theoretical and experimental load–indenter displacement curves: (a) simply supported; (b) fully fixed (Reid and Goudie, 1989) (reproduced with kind permission of John Wiley & Sons Inc.).

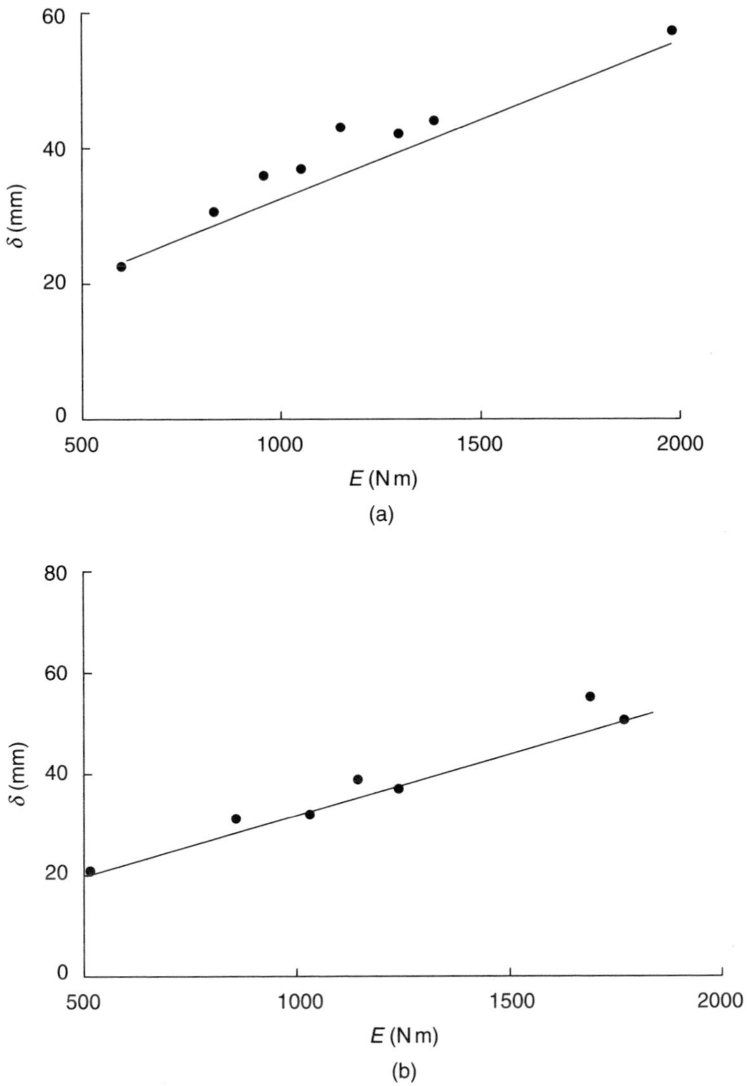

5.6 Comparison of tube deflection from theory (straight line) and impact experiments (dots): (a) impact at mid-span; (b) impact at one quarter span (Jones *et al.*, 1992) (reproduced with kind permission of the Council of the Institution of Mechanical Engineers).

deflection is slightly less if an average of yield stress and ultimate stress is taken as the flow stress.

Impact energy leading to tube material rupture was studied by Shen and Chen (1998). Denting of a tube by two wedge indenters was reported by

Watson *et al.* (1976) and Lu (1993b). Empirical formulae were given by Lu (1993b) and Ong and Lu (1996). In particular, for mild steel tubes with $L/D = 10$, $P = 3.78Y\delta^{0.47}h^{1.6}D^{-0.07}$, where δ is the indenter displacement. Kardaras and Lu (2000) conducted a finite element analysis of tube indentation by point loads.

5.3 Bending collapse of thin-walled members

5.3.1 Square and rectangular sections

Tubes of square and rectangular cross-sections are representative of thin-walled beams in vehicle and building structures. Their ***bending collapse*** behaviour is important in assessing the energy absorption of the structure as a whole. Bus rollover is such an example where most of the energy is absorbed by being at general plastic hinges in the frame structure. Kecman (1983) first produced a theoretical model for this problem and his analysis is described here, although other theoretical treatments have also been conducted recently (Wierzbicki *et al.* 1994a and 1994b; Kim and Reid, 2001a).

Figure 5.7 shows a photograph of an actual ***plastic hinge*** formed by bending a mild steel square hollow section (38 × 38 × 1.6 mm) tube as a 1 m long cantilever. The detailed bending mechanism involves initial bulging of the side webs, followed by a well-developed collapse mechanism with ***travelling hinge*** lines. At a much later stage (rotational angle $\theta = 25$–$30°$), the travelling hinge lines stop and additional hinge lines develop. The

5.7 Typical plastic hinge formed during bending of rectangular tube (Kecman, 1983) (reproduced with kind permission of Elsevier).

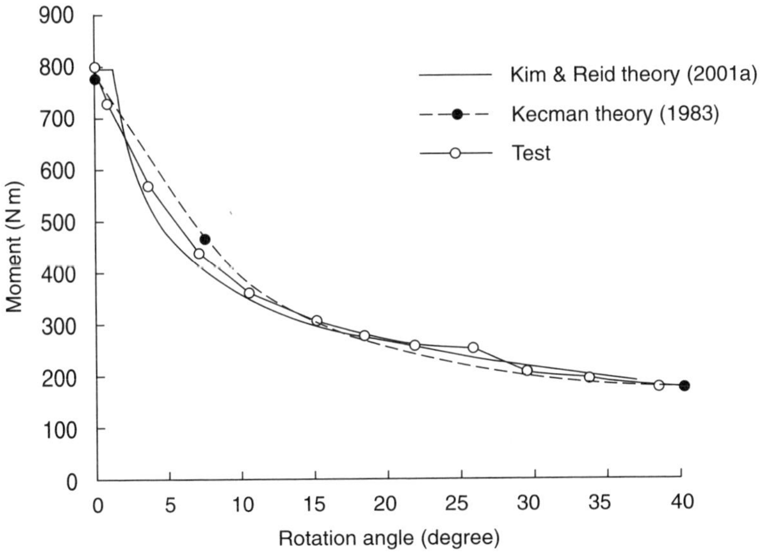

5.8 Moment–rotation curves from experiment and theories (Kim and Reid, 2001a) (reproduced with kind permission of Elsevier).

bending mechanism terminates when jamming occurs between the two buckled halves of the compression flange. Further bending initiates a secondary hinge. A typical moment–plastic rotation angle curve is shown in Fig. 5.8, for a rectangular mild steel tube (50.8 × 38.1 × 1.26 mm) (Y = 253 MPa, σ_u = 284 MPa; bending about the minor axis). The moment-carrying capacity decreases dramatically as the plastic hinge rotation angle increases. Note that in this plot the elastic deflection has been subtracted and hence the moment does not start from zero.

A theoretical treatment of such cross-sections was given by Kecman (1983). The maximum bending strength is governed by elastic buckling of the compression flange for thin-walled sections, or yielding of the material for relatively thick ones. After initial buckling of the flange, its load-carrying capacity reduces and an effective flange width concept is used. The equations for the ***maximum bending moment*** at the onset of collapse were thus obtained (Kecman and Suthurst, 1984). For a rectangular section of width a, depth b and thickness h, the critical stress of the compression flange is

$$\sigma_{cr} = 0.9E\left(\frac{h}{a}\right)^2\left(5.23 + 0.16\frac{a}{b}\right) \qquad [5.12]$$

where E is the material elastic modulus.

Hence, if $\sigma_{cr} < Y$ with Y being the material yield stress,

$$M_{\max} = Yhb^2 \frac{2a + b + a\left(0.7\frac{\sigma_{cr}}{Y} + 0.3\right)\left(3\frac{a}{b} + 2\right)}{3(a+b)} \qquad [5.13]$$

If $2Y \leq \sigma_{cr}$

$$M_{\max} = M_p = Yh\left[a(b-h) + \frac{1}{2}(b-2h)^2\right] \qquad [5.14]$$

If $Y \leq \sigma_{cr} < 2Y$,

$$M_{\max} = Yhb\left(a + \frac{b}{3}\right) + \frac{\sigma_{cr}}{Y}\left[M_p - Yhb\left(a + \frac{b}{3}\right)\right] \qquad [5.15]$$

The moment–rotation curve can be obtained from an idealised collapse mechanism, Fig. 5.9(a). Here, wall deformation occurs by bending along straight hinge lines only, and the wall is inextensible. During deformation, point A moves downwards; hence eight travelling (also known as rolling) hinges are involved, such as AG, AE, AK and AL, with the remaining hinges, such as KG, KL, LE, GB, BE and AJ, being stationary. The length of the stationary hinges does not change except for AB, which increases. From the geometry (Figs 5.9(b) and (c)), coordinates of point B are

$$x_B = H, \quad Y_B = b\cos\rho - \sqrt{b\sin\rho(2H - b\sin\rho)}, \quad z_B = 0 \qquad [5.16]$$

where $\rho = \theta/2$ and H is the half hinge length.

Because the length of the middle section remains the same, with $y_A = y_B$

$$b = z_A + \sqrt{y_A^2 + z_A^2} \qquad [5.17]$$

Solving for z_A:

$$z_A = b\sin^2\rho - H\sin\rho + \sqrt{b\sin\rho(2H - b\sin\rho)}\cos\rho \qquad [5.18]$$

Similarly, continuity in the longitudinal direction for a fibre originally parallel to the tube axis and passing through point A leads to

$$H^2 = [H - (b - z_A)\sin\rho]^2 + [y_A - (b - z_A\cos\rho)]^2 + z_A^2 \qquad [5.19]$$

Substituting Eqs [5.16–5.18] into [5.19] and recognising that Eq. [5.19] must hold for any values of θ and a/b, we obtain

$$2H = a \text{ or } 2H = b$$

Taking the smaller value between a and b (so as to minimise energy)

$$2H = \min\{a, b\} \qquad [5.20]$$

The above equation defines the overall size of the plastic hinge. The rotation corresponding to jamming when points G and E meet is

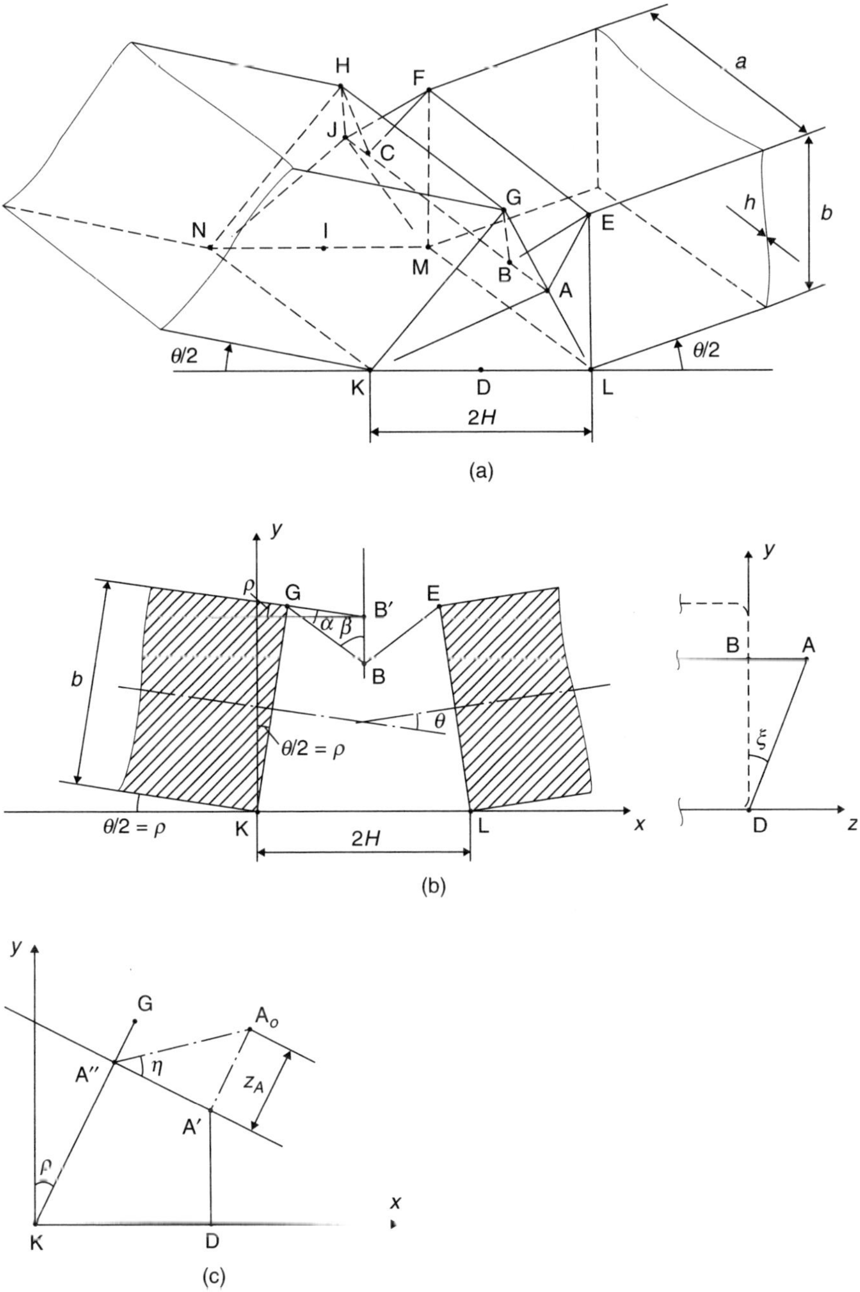

5.9 Idealised bending collapse mechanism for rectangular thin-walled tube (a) and corresponding geometry detail (b, c) (Kecman, 1983) (reproduced with kind permission of Elsevier).

$$\theta_J = 2\arcsin\left(\frac{H-0.5h}{b}\right) \quad [5.21]$$

The moment–rotation relationship is obtained by first evaluating the total energy absorbed by the hinge lines for a given hinge rotation $\theta\,(=2\rho)$. The angle of rotation at GH and EF is

$$\alpha = \pi - \frac{\pi}{2} - \rho - \beta = \frac{\pi}{2} - \rho - \arcsin\left(1 - \frac{b}{H}\sin\rho\right) \quad [5.22]$$

and the plastic bending energy absorbed is

$$W_1 = W_{EF+GH} = 2M_o a\left[\frac{\pi}{2} - \rho - \arcsin\left(1 - \frac{b}{H}\sin\rho\right)\right] \quad [5.23]$$

where $M_o = Yh^2/4$ is the fully plastic bending moment per unit length and H is the half hinge length.

For BC

$$W_2 = W_{BC} = M_o a\left[\pi - 2\arcsin\left(1 - \frac{b}{H}\sin\rho\right)\right] \quad [5.24]$$

For AB and CJ

$$\begin{aligned} W_3 &= W_{AB+CJ} \\ &= 2M_o\left[b\sin^2\rho - H\sin\rho + \sqrt{b\sin\rho(2H - b\sin\rho)}\cos\rho\right] \\ &\quad \times\left[\pi - 2\arcsin\left(1 - \frac{b}{H}\sin\rho\right)\right] \end{aligned} \quad [5.25]$$

Note that the length of AB is z_A and its angle of rotation is $\pi - 2\beta$. The energy absorbed for BG, BE, CH and CF does not change with θ, and is:

$$W_4 = W_{BG+BE+CH+CF} = 4M_o H\frac{\pi}{2} = 2M_o H\pi \quad [5.26]$$

Also

$$\begin{aligned} W_5 &= W_{GK+EL+HN+FM} \\ &= 4M_o b\arctan\left[\frac{z_A}{\sqrt{(H - x_{A''})^2 + (y_{A''} - y_B)^2}}\right] \end{aligned} \quad [5.27]$$

where $y_{A''}$ and $x_{A''}$ are given by (Fig. 5.9(c)):

$$y_{A''} = \frac{H\tan\rho + b\cos\rho - \sqrt{b\sin\rho(2H - b\sin\rho)}}{1 + \tan^2\rho} \quad [5.28]$$

$$x_{A''} = y_{A''}\tan\rho \quad [5.29]$$

The energy absorbed by the four top half travelling hinge lines is equal to the area swept multiplying by the mean curvature and M_0 (refer to Section 6.2.3), i.e.

$$W_6 = W_{GA+AE+CH+CF} = 4\frac{2M_o}{r}\frac{Hz_A}{2} - 4M_o\frac{H}{r}z_A \qquad [5.30]$$

where r is the radius of curvature and Kecman (1983) assumed, empirically

$$r = r(\theta) = \left(0.07 - \frac{\theta}{70}\right)H \qquad [5.31]$$

Similarly, for the four bottom half travelling hinge lines, assume that the curvature varies linearly along KA, i.e. at a distance l_K from K

$$r_{KA} = \frac{KA}{l_K}r \qquad [5.32]$$

Also, the rolled length is assumed as

$$l_r = \frac{l_K}{KA}z_A \qquad [5.33]$$

and hence for KA

$$W_{KA} = \int 2M_o\frac{l_r}{r_{KA}}\mathrm{d}l_K = 2M_o\int_0^{KA}\frac{l_K}{KA}z_A\frac{l_K}{KA\cdot r}\mathrm{d}l_K = \frac{2M_o z_A KA}{3r}$$

So

$$W_7 = W_{KA+LA+NJ+MJ} = \frac{8}{3}M_o\frac{z_A}{r}\sqrt{H^2 + y_B^2 + z_A^2} \qquad [5.34]$$

Finally, $\xi = \arctan(z_A/y_A)$ (Fig. 5.9(b)), and

$$W_8 = W_{KN+LM+KL+MN} = 2M_o\left[a\rho + 2H\arctan\left(\frac{z_A}{y_A}\right)\right] \qquad [5.35]$$

Therefore, the total energy absorbed by plastic bending along all the hinge lines is the sum of all the eight energy components:

$$W(\theta) = \sum_{i=1}^{8} W_i(\theta) \qquad [5.36]$$

The bending moment at any hinge rotation θ can be obtained numeri cally by taking a small increment $\Delta\theta$ and

$$M(\theta) = \frac{W(\theta + \Delta\theta) - W(\theta)}{\Delta\theta} \qquad [5.37]$$

Note that the above analysis is valid for a well-developed mechanism at a given hinge rotation θ prior to jamming. It is therefore not applicable at the initial collapse stage; the theoretical moment thus obtained is much higher than the experimental one. Kecman approximated the initial *M–θ* ***curve*** by drawing a straight line from M_{max} tangential to the numerically calculated *M–θ* curve according to the above model. Denoting the contact point as θ_T, the energy is then, for $0 < \theta \leq \theta_T$

$$W(\theta) = 0.5\left[M_{\max}\left(2-\frac{\theta}{\theta_T}\right)+\frac{\theta}{\theta_T}M(\theta_T)\right]\theta \qquad [5.38]$$

The *M–θ* curve after jamming was given, empirically, for $\theta > \theta_j$, as

$$M(\theta) = M(\theta_J) + 1.4[M_{\max} - M(\theta_J)](\theta - \theta_J) \qquad [5.39]$$

Theoretical results based on the above procedure agree well with the test results (Fig. 5.8). Note that for better agreement with the test results, the ultimate stress, instead of the yield stress, should be used in calculating M_o. This model has been improved by Kim and Reid (2001a) who used a toroidal surface at A, in order to make the mechanism kinematically admissible. Stretching is therefore invoked. The unknown parameters such as rolling radius were obtained completely theoretically by minimising the total energy. Their theoretical results are also plotted in Fig. 5.8, showing good agreement with the test results. Pure bending tests of steel tubes were also reported by Cimpoeru and Murray (1993) and Kim *et al.* (1997). ***Biaxial bending*** of the rectangular section was investigated by Brown and Tidbury (1983) and more recently by Kim and Reid (2001b). Further impact studies of box structures have been conducted by Zhou *et al.* (1990).

5.3.2 Circular tubular sections

Bending collapse of thin-walled beams of circular tubular cross-section exhibits a similar softening behaviour to that of rectangular tubular sections discussed above. The moment decreases rapidly with rotation (Fig. 5.10) for steel tubes with $D/h = 26\sim57$ (Mamalis *et al.*, 1989). Note that here the angle of rotation is the total angle including elastic rotation. A plug was used to clamp one end of the tube in these tests, which used a cantilever setup. As for the rectangular sections, here the compression side may deform inwards in the same manner as the denting of a circular tube discussed before. However, quite often a bulge forms on the compression face, which replaces a great part of the triangular region. Fracture may also occur in the tension face at the clamping end.

Mamalis *et al.* (1989) proposed a ***bending collapse mechanism*** (Fig. 5.11) which is almost identical to that of tube indentation (Fig. 5.4). The tube wall

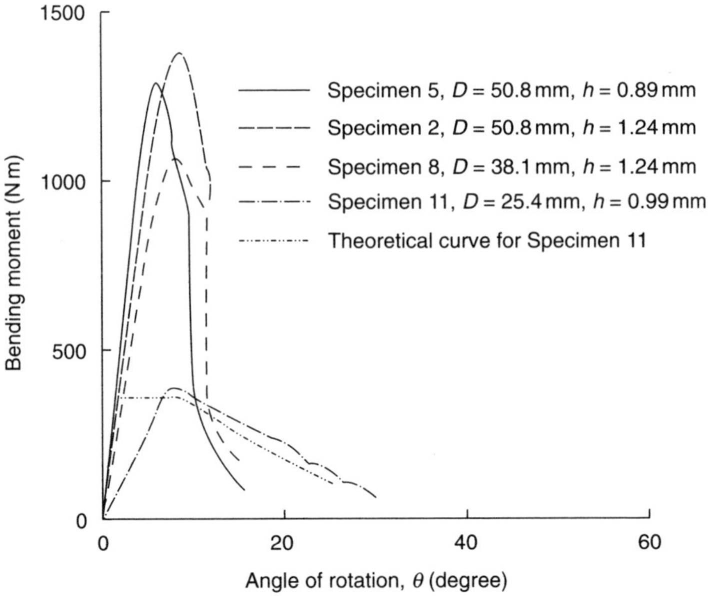

5.10 Moment–rotation curves from experiments and theory for bending of circular thin-walled tubes (reproduced with kind permission of the Council of the Institution of Mechanical Engineers).

was assumed to be inextensional and plastic bending energy was expressed as follows.

For flattening of triangular regions AEC and AFC

$$W_1 = 4M_o R\phi_o^2 \tag{5.40}$$

where ϕ_o^2 specifies the extent of the straight line at the central cross-section (Fig. 5.11(c)), and is related to the rotation θ. $M_o = Yh^2/4$ is the fully plastic bending moment per unit length, as before.

For flattening of the circular region

$$W_2 = 4M_o R(\pi - \phi_o)(\phi'_o - \phi_o) \tag{5.41}$$

For hinge line AC

$$W_3 = ACM_o(\pi - 2\alpha) = 2\phi_o RM_o(\pi - 2\alpha) \tag{5.42}$$

Finally, for oblique hinge lines AE, AF, CE and CF

$$W_4 - 4\int_0^{th} M_o \frac{x\phi'_o}{l}\,dx - 2M_o \frac{l_h^2}{l}\phi'_o \tag{5.43}$$

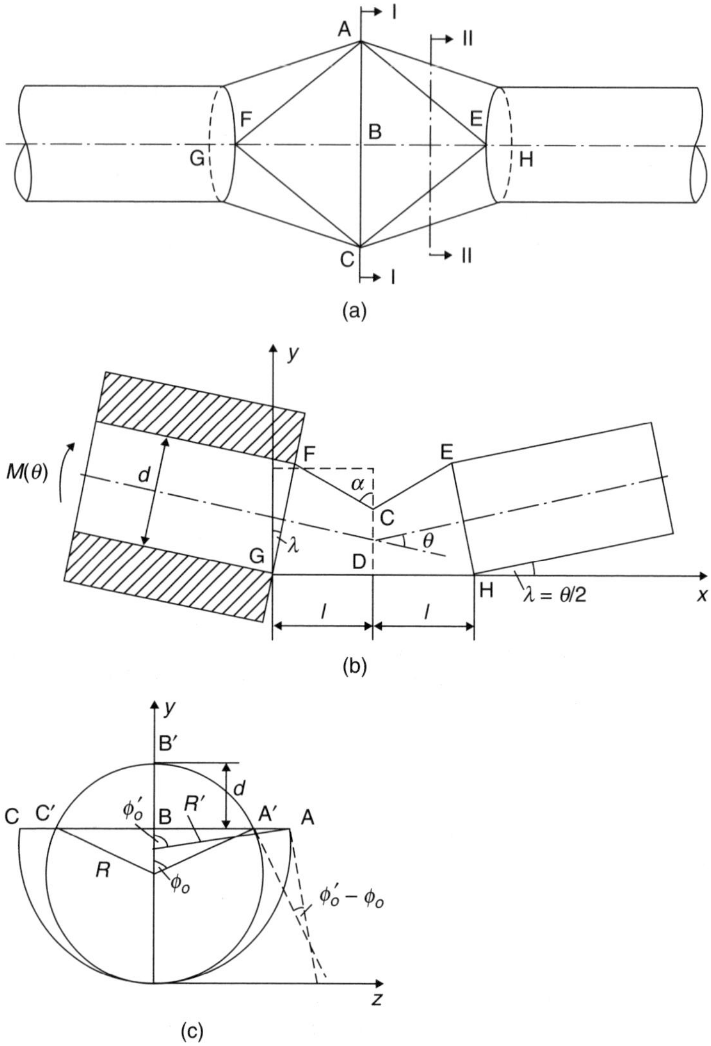

5.11 Assumed collapse mechanism of a thin-walled circular tube (a) and corresponding detail geometry (b, c) (reproduced with kind permission of the Council of the Institution of Mechanical Engineers).

The total energy is $W(\theta) = W_1 + W_2 + W_3 + W_4$, and the moment–rotation curve can be obtained in exactly the same way as for rectangular sections. This analysis appears to agree fairly well with the experimental data (Fig. 5.10).

Yu *et al.* (1993) investigated the large deformation mechanism and cross-sectional distortion of a circular tubular cantilever beam subjected to a concentrated force at the tip and produced an analysis of the post-collapse curves, together with experiments.

5.3.3 Bending collapse of channel sections

Plastic hinge lines develop as beams of channel section are subjected to bending, as was the case for rectangular tubular cross-sections. In principle, one can use the same approach as for rectangular and circular tubular sections presented above, in order to work out the load–deflection curve. Nevertheless, we note that this method of bending energy consideration seems less successful in the initial collapse stage before the mechanism is fully developed. Consideration of equilibrium for suitable strips containing the plastic hinge lines may overcome the above shortcoming of the energy method, as proposed by Murray (1983). Before we discuss the load–deflection curve for a channel section under bending from central loading, we will present the ***strip method***.

Pin-ended strut of a rectangular cross-section

Consider a strut with pinned ends (Fig. 5.12), which collapses with a central plastic hinge. This central plastic hinge forms as a result of both the axial force and the bending moment. Hence Eq. [2.22] applies as a general yield criterion for a solid rectangular section of depth h and width b. For a central deflection Δ, we have

$$P\Delta = M = M_p\left[1-\left(\frac{P}{N_p}\right)^2\right] \qquad [5.44]$$

Here as before, $M_p = Ybh^2/4$ and $N_p = Ybh$.

Solving Eq. [5.44] explicitly for P

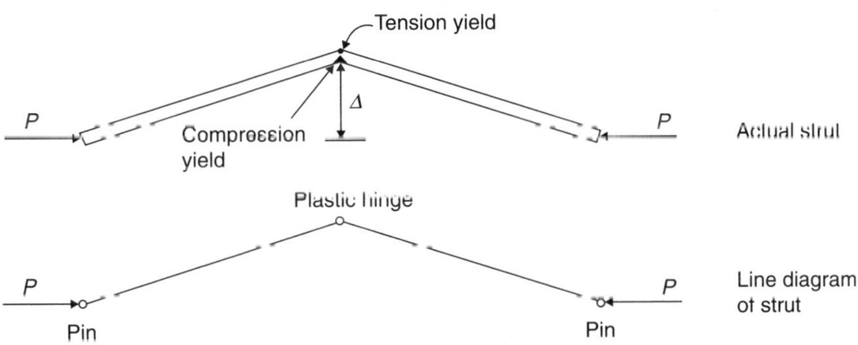

5.12 A strut with pinned ends.

$$\frac{P}{N_p} = -\frac{N_p \Delta}{2M_p} + \left[\left(\frac{N_p \Delta}{2M_p^2} \right)^2 + 1 \right]^{\frac{1}{2}} \qquad [5.45]$$

or

$$\frac{P}{N_p} = -\frac{2\Delta}{h} + \left[\left(\frac{2\Delta}{h} \right)^2 + 1 \right]^{\frac{1}{2}} \qquad [5.46]$$

The end shortening δ can be easily determined from the geometry. For a beam of initial length $2L$

$$\delta = 2\left[L - (L^2 - \Delta^2)^{\frac{1}{2}} \right] \qquad [5.47]$$

or, expanding the term in the bracket using the binomial theorem

$$\delta = \frac{\Delta^2}{L}\left[1 + \frac{1}{4}\left(\frac{\Delta}{L}\right)^2 + \frac{1}{8}\left(\frac{\Delta}{L}\right)^4 + \ldots \right] \qquad [5.48]$$

For small values of Δ/L (≤ 0.5)

$$\delta \approx \frac{\Delta^2}{L} \qquad [5.49]$$

The error thus introduced is less than 8%.

Equations [5.46] and [5.49] give a theoretical P–δ curve after eliminating Δ. By integrating this curve, the energy absorbed up to a particular load P_1 is

$$W_1 = \frac{M_p^2}{LN_p}\left[-\frac{8}{3} + \frac{2N_p}{P_1} + 2\frac{P_1^3}{3N_p^3} \right] \qquad [5.50]$$

For large deflections, the bending moment is dominant and P is small compared with the initial collapse load. The axial force effect can then be neglected in the yield condition and we simply have

$$P = \frac{M_p}{\Delta} \qquad [5.51]$$

This strut collapse problem will also be discussed in Section 7.2.2 in the context of inertia-sensitive structures.

Fixed-end strut of rectangular cross-section

The above analysis can be applied to the case where both ends of the strut are fixed. A plastic hinge develops at each end, in addition to the central one. Equations [5.44] and [5.46] become, respectively

$$P\Delta = 2M_p\left[1-\left(\frac{P}{N_p}\right)^2\right] \qquad [5.52]$$

and

$$\frac{P}{N_p} = -\frac{\Delta}{h}+\left[\left(\frac{\Delta}{h}\right)^2+1\right]^{\frac{1}{2}} \qquad [5.53]$$

The energy absorbed corresponding to any load P_1 is four times that of the pin-ended strut, and hence

$$E_2 = 4E_1 = \frac{4M_p^2}{LN_p}\left(-\frac{8}{3}+\frac{2N_p}{P_1}+\frac{2P_1^3}{3N_p^3}\right) \qquad [5.54]$$

Moment-carrying capacity of inclined plastic hinges

In the struts discussed above, the yield line of the central plastic hinge is perpendicular to the direction of the axial force P. When the yield line is inclined with respect to this force direction (hinge line AB in Fig. 5.13), the effective moment-carrying capacity perpendicular to the direction of thrust is (Murray, 1973)

$$M_p'' = M_p \sec^2 \beta \qquad [5.55]$$

where M_p is the moment-carrying capacity when the hinge is perpendicular to the axial force [Eq. 5.44]. This expression was later modified by Zhao and Hancock (1993), but we shall use the above equation in the following analysis.

Basic plastic collapse mechanisms

True plastic collapse mechanisms observed in experiments may be idealised into ones which are composed of several basic folding mechanisms, as summarised in Table 5.1 (Murray and Khoo, 1981). Here a positive hinge indicates rising up out of the plate plane and a negative hinge means deflecting down below the plane. Each basic mechanism can be represented using a

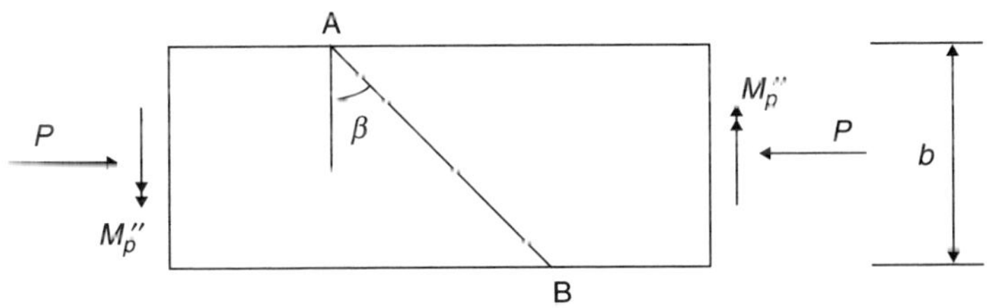

5.13 Plastic moment-carrying capacity of an inclined hinge AB.

Table 5.1 Basic folding mechanisms and force–displacement relationship (Murray and Khoo, 1981)

$K_1 = 1 + \sec^2\beta$; $K_2 = \sec^2\beta$; $K_3 = 2\sec^2\beta$

1	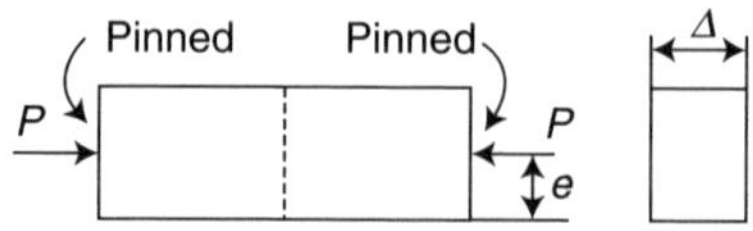	$P = YHb\left[\sqrt{\left(\frac{2\Delta}{H}\right)^2 + 1} - \frac{2\Delta}{H}\right]$ $e = b/2$
2	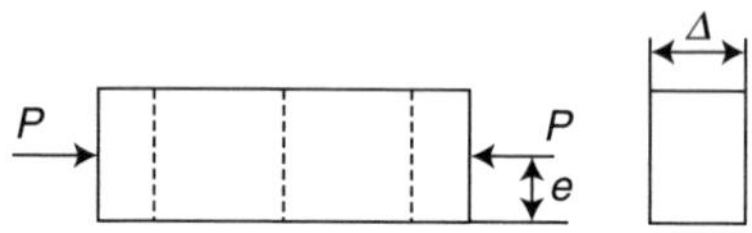	$P = YHb\left[\sqrt{\left(\frac{\Delta}{H}\right)^2 + 1} - \frac{\Delta}{H}\right]$ $e = b/2$
3	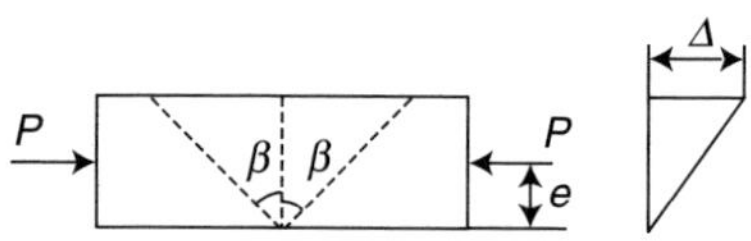	$P = \frac{YHb}{2}\left[\sqrt{\left(\frac{2\Delta}{K_1H}\right)^2 + 1} - \frac{2\Delta}{K_1H} + \frac{K_1H}{2\Delta}\ln\left\{\sqrt{\left(\frac{2\Delta}{K_1H}\right)^2 + 1} + \frac{2\Delta}{K_1H}\right\}\right]$ $Pe = \frac{YH^3b^2K_1^2}{12\Delta^2}\left[\left\{\left(\frac{2\Delta}{K_1H}\right)^2 + 1\right\}^{\frac{3}{2}} - 1 - \left(\frac{2\Delta}{K_1H}\right)^3\right]$
4	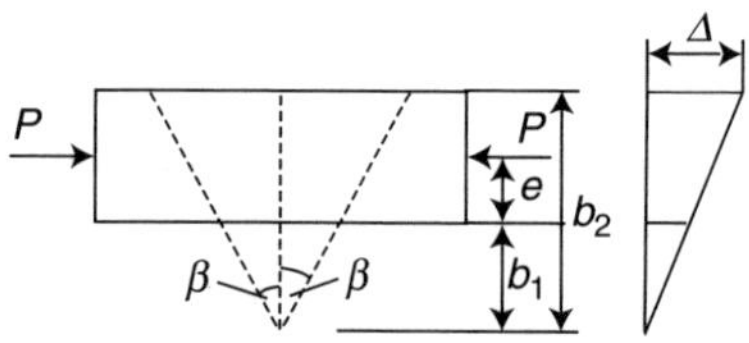	Obtain solution by using the difference between two Type 3 mechanisms $P = P_1 - P_2$ $Pe = P_1e_1 - P_2e_2$
5	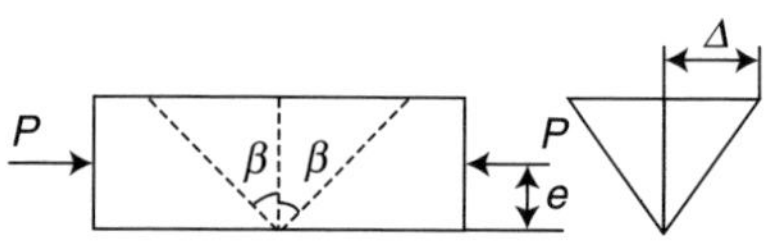End panels twist freely	Same equations as for Type 3 but replace K_1 by K_2
6	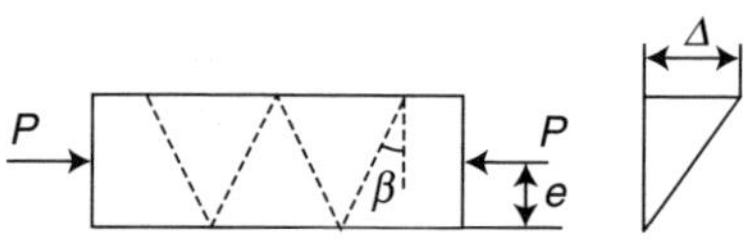All hinges inclined at β	Same equations as for Type 3 but replace K_1 by K_3
7	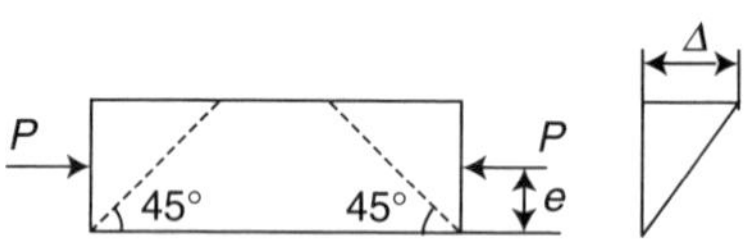	Same equations as for Type 5 but with $\beta = 45°$
8	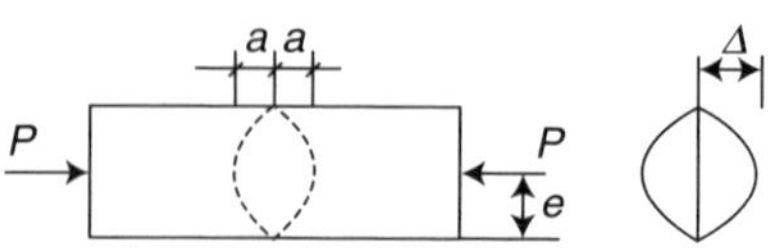End panels twist and bend freely	$P = \frac{YHb}{6}\left[1 - \frac{2\Delta}{H} + \sqrt{\left(\frac{2\Delta}{H}\right)^2 + 1} - \frac{6\Delta}{H(1 + 4a^2/b^2)} + 4\sqrt{\left\{\frac{3\Delta}{2H(1 + 4a^2/b^2)}\right\}^2 + 1}\right]$

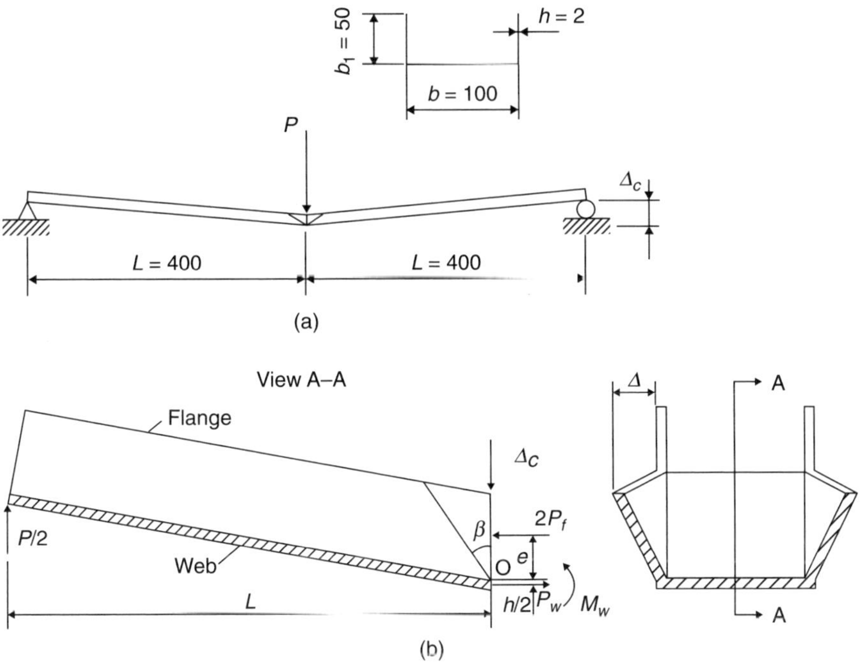

5.14 (a) A channel section beam under central loading; (b) details of equilibrium for one half of the beam (Murray, 1983).

series of finite strips (viz. the strut of width b discussed before). The relationship between force P and out-of-plane deflection Δ can be worked out accordingly, as per Table 5.1. The parameter β, which specifies the hinge inclination, is not well defined; usually different values of β are tried and the one which gives the lowest energy is chosen. In most cases, the result is not very sensitive to the value of β. The following example demonstrates this method of static analysis.

Channel beam with a central load P

Figure 5.14 shows a channel beam simply supported at both ends (Murray, 1983). Length $2L = 800$ mm, $b = 100$ mm, $b_1 = 50$ mm, $h = 2$ mm and $Y = 250$ MPa. The collapse mechanism developed is assumed to be as shown in Fig. 5.14(b). Equilibrium considerations lead to

$$2P_f = P_w \qquad [5.56]$$

where P_f is the compressive force in each flange and P_w is the tensile force in the web. Rotational equilibrium of one half of the beam about O gives

$$\frac{PL}{2} = M_w + 2P_f e + \frac{P_w h}{2} \tag{5.57}$$

The P_f–Δ (or $P_f e$–Δ) relationship for this mechanism is given in Table 5.1 (Mechanism 3). A numerical procedure is used to obtain the P–Δ_c curve, Δ_c is the central deflection. Assume a value of β and then for increasing values of Δ, calculate P_f from Table 5.1 and P_w from Eq. [5.56]. Using P_w, the reduced moment M_w is calculated from the yield condition, Eq. [2.22]. Substituting M_w and $P_f e$ (from Table 5.1) into Eq. [5.57], force P can be calculated. In this way, the P–Δ ***curve*** can be obtained. Repeat this process for a few values of β and take the one which gives the lowest load-carrying capacity. The axial shortening arising from Δ at the tip of the flange for half a beam is $\Delta^2/(2B_1 \tan\beta)$. The central deflection is therefore

$$\Delta_c = \frac{\Delta^2 L}{2B_1^2 \tan\beta} \tag{5.58}$$

The P–Δ_c curve thus obtained is shown in Fig. 5.15. Line 0E represents the elastic response. The 'best' value of β is 45°, but the force is not sensitive to β. The force corresponding to first yielding in the flange is 3.5 kN, and the force corresponding to a fully plastic hinge at the centre is 6.25 kN. Theoretical curves from this approach agree well with the four-point bending

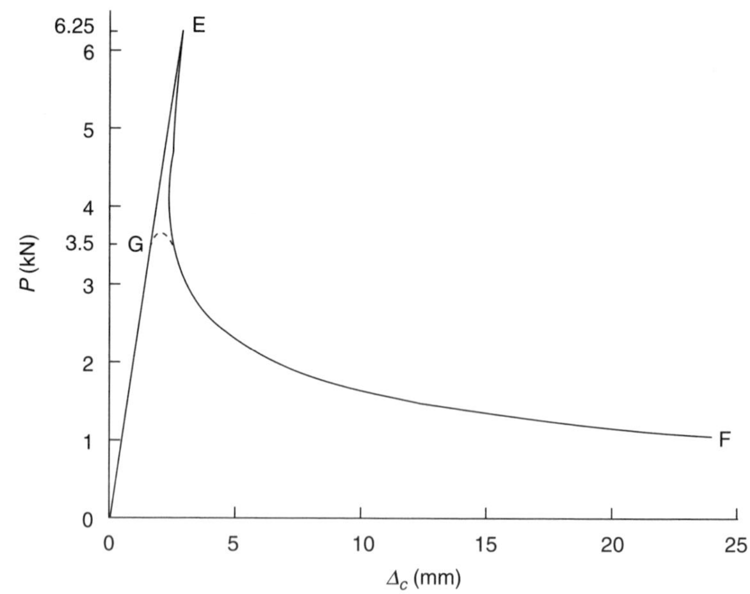

5.15 Theoretical load–central deflection curve for channel shown in Fig. 5.14(a).

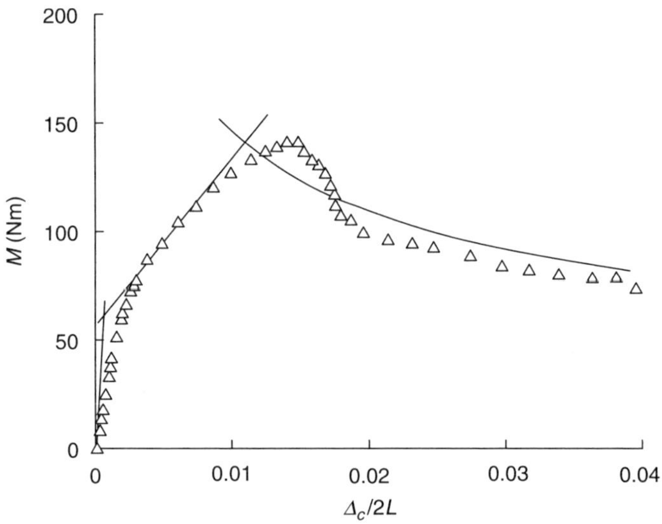

5.16 Moment against non-dimensionalised central deflection for channel tested by Fok *et al.* (1993): theoretical curves and experimental results (triangles) (reproduced with kind permission of the Council of the Institution of Mechanical Engineers).

test results (e.g. Fig. 5.16) by Fok *et al.* (1993). However, unlike Fig. 5.15, a post-buckling stage exists before plastic collapse in this case ($b_1 = 51$ mm, $b = 52$ mm, $h = 0.88$ mm, $2L = 660$ mm, $Y = 289$ MPa).

5.3.4 Bending of angle sections

When a thin angle section is subjected to bending, localised plastic deformation with yield lines is likely to occur, similar to the channel section discussed above. Nevertheless, for thick or moderately thick angle sections, deformation is global and no localised hinge lines occur. Yu and Teh (1997) have demonstrated this for aluminium alloy angle beams of equal flange length w = 25.4~50.8 mm and thickness h = 1.59~3.18 mm. Four-point bending tests showed a softening moment–average curvature relationship (Fig. 5.17). Two major plastic energy dissipation mechanisms exist: longitudinal bending and opening-up of the cross-section (Fig. 5.18).

Assume a rigid-linear hardening moment curvature relationship for the cross-section [Eq. 4.32]. Here the fully plastic bending moment is $M_p = (\sqrt{2}/4)Yhw^2$. Let the plastic segment length be λ. From the geometry of the bent segment (Fig. 5.18(b))

$$c = \frac{\lambda^2}{8R} - \frac{1}{8}\lambda^2\kappa \qquad [5.59]$$

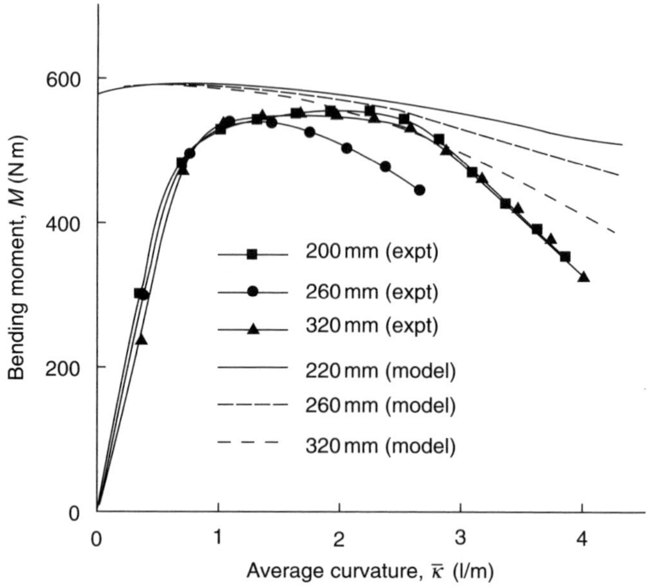

5.17 Moment versus average curvature of angle section under pure bending for three different values of *b*, the spacing between the two applied loads (Yu and Teh, 1997) (reproduced with kind permission of Elsevier).

where c denotes the height of the circular arc formed by the bent centroidal line, R and κ are radius of curvature and curvature, respectively, of the bent centroidal line of the plastic segment. For a change of inclination angle ϕ (Fig. 5.18(a))

$$\kappa = \frac{\phi}{\lambda} \qquad [5.60]$$

Thus, Eq. [5.59] can be rewritten as

$$c = \frac{1}{8}\lambda^2\kappa = \frac{1}{8}\lambda\phi \qquad [5.61]$$

The opening-up of the cross-section is shown in Fig. 5.18(c), with one flange moving from original position EF to E′F′. Assume that the flange remains straight and F′ lies on the line EF. The value of c is

$$c = d - d_1 - d_2 = \frac{w}{\sqrt{2}}\sin\left(\frac{\pi}{4}+\frac{\Delta\theta}{2}\right) - \frac{w}{2\sqrt{2}} - \frac{w}{2\sqrt{2}}\cos\left(\frac{\pi}{4}+\frac{\Delta\theta}{2}\right) = \frac{w}{2\sqrt{2}}\left(3\sin\frac{\Delta\theta}{2}+\cos\frac{\Delta\theta}{2}-1\right) \qquad [5.62]$$

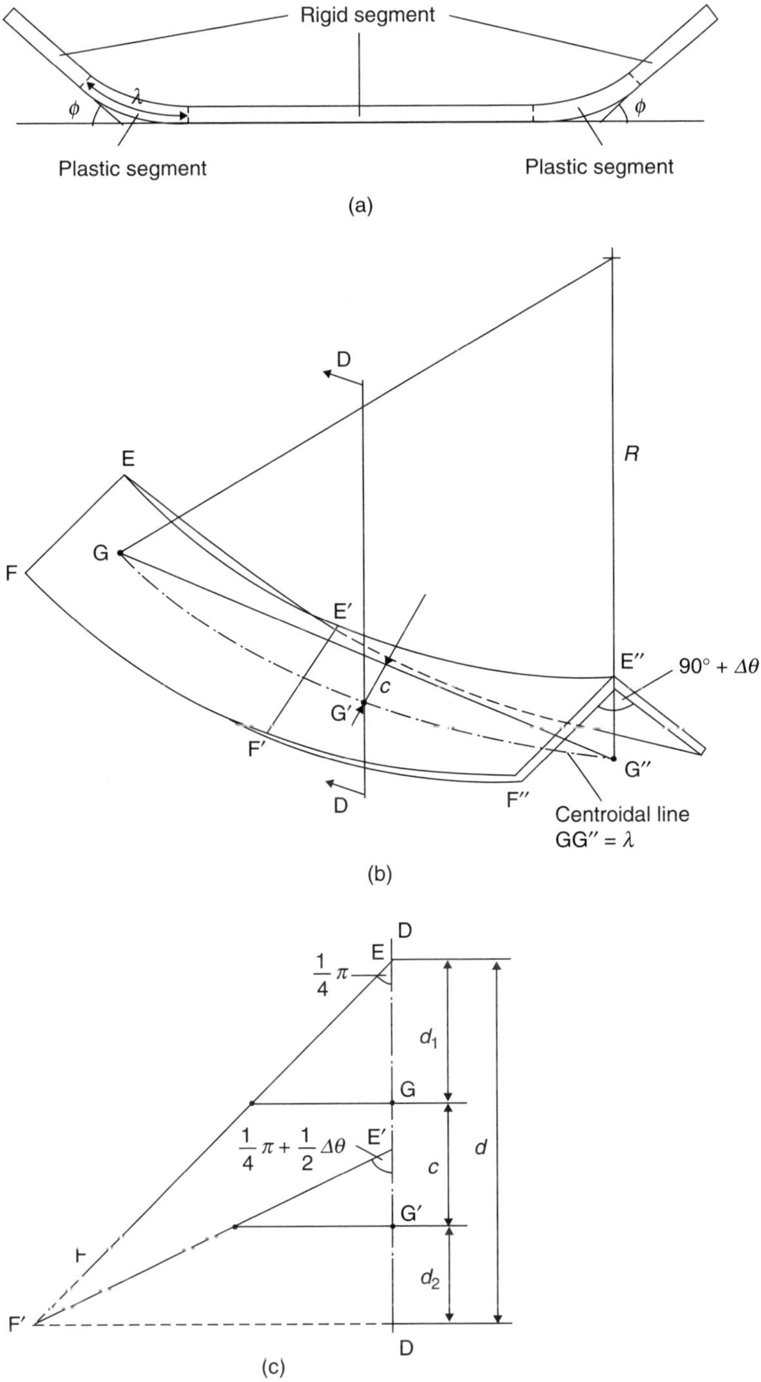

5.10 Plastic deformation mechanism assumed by Yu and Teh (1997) for angle section: (a) overall section of rigid and plastic bending segments; (b) details of a bending segment; (c) details of opening-up of the cross section (reproduced with kind permission of Elsevier).

Therefore, for small values of $\Delta\theta$, we have

$$2\sqrt{2}\frac{c}{w} \approx \frac{3}{2}(\Delta\theta) - \frac{1}{8}(\Delta\theta)^2 \qquad [5.63]$$

Combining Eqs [5.62] and [5.63]

$$\Delta\theta \approx \frac{1}{3\sqrt{2}}\frac{\lambda^2\kappa}{w} + \frac{1}{12}(\Delta\theta)^2 \approx \frac{1}{3\sqrt{2}}\frac{\lambda^2\kappa}{w} + \frac{1}{216}\frac{\lambda^4\kappa^2}{w^2}$$
$$\approx \frac{1}{3\sqrt{2}}\frac{\lambda\phi}{w} + \frac{1}{216}\frac{\lambda^2\phi^2}{w^2} \qquad [5.64]$$

Here ϕ specifies the deformation stage and λ is still unknown. The plastic energy for longitudinal bending of the segment is

$$W_1 = M_p\left[1 - \frac{1}{3}(\Delta\theta)\right]\kappa + \frac{1}{2}E_pI\left[1 - \frac{2}{3}(\Delta\theta)\right]\kappa^2\lambda$$
$$= M_p\left[1 - \frac{1}{3}(\Delta\theta)\right]\phi + \frac{1}{2}E_pI\left[1 - \frac{2}{3}(\Delta\theta)\right]\frac{\phi^2}{\lambda} \qquad [5.65]$$

The effect of $\Delta\theta$ on M_p and I is taken into account here by regarding the average value $(\Delta\theta)_{ave.} = (2/3)(\Delta\theta)$. Opening up of the flange dissipates plastic work, as follows

$$W_2 = M_o(\Delta\theta)_{ave.}\lambda = \frac{1}{6}Yh^2\Delta\theta\lambda \qquad [5.66]$$

Here $M_o = Yh^2/4$ is the plastic bending moment per unit width of the flange. Neglecting the torsion of the flange due to the variable angular distortion along the longitudinal direction, the total plastic energy dissipation is

$$W = W_1 + W_2 \qquad [5.67]$$

For a given value of ϕ, $\Delta\theta$ is related to λ only (Eq. [5.64]). Hence, W is a function of λ only. The 'optimum' length λ is found by minimising the total energy with respect to λ

$$\frac{\partial W}{\partial\lambda} = \frac{\partial}{\partial\lambda}(W_1 + W_2) = 0$$

which leads to

$$\frac{1}{12}\frac{h}{w}\left(\frac{\lambda}{w}\right)^4 + \left(\frac{2\sqrt{2}t}{w\phi} - \frac{\sqrt{2}\phi}{36}\right)\left(\frac{\lambda}{w}\right)^3$$
$$-\left(1 + \frac{\phi^2}{216}\frac{E_p}{Y}\right)\left(\frac{\lambda}{w}\right)^2 = \frac{3E_p}{Y} \qquad [5.68]$$

For every increasing value of ϕ, the corresponding value of λ can be solved numerically, which leads to curvature κ (Eq. [5.60]) and $\Delta\theta$ (Eq. [5.64]). The moment capacity of the new cross-section can be evaluated accordingly. This theoretical analysis broadly agrees with the experimental results (Fig. 5.17).

5.4 Other loading systems and comments

Similar to circular tubes, square tubes have been crushed laterally between rigid plates by Gupta and Ray (1998). The effect of ***foam-filling*** was investigated in their studies. Also, crossed layers of square tubes have been subjected to lateral compression (Gupta and Sinha, 1990a and 1990b).

Energy absorption of square tubes under ***torsional crushing*** has been reported (Santosa and Wierzbicki, 1997; Chen *et al.*, 2001). Circular tubes subjected to ***combined torsion and bending*** have been studied (Reddy *et al.*, 1996).

Thin-walled tubes under lateral loading can undergo local deformation such as indentation, followed by global plastic bending/stretching. A quasi-static approach through consideration of equilibrium and yield conditions seems more suitable in the initial bending of tubes. However, for a fully developed plastic bending deformation with hinge lines, the energy method can be applied successfully. Dynamic indentation of tubes has been reported, as briefly mentioned in this chapter. Nevertheless, there seems to be little study on the dynamic bending of thin-walled tubes.

6

Axial crushing of thin-walled members

This chapter describes analyses and experiments with thin-walled members under axial loading. Various analytical models are presented, and the strain-rate effect is discussed. A large number of experimental data are given in a dimensionless form. The members discussed are circular, square/rectangular and hat sections.

6.1 Circular tubes

6.1.1 Axial collapse modes and typical force–displacement curves

When a circular thin-walled tube is crushed axially, it collapses either axisymmetrically or non-symmetrically, depending primarily on the ratio of diameter and thickness (D/h). The axisymmetrical mode is often known as the ***ring mode*** or concertina mode, while the non-symmetrical mode is referred to as ***diamond mode***. Examples of these are shown in Figs 6.1(a) and (b), respectively. The diamond mode is characterised by the number of lobes, which can vary from two to five for most practical tubes. For certain values of D/h, a tube may start to collapse with the ring mode and then switch to the diamond mode, hence exhibiting a ***mixed mode***, see Fig. 6.1(c). Based on a large number of experiments with tubes of various dimensions, a mode classification chart can be established for a given material (Andrews *et al*., 1983). Figure 6.2 is such a plot for aluminium tubes (Guillow *et al*., 2001). Broadly speaking, the diamond mode occurs when D/h is greater than 80. For D/h less than 50, the ring mode is present for L/h less than 2 and a mixed mode for L/h larger than 2. For long tubes, Euler-type buckling takes place.

A typical force–displacement curve is shown in Fig. 6.3, which is for an aluminium tube of $D = 97$ mm, $L = 196$ mm and $h = 1.0$ mm. The tube collapses axisymmetrically. The axial force reaches an initial peak, followed by a sharp drop and then fluctuations. These fluctuations are a result of formation of the successive folding; each subsequent peak corresponds to the

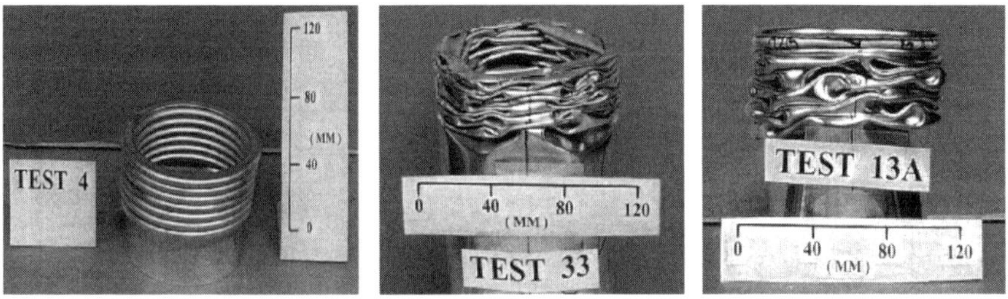

(a) (b) (c)

6.1 Collapse modes for circular tubes under axial loading: (a) ring mode; (b) diamond mode; and (c) mixed mode (reproduced with kind permission of Elsevier).

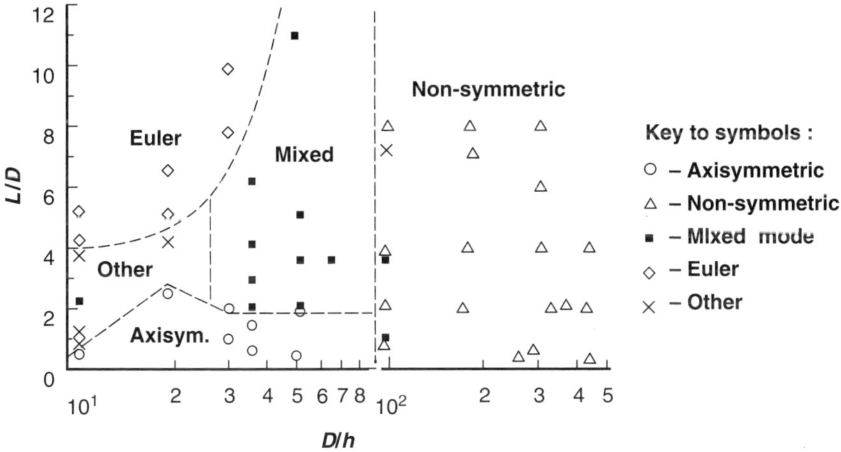

6.2 Mode classification chart for circular aluminium tubes (reproduced with kind permission of Elsevier).

onset of a folding process. Sometimes, however, there is a secondary peak in between the two successive peaks. The energy absorbed is simply the area under this curve. For practical purposes, the average force is often worked out as an indication of energy-absorption capacity. The non-symmetric mode exhibits similar characteristics in the force–displacement curves.

6.1.2 Theoretical models

Alexander model for ring mode

Alexander (1960) was the first to provide a theoretical model for axial crushing of a circular tube for the ring mode. The model is shown in Fig. 6.4. During formation of a single fold, three circumferential plastic hinges

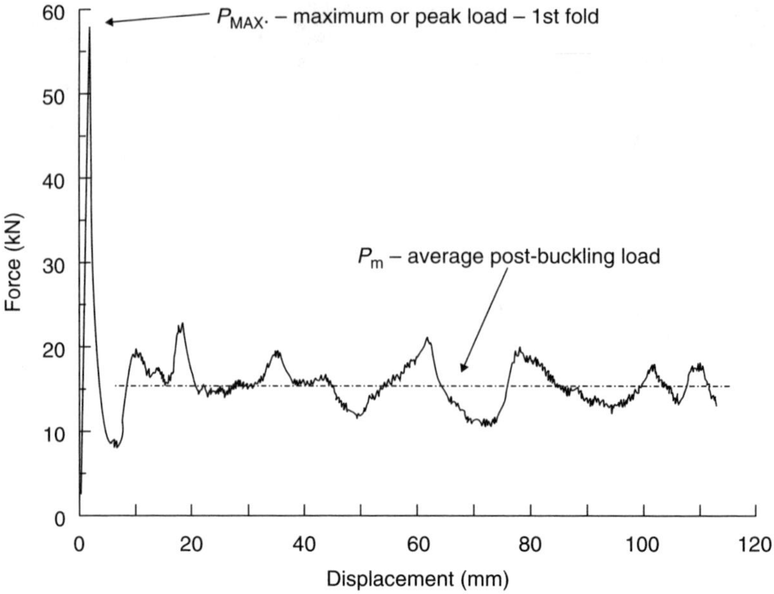

6.3 Typical force–displacement curve (reproduced with kind permission of Elsevier).

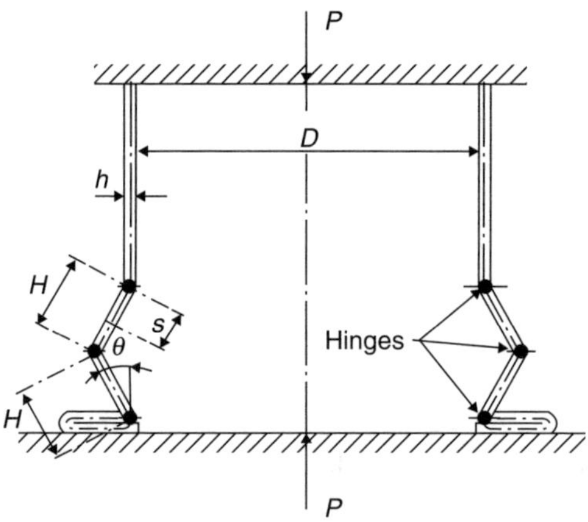

6.4 A simple theoretical model for axisymmetric collapse.

occur. Assuming that the fold goes completely outwards, all the material between the hinges experiences circumferential tensile strain. The external work done is dissipated by plastic bending of the three hinges and circumferential stretching of the materials in between.

In the following analysis, the material is assumed to be rigid, perfectly plastic. Further, there is no interaction between bending and stretching in the yielding criterion; hence, the material yields either by bending only or stretching only. For a complete collapse of a single fold, the plastic bending energy is

$$W_b = 2M_o \pi D \frac{\pi}{2} + 2M_o \int_0^{\pi/2} \pi(D + 2H\sin\theta)\mathrm{d}\theta$$

or

$$W_b = 2\pi M_o(\pi D + 2H) \qquad [6.1]$$

where H is the ***half-length of the fold*** and D is the tube diameter. M_o is the fully plastic bending moment per unit width as before.

The corresponding stretching energy is

$$W_s = 2\int_0^H Y\pi Dh \ \ln[(D + 2s\sin\theta)/D]\mathrm{d}s$$

where Y is the yield stress.

When $\theta = \pi/2$

$$W_s \approx 2\pi YhH^2 \qquad [6.2]$$

This equation can be obtained also by considering the change of the area between the three hinges, $\{= 2[\pi(D + 2H)^2/4 - \pi D^2/4] - 2\pi DH = 2\pi H^2\}$, and then multiplying it by Yh, the yielding membrane force per unit length. From the energy balance, the external work has to be dissipated by plastic energies in bending and stretching. Consequently

$$P_m 2h = W_b + W_s \qquad [6.3]$$

where P_m is the average external force over a complete collapse of the fold. Substituting Eqs [6.1] and [6.2] into Eq. [6.3], we have

$$\frac{P_m}{Y} = \frac{\pi h^2}{\sqrt{3}}\left(\frac{\pi D}{2H} + 1\right) + \pi Hh \qquad [6.4]$$

The unknown length H is determined by invoking the idea that the value of H is such that the external force P_m is minimum. Hence, let $\partial P_m/\partial H = 0$ to give

$$H = \sqrt{\left(\frac{\pi}{2\sqrt{3}}\right)}\sqrt{Dh} \approx 0.95\sqrt{Dh} \qquad [6.5]$$

Substituting Eq. [6.5] in Eq. [6.4]

$$\frac{P_m}{Y} \approx 6h\sqrt{Dh} + 1.8h^2 \qquad [6.6]$$

Remember that in the above analysis the material is assumed to deform completely outwards. If the material deforms inwards, a similar analysis leads to

$$\frac{P_m}{Y} \approx 6h\sqrt{Dh} - 1.8h^2 \tag{6.7}$$

In practice, as Alexander argued, the material deforms partially inwards and partially outwards. Hence, an average of Eqs [6.6] and [6.7] can be taken:

$$P_m \approx 6Yh\sqrt{Dh} \tag{6.8}$$

This completes the Alexander analysis of axisymmetric collapse of circular tubes, developed in 1960. The model is extremely simple, but it does capture most of the main features observed in experiments. Several modifications of this model have been presented. Johnson (1972) modified the expression for the stretching energy on the grounds that the circumferential strain varies along s.

It was recognised that the deforming tube wall bends in the meridian direction instead of the straight line (Abramowicz, 1983; Abramowicz and Jones, 1984b and 1986). In their modified model, two arcs join together to represent the deformed tube wall. This leads to an ***effective crush length*** δ_e which is smaller than $2H$

$$\frac{\delta_e}{2H} = 0.86 - 0.52\left(\frac{h}{D}\right)^{\frac{1}{2}} \tag{6.9}$$

Consequently, a slightly higher average force than Eq. [6.8] is obtained after assuming that H remains the same

$$P_m = 8.91Yh\sqrt{Dh}\left(1 - 0.61\sqrt{\frac{h}{D}}\right) \tag{6.10}$$

Grzebieta (1990) further modified the meridian profile, but adopted an equilibrium approach. Thus, the force–displacement curve can be worked out, not just the average force. To account for the fact that the tube wall deforms both inwards and outwards, Wierzbicki *et al.* (1992) introduced a parameter known as the ***eccentricity factor***, which defines the outward portion over the whole length H. The value of this parameter is about 0.65 based on experiments. In this way, the occurrence and position of a second peak within each fold can be predicted. This work has been further refined by Singace *et al.* (1995) and Singace and El-Sokby (1996).

Effects of strain-rate and inertia

In the dynamic case, the strain-rate effect can be approximately taken into account as follows. As discussed in Section 2.4.2, this effect plays a role by enhancing the yield stress of the material. Based on the Cowper-Symonds relation, Eq. [2.73], Eq. [6.8] can be rewritten as

$$P_m \approx 6Yh\sqrt{Dh}\left[1+(\dot{\varepsilon}/B)^{1/q}\right] \quad [6.11]$$

where B and q are constants for the tube material and $\dot{\varepsilon}$ is the strain rate. Their typical values are given in Table 2.1.

The key is to estimate the strain-rate over the dynamic collapse process. Here we present a simple estimate for the average circumferential strain-rate, which is assumed to be representative of the problem. The mean strain in a completely flattened fold of the circular tube is

$$\varepsilon_\theta \approx H/D \quad [6.12]$$

Assume that the tube starts to deform with an initial velocity and that this velocity decreases linearly with time. This corresponds to a constant deceleration with a constant external axial load. The total time to deform one fold completely is

$$T = 2H/V_o \quad [6.13]$$

Therefore the average strain-rate is

$$\dot{\varepsilon}_\theta = \varepsilon_\theta/T = V_o/2D \quad [6.14]$$

Substituting this into Eq. [6.11] results in

$$P_m = 6Yh\sqrt{Dh}\left[1+(V_o/2D)^{1/q}\right] \quad [6.15]$$

Note that the contribution of the second term is not as much as one might expect because the value of q is usually large.

The inertia effect of axial collapse of tubes can be large, similar to the type II inertia-sensitive structures discussed in Section 7.2. Detailed analysis can be found in Karagiozova *et al.* (2000).

Theories for non-symmetric modes

Theoretical models for the diamond mode are less successful than those for the ring mode. Most of the models involve bending of triangularised elements about hinge lines with the mid-surface being inextensional. Pugsley and Macaulay (1960) were among the first researchers to consider the diamond mode. They proposed

$$\frac{P}{Y\pi Dh} = 10\frac{h}{D} + 0.13 \quad [6.16]$$

The constants were determined to best fit the experiments. Johnson *et al.* (1977b) attempted to develop a theory for the diamond mode based on experiments with PVC tubes. From the actual geometry of folding, the

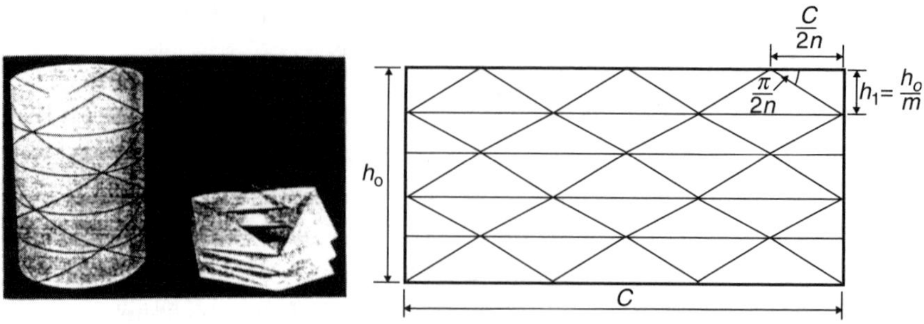

6.5 A theoretical collapse model for non-symmetric mode; $n = 3$ (reproduced with kind permission of the Council of the Institution of Mechanical Engineers).

arrangement of hinge lines can be worked out for a given number of lobes. Figure 6.5 shows such an arrangement for three lobes. The external work is dissipated only by plastic bending of elements about hinge lines together with flattening of the initially curved elements. For long tubes, the calculated force is

$$\frac{P_m}{2\pi M_o} \approx 1 + n\,\mathrm{cosec}\left(\frac{\pi}{2n}\right) + n\cot\left(\frac{\pi}{2n}\right) \qquad [6.17]$$

where n is the number of circumferential lobes. This formula requires a prior knowledge of n and there is no established method of determining this.

Further theoretical studies have been conducted by Singace (1999). An eccentricity factor was introduced in the same way as for the ring mode. The equation developed is

$$\frac{P_m}{M_o} = -\frac{\pi}{3}n + \frac{2\pi^2}{n}\tan\left(\frac{\pi}{2n}\right)\frac{D}{h} \qquad [6.18]$$

Experiments on axial crushing of circular tubes

Experimental results on axial crushing of circular tubes are extensive. Most of the researchers who proposed theoretical models conducted experiments in an attempt to validate the models. However, the range of D/h used was usually very limited. The most recent work is by Guillow *et al.* (2001) and it covers a sufficiently large range of D/h and L/D in a single testing program. For aluminium tubes, the average force is plotted against D/h in Fig. 6.6.

Figure 6.6 is a double logarithmic plot. It is surprising to note that all the points, regardless of the collapse mode, fall into a single curve, which may

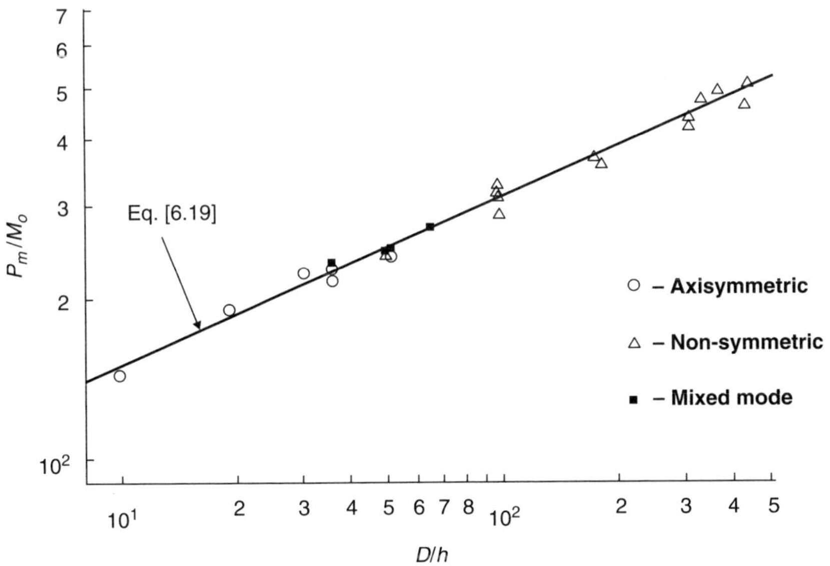

6.6 Dimensionless plot of average force against *D/h* for aluminium tubes (reproduced with kind permission of Elsevier).

be approximated using a straight line. Hence an empirical equation emerges as

$$\frac{P_m}{M_o} = 72.3\left(\frac{D}{h}\right)^{0.3} \qquad [6.19]$$

Recall the theoretical analysis presented above for the ring mode. The value of P_m/M_o is largely proportional to $\sqrt{D/h}$ and dependent upon n. Experimental results in Fig. 6.6 clearly defy these theoretical observations. Huang and Lu (2003) have explained this by proposing a model with an effective hinge arc length, the value of which is $3h$ to $5h$.

Structural effectiveness and solidity ratio

To facilitate the presentation of test results, two important parameters are introduced here, namely ***structural effectiveness*** and ***solidity ratio***. The structural effectiveness is defined as

$$\eta = \frac{P_m}{AY} \qquad [6.20]$$

where A is the net cross-sectional area of the thin-walled tube. Hence, AY represents the axial squash load and η is always less than 1. For a circular tube, therefore

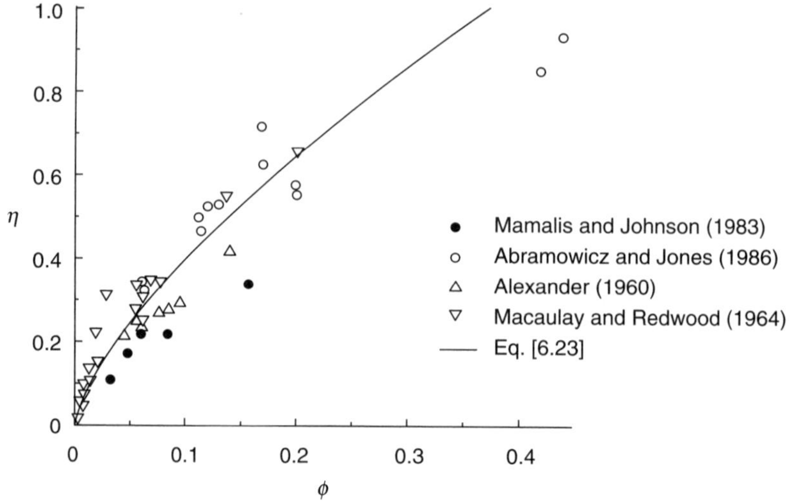

6.7 Plot of structural effectiveness against solidity ratio for circular tubes.

$$\eta = P_m / \pi D h Y \qquad [6.21]$$

The solidity ratio is defined as

$$\phi = A / A_1 \qquad [6.22]$$

Here A_1 is the enclosed total area of the section and is $\pi D^2/4$ for circular tubes. Obviously, $\phi < 1$.

Experiments for thin-walled tubes can be summarised using these two parameters. Figure 6.7 shows a plot for circular tubes. It is clear that empirically for circular tubes

$$\eta = 2\phi^{0.7} \qquad [6.23]$$

The empirical equation [6.19] can be recast into the form

$$\eta = 5.7\phi^{0.7} \qquad [6.24]$$

This has the same value of power 0.7 as in Eq. [6.23], but with a much higher value of coefficient 5.7. Wierzbicki and Abramowicz (1983) obtained a value of 5.15, which is close to the present one.

Dynamic tests of circular tubes lead to a higher average load compared with the static tests (Abramowicz and Jones, 1984b). Figure 6.8 shows the ratio of the dynamic average load, P_m^d, to the static one, P_m^s. Equation [6.15] is also plotted with $D = 6844s^{-1}$ and $q = 3.91$ for the mild steel's ultimate stress.

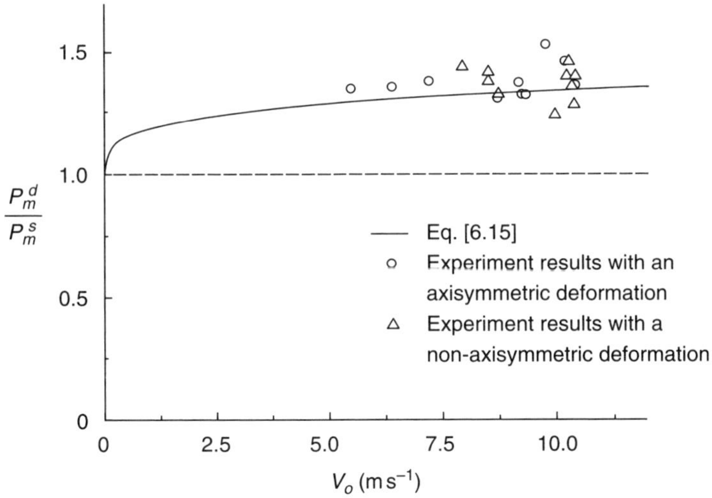

6.8 Effect of the impact velocity on the ratio between the average dynamic and static loads for circular tubes.

6.2 Square tubes

6.2.1 Axial collapse modes and typical force–displacement curves

Thin-walled square tubes are often subjected to axial loads. They are representative of a number of structural components in, for example, cars, railway coaches and ships. Their collapse modes are very different from those for circular tubes, but the general characteristics of the force–displacement curves are similar. This is because both square and circular tubes undergo progressive collapse when subjected to axial loading.

A typical view of the fully crushed square box column is shown in Fig. 6.9(a), which is for an aluminium tube with $c/h = 23$; here c is the side length and h is the thickness. The tube wall undergoes severe inward and outward plastic bending, with possible stretching.

Note that when the tube is thin, a non-compact collapse mode may occur. In this case, the folds are not continuous, as for compact sections, and they are separated by slightly curved panels, see Fig. 6.9(b) for $c/h = 100$ (Reid *et al.*, 1986). This mode may be relatively unstable globally with the tendency to Euler-type buckling, which is an undesirable energy-dissipating mechanism. A typical force–displacement curve is plotted in Fig. 6.10. It is clear that the force falls sharply after the initial peak and then it fluctuates periodically, corresponding to the formation and complete collapse of folds one by one.

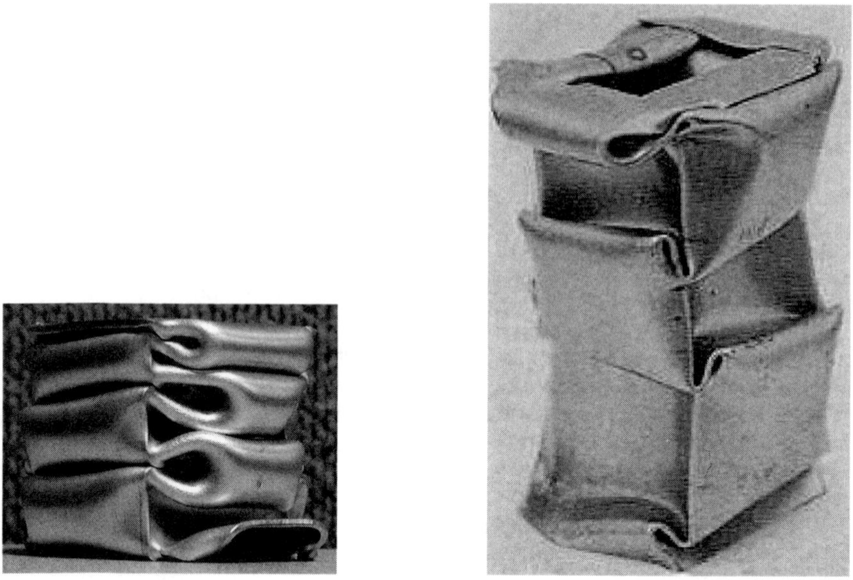

(a) (b)

6.9 Collapse modes for square tubes: (a) compact mode (aluminium tube $c/h = 23$); (b) non-compact mode ($c/h = 100$) (part(b) reproduced with kind permission of Elsevier).

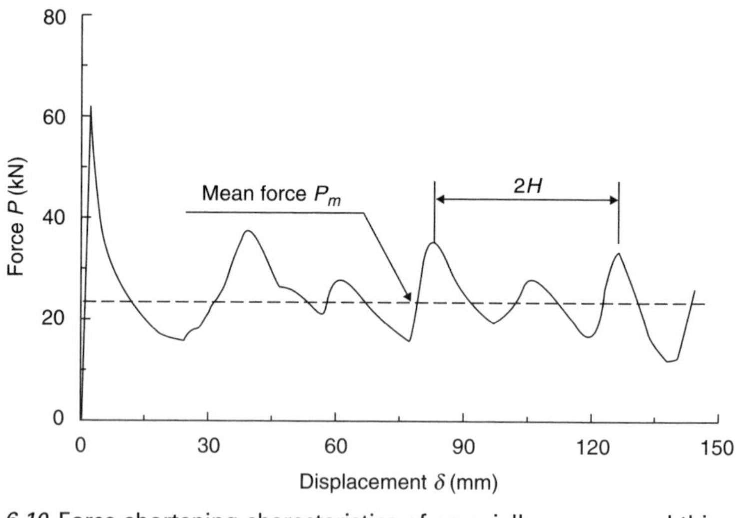

6.10 Force-shortening characteristics of an axially compressed thin-walled aluminium column ($c = 51.0$ mm, $h = 2.19$ mm).

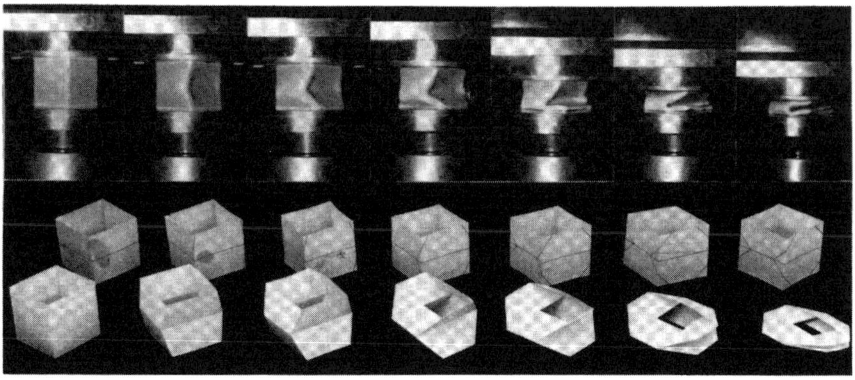

6.11 PVC specimens with $L/c = 1$ and $h/c = 0.034$ showing various stages of the deformation process corresponding to a series of paper models (Meng *et al.*,1983) (reproduced with kind permission of Elsevier).

In order to understand further the axial collapse mechanism of square tubes, Meng *et al.* (1983) used a series of PVC tubes and successive deformation events were captured, see Fig. 6.11. Included are photographs of the corresponding paper models with stationary plastic hinges. But globally, during the whole deformation, the initially vertical hinges at the four corners gradually become inclined, as marked in the undeformed paper model. The final inclination is about $\pi/4$. The general features are similar for tubes with a range of L/c (L is the tube length) and h/c ratios.

6.2.2 Idealisation of collapse mechanism

Based on the observations of the collapse process, we may summarise a typical folding sequence in Fig. 6.12, which shows a deforming stage for one quarter of a square section. This idealised mechanism and the subsequent analysis is due to Wierzbicki and Abramowicz (1983). Globally, this element consists of two sorts of hinges: fixed horizontal hinges (AC and CD) and inclined travelling hinges (KC and CG). The travelling hinge KC originates from the vertical corner K′C′ and the inclination angle increases as the deformation progresses.

The initial geometry of this element is defined by the total height of this element $2H$, being analogous to the length of fold for circular tubes. For a general case, let $2\psi_o$ be the angle between two adjacent plates as viewed along the tube axis, and c be the side length of AC′ and C′D. Both $2\psi_o$ and

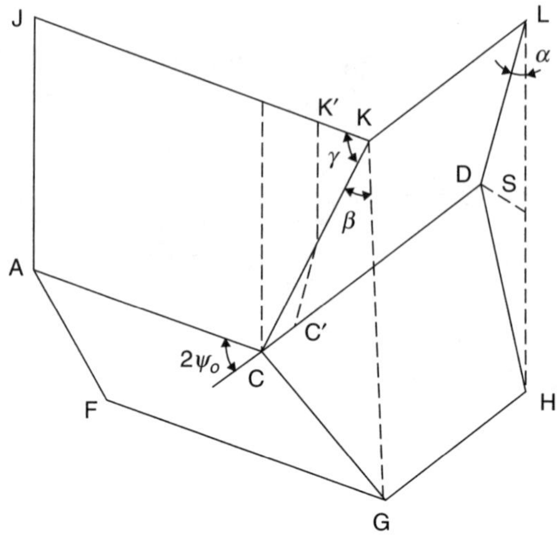

6.12 A quarter of a square section during folding (Wierzbicki and Abramowicz, 1983).

c are assumed to be constant during deformation. For square tubes, $2\psi_o$ is $\pi/2$ and AC = CD = c. But, for the analysis to be readily applicable to other cases which we shall see later, we keep these two parameters.

The status of this collapse element can be described by one of the parameters: crushing distance δ, angle of rotation of panel KLDC, α, and the horizontal distance at D, S. Of course, they are related

$$\delta = 2H(1 - \cos\alpha) \qquad [6.25]$$

$$S = H\sin\alpha \qquad [6.26]$$

and

$$\tan\gamma = \frac{\tan\psi_o}{\sin\alpha} \qquad [6.27]$$

$$\tan\beta = \frac{\tan\alpha}{\sin\psi_o} \qquad [6.28]$$

The relationship between the velocities is, by differentiating the above equation

$$\dot{\delta} = 2H\sin\alpha\dot{\alpha} \qquad [6.29]$$

and the horizontal velocity of the point D is

$$V = \dot{S} = H\cos\alpha\dot{\alpha} \qquad [6.30]$$

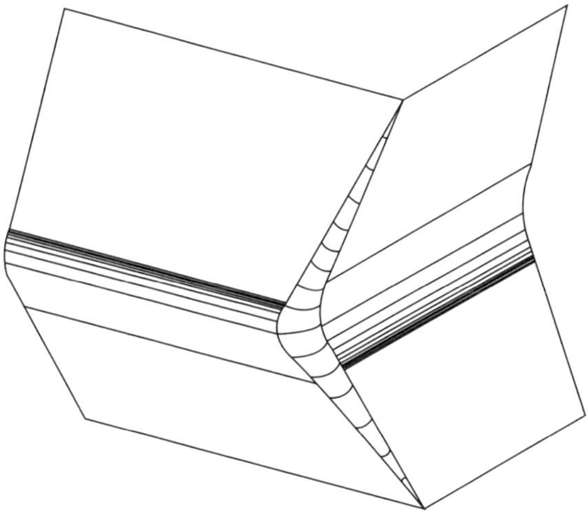

6.13 A more realistic kinematically admissible folding mechanism (Wierzbicki and Abramowicz, 1983).

6.2.3 Details of the plastic zones

In the idealised deformation mode in Fig. 6.12, all the plastic deformation occurs within localised plastic hinges. This would be acceptable if there were no propagation of plastic deformation. However, in this case, we mentioned that the plastic hinge KC is travelling as deformation proceeds, originally from position K′C′. A localised hinge with infinite curvature would absorb an infinite amount of plastic energy when it travels, as will be seen later. The same can be said about point C, which moves from its original position C′.

We therefore need a more realistic model which is kinematically admissible. Figure 6.13 shows such a model, obtained by extending the plastic deformation into a plastic zone instead of concentrated hinges. In this model, plastic deformation occurs only in shaded regions. Thus, during deformation the four plane trapezoidal plates move as rigid bodies. Two cylindrical surfaces are bounded by two straight hinge lines which propagate in opposite directions, leading to a wider zone. Two adjacent trapezoidal plates are connected via a conical surface bounded by two straight lines. As KC in Fig. 6.12 moves, one straight line imparts a curvature to an originally flat sheet (part of JKCA) and the other removes this curvature so that the curved sheet bends back to flat, joining KLCD. Finally, the four active deforming zones are connected by a section of a ***toroidal shell***. This doubly curved surface has a non-zero Gaussian curvature (defined as the product of two principal curvatures), while the cylindrical segments before and after passing this toroidal shell have a zero Gaussian curvature. There is therefore a change of Gaussian curvature when the material deforms into this shell and then back to a cylindrical one, and there must be in-plane stretching (Calladine, 1983b).

Our next task is to evaluate energy dissipation for each of the four types of plastic zone. Those for the two cylindrical shells are straightforward to perform, similar to the analysis in the previous chapters. The energies dissipated in the two travelling conical zones and in the toroidal shell are less so, and this issue will be dealt with next.

Energy dissipation in a travelling hinge

Consider a strip shown in Fig. 6.14, with a travelling hinge defined by an arc AB of radius *r*. Suppose this hinge moves by a distance Δs into a new position, but the radius remains unchanged. For the sake of convenience, we assume that Δs is sufficiently large that the whole arc AB is unbent into a straight segment $A'B'$. The energy for unbending AB is then

$$W_{AB} = \overline{AB}\frac{1}{r}M_p = M_p(\pi - \beta) \qquad [6.31]$$

where M_p is the fully plastic bending moment. Segment BC has been first bent into an arc of radius *r* and then unbent to a flat strip $B'C'$. The energy required by this process is

$$W_{BC} = \overline{BC}\frac{1}{r}M_p\,2 \qquad [6.32]$$

Similarly, for CD

$$W_{CD} = \overline{CD}\frac{1}{r}M_p \qquad [6.33]$$

and for DE

$$W_{DE} = \overline{DE}\frac{1}{r}M_p \qquad [6.34]$$

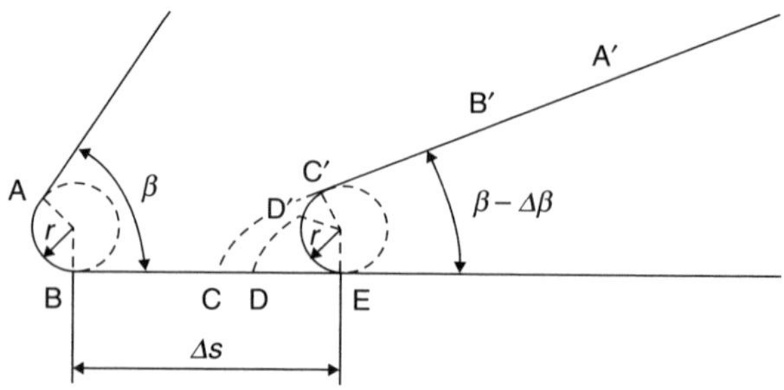

6.14 A strip demonstrating a travelling hinge.

Therefore, the total bending energy for this plastic hinge to travel a distance Δs is

$$W = W_{AB} + W_{BC} + W_{CD} + W_{DE}$$
$$= \left(\overline{AB} + 2\overline{BC} + \overline{CD} + \overline{DE}\right) \cdot \frac{1}{r} \cdot M_p$$
$$= \frac{1}{r} M_p \left[\overline{AB} + 2\left(\Delta s - \frac{\overline{AB} + \overline{CD}}{2} \right) + \overline{CE} \right]$$
$$= 2 \frac{1}{r} M_p \Delta s \qquad [6.35]$$

Equation [6.35] was also given by Meng *et al.* (1983). This equation demonstrates that the energy absorbed by a travelling hinge is directly proportional to the distance travelled and inversely proportional to the radius r of the hinge. This explains why a sharp crease (of a radius zero) could not travel – to do so would require an infinite amount of external work. The process of a travelling hinge may be understood in another way: it can be regarded as material being pushed, in an opposite direction, through an anvil of radius r. Thus, the energy is absorbed by bending and then unbending a strip of distance Δs, leading immediately to Eq. [6.35].

Energy dissipation in a sheet passing over a toroidal surface

For the toroidal segment shown in Fig. 6.15, a generic point within this surface may be described by two coordinates (θ, ϕ). Here θ denotes the meridian coordinate (Fig. 6.15(c)) and ϕ is along the circumferential direction (Fig. 6.15(b)). The limits of θ and ϕ are

$$\frac{\pi}{2} - \psi \leq \theta \leq \frac{\pi}{2} + \psi \qquad [6.36]$$

$$-\beta \leq \phi \leq \beta \qquad [6.37]$$

Also

$$r = b\cos\theta + a \qquad [6.38]$$

where b is the radius in the meridian direction (Fig. 6.15). When a material is forced to pass outward over this toroidal surface, there is a circumferential strain and its increment corresponding to a tangential velocity v_t is

$$\dot{\varepsilon}_{\phi} = \frac{v_t \sin\theta}{r} = \frac{H\cos\alpha\dot{\alpha}}{\tan\psi_o} \frac{\sin\theta}{b\cos\theta + a} \qquad [6.39]$$

Here, $v_t \sin\theta$ is the horizontal component of v_t, and

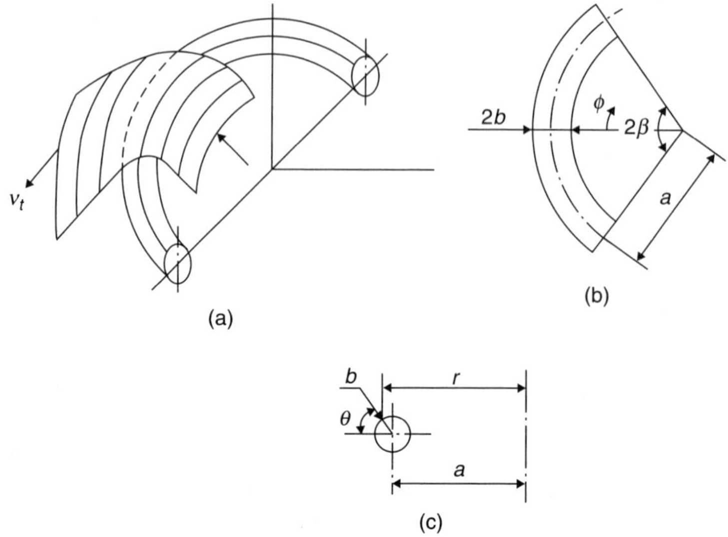

6.15 Plastic deformation of a toroidal surface: a 3D sketch (a) and definitions of various parameters (b,c) (Wierzbicki and Abramowicz, 1983).

$$v_t = \frac{\dot{s}}{\tan\psi_o} = \frac{H\cos\alpha\dot{\alpha}}{\tan\psi_o} \tag{6.40}$$

In the toroidal region, the main plastic flow occurs in terms of $\dot{\varepsilon}_\phi$. Though there is a non-zero curvature change in circumferential direction, the corresponding bending energy can be shown to be zero (resulting from the fact that yielding occurs in circumferential tension with a fully yielding membrane force $N_o = Yh$). Hence, the plastic dissipation rate is

$$\dot{W}_1 = \int_s N_o\dot{\varepsilon}_\phi \mathrm{d}s = \int N_o\dot{\varepsilon}_\phi r\mathrm{d}\phi b\mathrm{d}\theta \tag{6.41}$$

Substituting Eqs [6.38], [6.39] and [6.40] into Eq. [6.41], we have

$$\dot{W}_1 = 4N_o bH\frac{\pi}{(\pi-2\psi_o)\tan\psi_o}$$
$$\cos\alpha\left[\cos\psi_o - \cos\left(\psi_o + \frac{\pi-2\psi_o}{\pi}\beta\right)\right]\dot{\alpha} \tag{6.42}$$

Note that the angle ψ is assumed to increase linearly with the coordinate ϕ from ψ_o to $\pi/2$ as

$$\psi = \psi_o + \frac{\pi-2\psi_o}{\pi}\phi$$

Integrating $\dot{W}_1$ with respect to α, we obtain

$$W_1 = 4N_o bH\mathrm{I}_1(\psi_o) = 16M_o \frac{Hb}{h}\mathrm{I}_1(\psi_o) \quad [6.43]$$

where

$$\mathrm{I}_1(\psi_o) = \frac{\pi}{(\pi - 2\psi_o)\tan\psi_o}\int_0^{\frac{\pi}{2}} \cos\alpha \left\{\sin\psi_o \sin\left(\frac{\pi - 2\psi_o}{\pi}\right)\beta + \cos\psi_o\left[1 - \cos\left(\frac{\pi - 2\psi_o}{\pi}\right)\beta\right]\right\}\mathrm{d}\alpha \quad [6.44]$$

Since β is a function of α, $\mathrm{I}_1(\psi_o)$ can be evaluated, e.g. $\mathrm{I}_1(\pi/4) = 0.58$ for $\psi_o = \pi/4$, and $\mathrm{I}_1(\pi/6) = 1.05$ for $\psi_o = \pi/6$.

Energy dissipation in plastic hinges

The plastic energy dissipated in the fixed horizontal hinges AC and CD is

$$\dot{W}_2 = 2M_o c\dot{\alpha} \quad [6.45]$$

or

$$W_2 = 2\int_0^{\frac{\pi}{2}} M_o c\,\mathrm{d}\alpha = \pi M_o c \quad [6.46]$$

Finally, the inclined hinges have a total length

$$L = \frac{2H}{\sin\gamma} \quad [6.47]$$

Hence

$$\dot{W}_3 = 2M_o L\frac{v_t}{b} = 4M_o \frac{H^2}{b}\frac{1}{\tan\psi_o}\frac{\cos\alpha}{\sin\gamma} \quad [6.48]$$

$$W_3 = 4M_o \mathrm{I}_3(\psi_o)H^2/b \quad [6.49]$$

where

$$\mathrm{I}_3(\psi_o) = \frac{1}{\tan\psi_o}\int_0^{\frac{\pi}{2}} \frac{\cos\alpha}{\sin\gamma}\mathrm{d}\alpha$$

Hence, $\mathrm{I}_3(\pi/4) = 1.11$ and $\mathrm{I}_3(\pi/6) = 2.39$. The rate of external work done is

$$\dot{W}_{ext} = P\delta = 2PH\sin\alpha\dot{\alpha} \quad [6.50]$$

or

$$\dot{W}_{ext} = 2PH \quad [6.51]$$

The work balance therefore requires

$$2P_m H = W_1 + W_2 + W_3 \quad [6.52]$$

where P_m is the average load. It can be seen that, on substituting the expressions for the three energy components (Eqs [6.43], [6.46] and [6.49]), the average load is of the following form

$$\frac{P_m}{M_o} = A_1 \frac{b}{h} + A_2 \frac{c}{H} + A_3 \frac{H}{b} \tag{6.53}$$

where A_1, A_2 and A_3 are appropriate functions. The only two unknown parameters, the radius b and the height of half a fold H, can be determined by letting

$$\frac{\partial P_m}{\partial H} = 0, \quad \frac{\partial P_m}{\partial b} = 0 \tag{6.54}$$

This leads to

$$b = \sqrt[3]{A_1 A_3 / A_1^2}\,\sqrt[3]{ch^2} \tag{6.55}$$

$$H = \sqrt[3]{A_2^2 / A_1 A_3}\,\sqrt[3]{c^2 h} \tag{6.56}$$

Substituting them back into Eq. [6.53], we have

$$\frac{P_m}{M_o} = 3\sqrt[3]{A_1 A_2 A_3}\,\sqrt[3]{c/h} \tag{6.57}$$

This indicates that total plastic energy comprises equal contributions from all the three major mechanisms of energy dissipation.

For a square or rectangular section of $c_1 \times d$, we take $c = \frac{1}{2}(c_1 + d)$. $I_1 = 0.58$, $I_3 = 1.11$. Also, because the top and bottom horizontal hinges (JK, KL, FG and GH) occur, W_2 has to be doubled. The corresponding energy balance gives

$$2HP_m = M_o\left(64 I_1 \frac{bH}{h} + 8\pi c + 16 I_3 \frac{H^2}{b}\right) \tag{6.58}$$

Hence, $A_1 = 32I_1 = 18.56$, $A_2 = 4\pi$, $A_3 = 8I_3 = 8.91$. Consequently

$$H = 0.983\sqrt[3]{hc^2}, \quad b = 0.687\sqrt[3]{h^2 c} \tag{6.59}$$

and

$$\frac{P_m}{M_o} = 38.27\sqrt[3]{\frac{c}{h}} \tag{6.60}$$

For a square tube, $c_1 = d = c$ and we have

$$P_m = 9.56 Y h^{\frac{5}{3}} c^{\frac{1}{3}} \tag{6.61}$$

The fact that P_m is proportional to $h^{\frac{5}{3}}$ is a reflection of the energy contribution of bending and stretching, in this case, 2:1. We know that for a bending-only deformation, the force is proportional to h^2 while it is proportional to h for a membrane-only deformation. When both are present, the force is proportional to h raised to a power between 1 and 2.

Using the structural effectiveness η and solidity ratio ϕ introduced earlier, we have, for square sections

$$\eta = 0.948\phi^{\frac{2}{3}} \tag{6.62}$$

In the above model, only the circumference c plays a role in the calculation of P_m and H. The aspect ratio is immaterial. This has been partially verified by experiment: the observed fold length is indeed independent of the aspect ratio (Aya and Takahashi, 1974).

6.2.4 Comparison with experiments

Experimental results are plotted in Fig. 6.16, in terms of η and ϕ (Wierzbicki and Abramowicz, 1983). Note that the ultimate stress was used in converting the average force P_m to η. The theoretical prediction, Eq. [6.62], does agree well with the experiments. The best fitted curve by Magee and Thornton (1978) gives $\eta = 1.4\phi^{0.8}$.

Compared with the results for circular tubes, square or rectangular ones do not seem as efficient in absorbing energy. Similar to the case of circular tubes, the effective crushing distance is smaller than $2H$ in practice. It is shown to be (Abramowicz and Jones, 1984a and 1986)

$$\delta_e / 2H = 0.73 \tag{6.63}$$

This leads to a modified average force

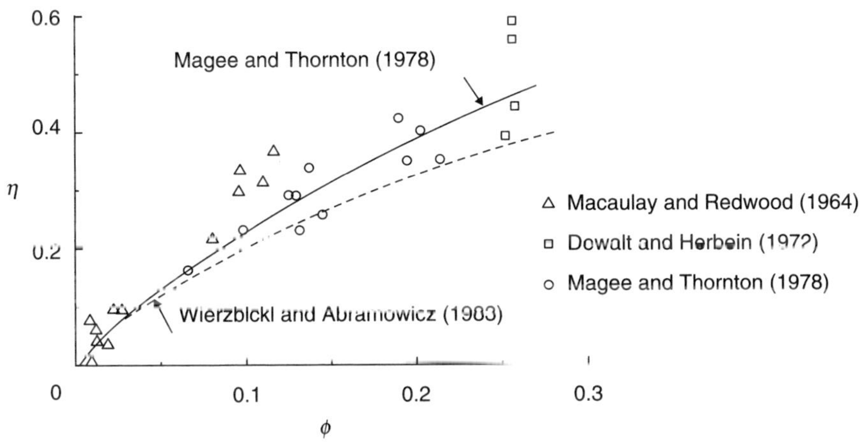

6.16 Comparison of theory with experiments for square and rectangular sections (Wierzbicki and Abramowicz, 1983).

$$\frac{P_m}{M_o} = 52.42\left(\frac{c}{h}\right)^{\frac{1}{3}} \qquad [6.64]$$

or

$$\eta = 1.3\phi^{\frac{2}{3}} \qquad [6.65]$$

This would bring the theoretical curve in Fig. 6.16 a little higher.

6.2.5 Dynamic effect

As with circular tubes, the exact value of strain-rate is difficult to calculate in this case. However, an estimate is made (Abramowicz and Jones, 1984a) as

$$\dot{\varepsilon} = 0.33\frac{V_o}{c} \qquad [6.66]$$

Therefore, this strain-rate will enhance the yield stress and, by incorporating Eq. [6.66], Eq. [6.64] becomes

$$\begin{aligned} P_m/M_o &= 52.42(c/h)^{\frac{1}{3}}\left[1+(\dot{\varepsilon}/D)^{\frac{1}{q}}\right] \\ &= 52.42(c/h)^{\frac{1}{3}}\left[1+(0.33V_o/cD)^{\frac{1}{q}}\right] \end{aligned} \qquad [6.67]$$

or

$$\eta = 1.3\phi^{\frac{2}{3}}\left[1+(0.33V_o/cD)^{\frac{1}{q}}\right] \qquad [6.68]$$

The term inside the square bracket represents the enhancement of the dynamic average force over its static counterpart. Figure 6.17 shows experimental results for mild steel square tubes, together with Eq. [6.67] with $D = 6844s^{-1}$ and $q = 3.91$ (Abramowicz and Jones, 1984a). It is evident that Eq. [6.67] underestimates the dynamic load. The inertia effect would be responsible for the large discrepancy, as discussed for circular tubes.

6.3 Top-hat and double-hat sections

Under axial compression, tubes of hat sections behave similarly to square and rectangular boxes. Figure 6.18 shows compression of a ***top-hat structure*** and a ***double-hat structure*** (White *et al.*, 1999). Force–displacement curves (not shown) largely resemble those for other sections. A large

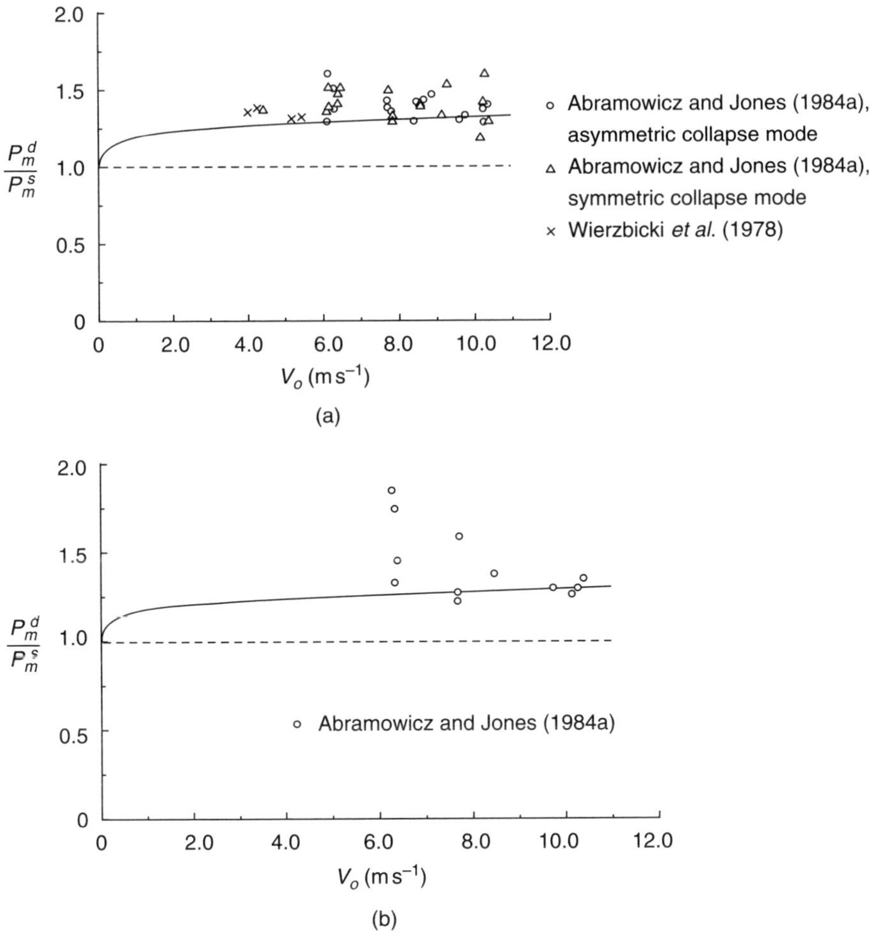

6.17 Ratio of average dynamic and static forces versus impact velocity: (a) $c = 37.07$ mm, $h = 1.152$ mm; (b) $c = 49.31$ mm, $h = 1.63$ mm (Abramowicz and Jones, 1984a).

number of test points are plotted in terms of η and ϕ (Fig. 6.19), together with an empirical equation (White *et al.*, 1999)

$$\eta = 0.57\phi^{0.63} \quad [6.69]$$

Theoretical analysis can be performed following the procedure described for square sections. Let $L = 2a + 2b + 4f$, where a, b and f are defined in Fig. 6.20. The following two results are obtained.

For a top-hat section

$$\frac{P_m}{M_o} = 32.89\left(\frac{L}{h}\right)^{\frac{1}{3}} \quad [6.70]$$

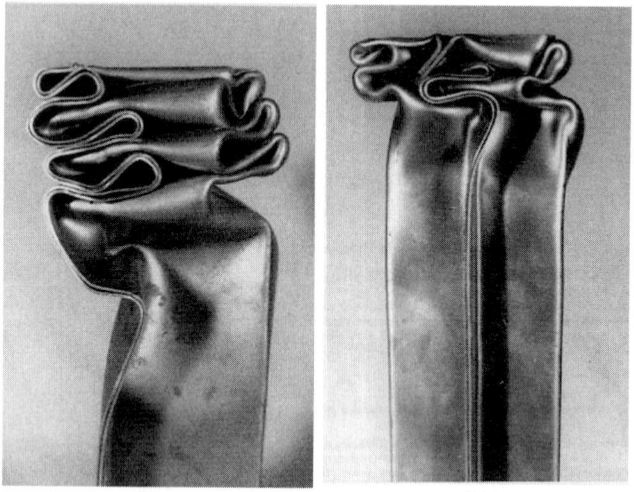

6.18 Photograph of a compressed top-hat section (a) and double-hat section (b) (White and Jones, 1999) (reproduced with kind permission of Elsevier).

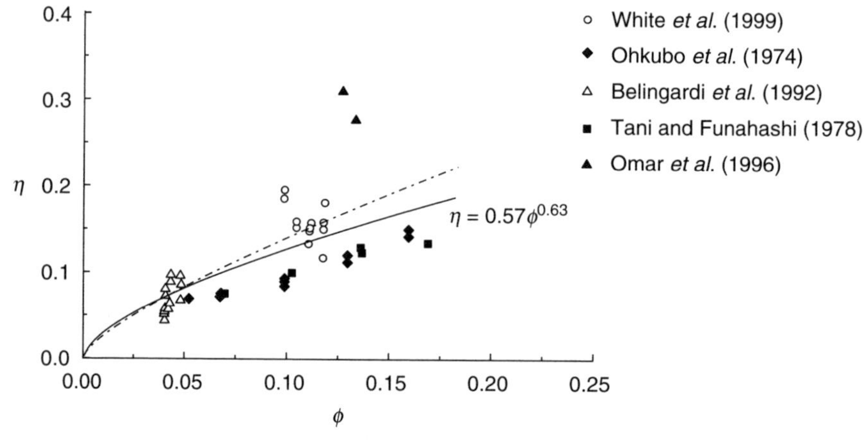

6.19 Plot of η versus ϕ for hat sections (White *et al.*, 1999) (reproduced with kind permission of Elsevier).

or if the ultimate tensile stress σ_u is used for a strain-hardening material

$$\frac{P_m}{M_u} = 35.55\left(\frac{L}{h}\right)^{0.29} \quad [6.71]$$

where $M_u = (\sigma_u h^2)/4$.

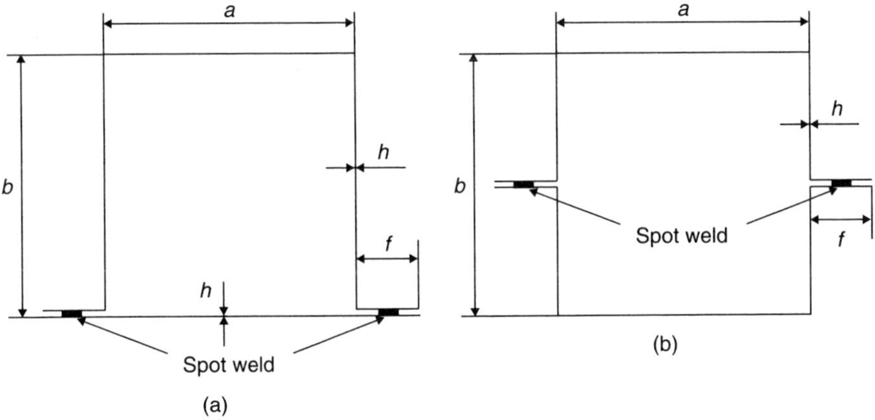

6.20 Sketch of a top-hat and double-hat sections showing several parameters.

Similarly, for a double-hat section

$$\frac{P_m}{M_o} = 52.20\left(\frac{L}{h}\right)^{\frac{1}{3}} \qquad [6.72]$$

and

$$\frac{P_m}{M_u} = 58.15\left(\frac{L}{h}\right)^{0.29} \qquad [6.73]$$

Test points fall within the bounds given by the two formulae, for each type of hat section.

6.4 Effect of foam filling

Thin-walled tubes can be filled with other materials in order to enhance their crashworthiness (Thornton, 1980). One commonly used in-filler is foam, polyurethane or metal foams. Foams have excellent energy-absorption behaviour, as discussed in Chapter 10, with an almost constant stress plateau and long stroke. Besides, the interaction between the foam and tube walls provides additional enhancement in energy dissipation. Two kinds of tubes are studied – those with circular and those with square sections.

Figure 6.21 shows load displacement curves for a typical circular aluminium tube filled with polyurethane foam (D = 97 mm, h = 1.0 mm). As expected, compared with an empty tube, all three in-filled tubes have a higher load and dense foams produce larger forces. Similar results for square tubes are shown in Fig. 6.22(a), where c = 75 mm, h = 0.76 mm.

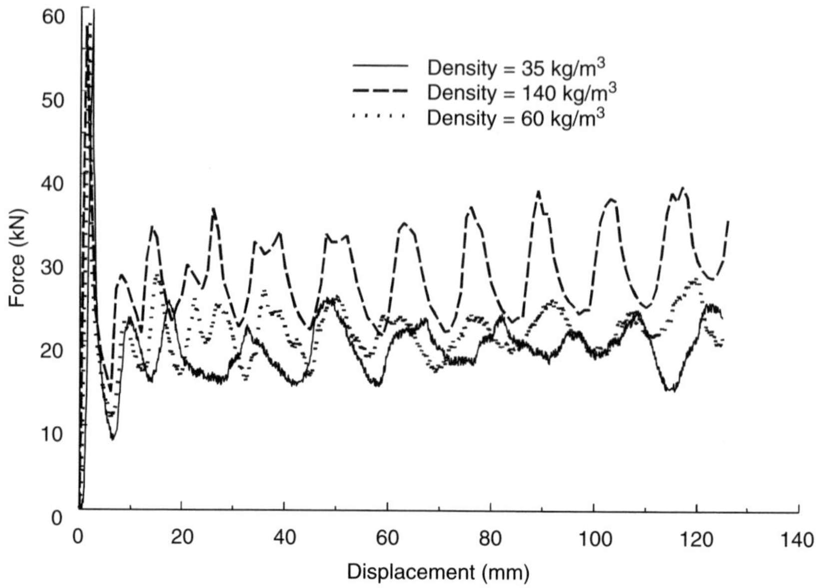

6.21 Force–displacement curves for a circular aluminium tube filled with polyurethane foam with three densities (Guillow *et al.*, 2001) (reproduced with kind permission of Elsevier).

The interaction between the in-filled foam and tube walls can be understood as follows. The foam provides constraint when a tube wall buckles inwardly – similar to a compression-only spring. In the case of weak or non-bonding between the foam and tube, the tube wall can bend freely outward without any action from the foam. This constraint of foam has two consequences. First, depending upon the level of plateau stress, the foam may change the tube collapse mode: for circular tubes, from the diamond mode of an empty tube to the ring mode, and for square tubes, from non-compact to compact mode, see Fig. 6.22(b). Second, even if the collapse mode may appear the same, ring or compact mode, the plastic fold length decreases with the presence of foam; so does the proportion of inward bending – a very strong foam would completely prohibit any inward bending.

When foams are compressed, densification occurs after the plateau stage and the stress increases rapidly with the strain. This corresponding strain, the locking strain, is the limit of a compression stroke achievable by the tube and this reduces the effective length of each fold, leading to a higher average force.

Previous theoretical models for empty tubes can be modified to account for the effect of in-filled foams. For circular tubes with an axisymmetric collapse mode, we assume that the tube wall moves only outwards and hence the model in Fig. 6.4 is applicable. We note that the collapse of one fold

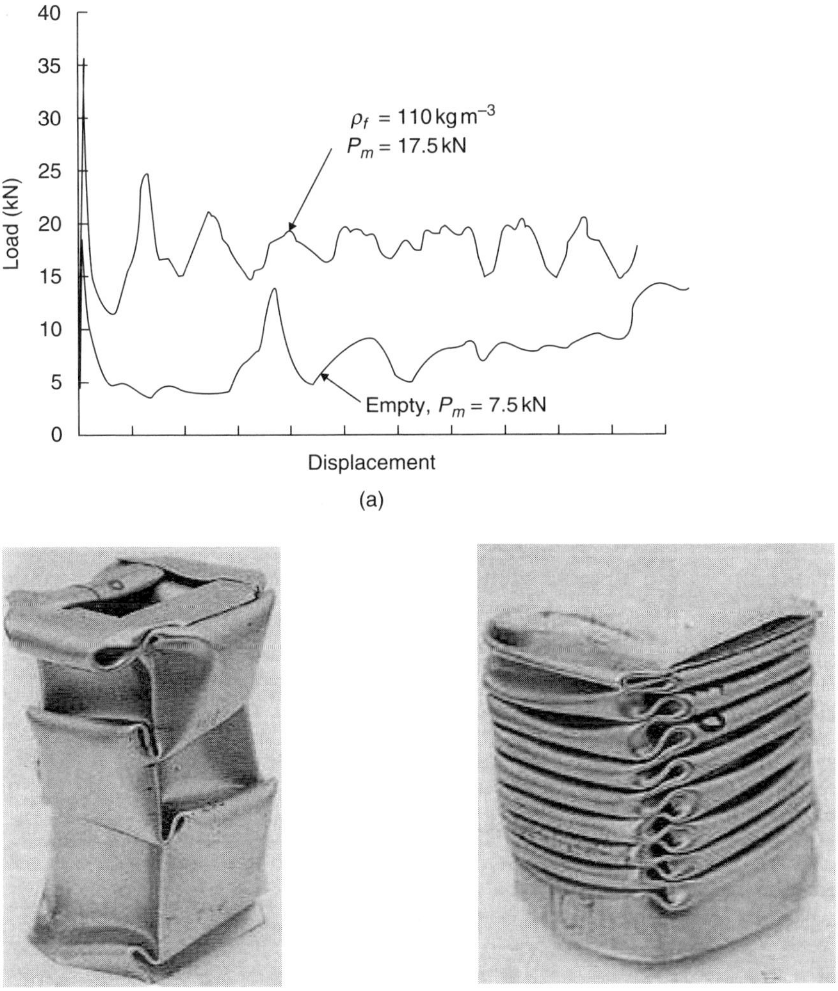

(a)

(b)

6.22 Comparison of an empty and a foam-filled square tube: (a) force–displacement curves; (b) crushing modes ($c = 75$ mm, $h = 0.76$ mm) (Reid *et al.*, 1986).

stops when the overall axial strain reaches the locking strain of the in-filled foam, ε_l. The nominal axial strain of a tube is

$$\varepsilon_l = 1 - \cos\theta_o \tag{6.74}$$

Hence

$$\theta_o = \cos^{-1}(1 - \varepsilon_l) \tag{6.75}$$

This defines the complete position of a collapsing fold. The locking strain of foam is related to its relative density, ρ^*/ρ_s, where ρ^* is the density of foam and ρ_s is the density of the solid from which the foam walls are made. If we take the strain as corresponding to a stress three times that of its plateau stress, $3\sigma_p$, σ_p being the plateau stress, then the locking strain is approximated as

$$\varepsilon_l = 1 - 3\rho^*/\rho_s \qquad [6.76]$$

Substituting this equation into Eq. [6.75], we have

$$\theta_o = \cos^{-1}(3\rho^*/\rho_s) \qquad [6.77]$$

In Eqs [6.1] and [6.2], using θ_0 instead of $\pi/2$, the modified average force for the tube alone is

$$P_{mt}(\theta_o) = 2\pi M_o\left[\frac{D(\theta_o + 2\sin\theta_o)}{H(1-\cos\theta_o)} + 1\right] \qquad [6.78]$$

The presence of foam reduces the fold length $2H$ slightly. But, if we assume that it is the same as an empty tube, or $H \approx \sqrt{Dh}$ (Eq. [6.5]), we obtain

$$P_{mt} = 2\pi M_o\left[\sqrt{\frac{D}{h}}\,\frac{\cos^{-1}(3\rho^*/\rho_s) + 2\sin(\cos^{-1}(3\rho^*/\rho_s))}{1-3\rho^*/\rho_s} + 1\right] \qquad [6.79]$$

By ignoring any possible increase in the cross-sectional area and from Eq. [10.35] of the relationship between the plateau stress and the relative density, the crushing force for the foam is

$$P_f = \sigma_p \frac{\pi D^2}{4} = 0.3Y_s(\rho^*/\rho_s)^{1.5}\frac{\pi D^2}{4} \qquad [6.80]$$

where Y_s is the yield stress of the solid cell wall of the foam. Hence, the average force for a foam-filled circular tube is

$$P_m = P_{mt} + P_f \qquad [6.81]$$

This analysis is due to Reddy and Wall (1988), and Eq. [6.81] agrees fairly well with their experiments (Fig. 6.23). Note that the force for this tube alone, P_{mt}, is also plotted and is shown to increase slightly with ρ^*. This reflects the interaction between the foam and tube walls discussed earlier. The increase of P_{mt} with ρ^* would be larger if the foam is assumed to lock earlier, say, when the stress is $2\sigma_p$. The choice of this lock strain is a little arbitrary here.

Similar theoretical analysis can be performed for square or rectangular tubes (Reid *et al.*, 1986), based on a modification of the analysis presented before for an empty square tube. Two empirical equations are worth

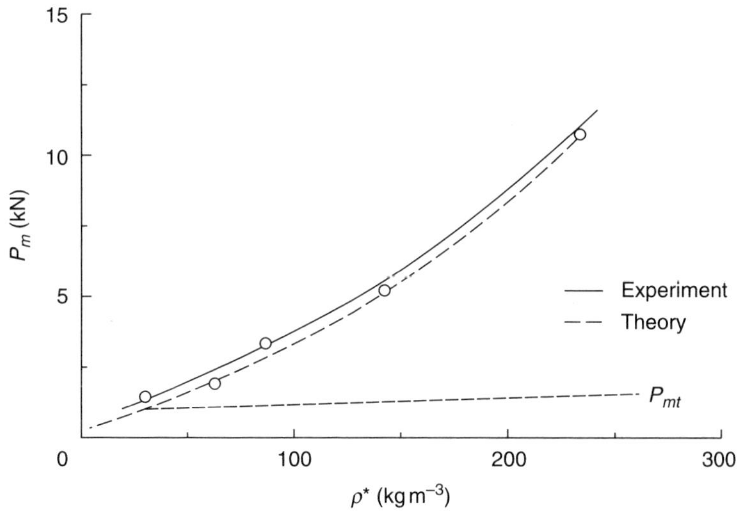

6.23 Average compression force versus foam density for a circular section (Reddy and Wall, 1988).

mentioning. Based on a numerical analysis, it is found that (Santosa and Wierzbicki, 1998b)

$$\begin{aligned} P_m &= P_{mt} + 2\sigma_p c^2 \\ &= 14Yc^{\frac{1}{3}}h^{\frac{5}{3}} + 2\sigma_p c^2 \end{aligned} \quad [6.82]$$

where c is the side length of a square tube and P_m here is the average force for an empty tube. To incorporate the strain-hardening effect, the equivalent flow stress is taken

$$\sigma_w = 2.23^n \frac{\sigma_u}{n+1}\left(\frac{2}{n+2}\right)^{\frac{2}{3}}\left(\frac{h}{c}\right)^{\frac{4n}{9}} \quad [6.83]$$

for a material having stress–strain relation

$$\sigma = \sigma_u\left(\frac{\varepsilon}{\varepsilon_u}\right)^n \quad [6.84]$$

In Eq. [6.82], if the coefficient of the second term being 2, instead of 1, this indicates the enhancement of force as a result of foam–tube wall interaction. It appears that this interaction is more significant than is the case with circular tubes (Fig. 6.23).

Equation [6.84] is verified for very low strength foams ($\sigma_p < 1.48$ MPa). For tubes filled with aluminium foams with σ_p in the range 1–12 MPa, another empirical formula was proposed by Hanssen *et al.* (1999):

$$P_m = 11.3Yc^{\frac{1}{3}}h^{\frac{5}{3}} + \sigma_p c^2 + 5\sqrt{\sigma_p Y}ch \qquad [6.85]$$

This equation correlates very well with a large number of experiments conducted. The crushable length (maximum stroke) is reduced by the foam and a new maximum stroke, $d_{\max}$, is given by

$$\frac{d_{\max}}{L} = 0.73 - \frac{c}{28h}\sqrt{\frac{\sigma_p}{Y}} \geq 0.55 \qquad [6.86]$$

where L is the tube length. The effect of strain-rate on P_{mt} can be taken into account in the same way as in Section 6.1.2. Similarly, the enhancement of the plateau stress of foam due to the strain-rate effect can be considered. To estimate the value of the strain-rate, assume that the foam is compressed uniformly over a fold of length $2H$. For an impact velocity V_o, the initial strain-rate is the $V_o/2H$. The average strain-rate is then

$$\dot{\varepsilon} = \frac{V_o}{4H} \qquad [6.87]$$

For $V_o = 10$ m/s and $2H = 2.5 \sim 3.5$ mm, the calculated strain-rate is between 100 and 200 s^{-1}. This could lead to an increase of 50 % in plateau stress for polyurethane foams.

6.5 Further remarks

In all of the above analyses of tubes under axial crushing, we are concerned only with the expressions for the average crushing force, because this is the most important parameter in evaluating the energy-absorption capacity of these tubes under such loading. Using the theoretical approaches described, a detailed force–displacement relationship can be worked out, although it would be slightly more complicated and hence is not presented here. Another parameter is the first, usually maximum, peak force. The level of such a force plays a very important role in designing energy-absorbing devices, because it dictates the maximum deceleration that a striking object experiences. Ideally, this peak force should not be excessive and it should be close to the average force. In practice, this level of force is reduced by introducing some trigging mechanisms in the tube. This includes chamfering one end of a tube to a small thickness, or slightly pre-bending this end of the tube. Theoretical analysis of the first peak load is not practical here as it is affected, for a given tube material and dimensions, by the initial imperfections over the tube and the detailed trigging mechanism. It should also be noted that the end-constraint may change the collapse mode of a circular tube (Singace and El-Sobky, 2001), although the level of axial force should be similar. Another type of tube behaviour under axial compression, tube inversion, will be discussed in Section 9.1.

Other forms of tubular members may be employed as energy absorbers. These include, square and rectangular tapered tubes (Reid and Reddy, 1986a and 1986b), frusta (Mamalis and Johnson, 1983), polygonal cylinders (Mamalis *et al.*, 1991a) and corrugated tubes (Singace and El-Sobky, 1997). Crushing of axially stiffened cylindrical shells and square tubes has been conducted (Birch and Jones, 1990; Jones and Birch, 1990). Interested readers should refer to these references.

Foam-filling of thin-walled tubes is effective in enhancing the energy-absorption performance. This enhancement is more significant as the plateau stress of the foam increases. There is, however, a limit to the maximum value of this plateau stress: if the foam is dense and hence has a high level of plateau stress, a tube may have a strong tendency to undergo Euler-type buckling, greatly reducing the energy absorption. Besides, such foam may cause ductile tearing of tube walls owing to the excessive tensile strain. If assessed in terms of the specific energy, energy absorbed over the mass, there exists an optimum density value for a given type of foam. In addition to polyurethane and aluminium foams, other materials can be used as in-fillers, such as wood (Reddy and Al-Hassani, 1993) and wood sawdust (Singace, 2000).

7
Impact on structures and inertia-sensitivity

This chapter first examines how to model the impact-contact between a rigid projectile and the surface of an elastic-plastic structure and how this local behaviour interacts with the global deformation of the structure. Then, a particular type of energy-absorbing structure, namely type II, is analysed in order to explore why the deformation of such structures is sensitive to the impact velocity, or to the inertia of both the striker and the structure itself.

7.1 Local deformation of structures due to impact

7.1.1 Kinetics of a direct central collision between two bodies

A description of a central collision process

In Section 2.4.2, the simplest case of central collision of two bodies was discussed. Now suppose that the centres of mass of bodies B_1 and B_2 travel along the same straight line with initial velocities v_{1o} and v_{2o}, respectively, before collision (Fig. 7.1(a)). To make a collision happen, it is assumed that $v_{1o} > v_{2o}$. Let F be the magnitude of the interactive forces acting on each other during the collision (Fig. 7.1(b)). Further assume that the contacting surfaces of the two bodies are oriented so that F is parallel to the line along which they travel and is directed toward their centres of mass. This condition, called ***direct central collision***, means that the two bodies will continue to travel along the same straight line after the collision (Fig. 7.1(c)).

Because the period of collision is usually very brief, the body forces (e.g. gravity) or other external forces applied to the two bodies would do no work during it. By neglecting the effect of the external forces, the linear momentum of the system, which consists of the two bodies involved, should be conserved; that is

$$m_1 v_{1o} + m_2 v_{2o} = m_1 v_{1f} + m_2 v_{2f} \qquad [7.1]$$

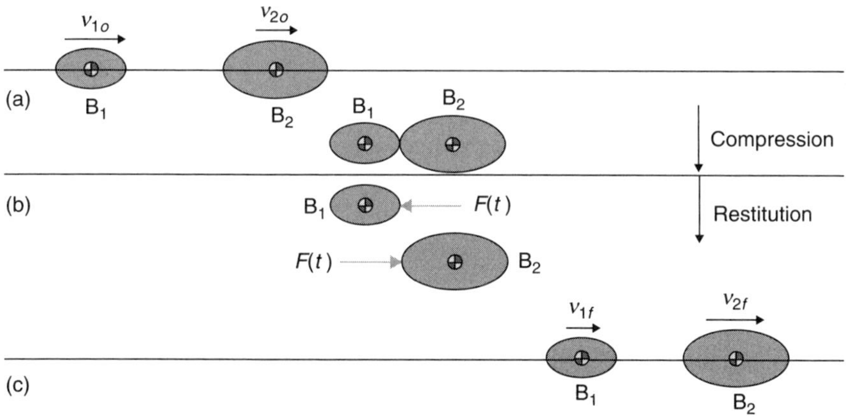

7.1 The process of a central collision of two bodies.

where m_1 and m_2 are the masses of bodies B_1 and B_2, respectively and v_{1f} and v_{2f} denote the velocities after the collision of bodies B_1 and B_2, respectively. Obviously, however, these two unknown velocities cannot be determined by a single equation (Eq. [7.1]) alone. Therefore, the collision process has to be considered in more detail.

Let t_o be the time instant at which B_1 and B_2 first come into contact. As a result of the collision, they will deform and their centres of mass will continue to approach each other, and this phase is called the ***compression phase*** of the collision. At time instant t_c, their centres of mass will reach their nearest proximity. At this instant, the relative velocity of the two centres of mass must be zero, so they possess the same velocity, denoted by v_c. The bodies then begin to move apart and separate from each other at time instant t_f. The period from t_c to t_f is called the ***restitution phase*** of the collision.

Analysis of the compression phase

In general, the impulse, $p(t)$, imparted to the two bodies is the integration of the contact force $F(t)$ over a time period starting from the initial instant of collision, t_o; that is

$$p(t) = \int_{t_o}^{t} F(t)\,\mathrm{d}t \qquad [7.2]$$

Applying the theorem of impulse and momentum to bodies B_1 and B_2, respectively, gives

$$-\mathrm{d}p = m_1 \mathrm{d}v_1 \qquad \mathrm{d}p = m_2 \mathrm{d}v_2 \qquad [7.3]$$

Therefore, the velocities of bodies B_1 and B_2 can be written as

$$v_1(t) = v_{1o} - \frac{p(t)}{m_1} \qquad v_2(t) = v_{2o} + \frac{p(t)}{m_2} \tag{7.4}$$

If the ***relative velocity*** between the two bodies is defined as

$$v^*(t) \equiv v_1(t) - v_2(t) \tag{7.5}$$

then it is found that

$$\mathrm{d}v^* = \mathrm{d}v_1 - \mathrm{d}v_2 = -\frac{\mathrm{d}p}{m_1} - \frac{\mathrm{d}p}{m_2} = -\left(\frac{1}{m_1} + \frac{1}{m_2}\right)\mathrm{d}p = -\frac{\mathrm{d}p}{m^*} \tag{7.6}$$

where $m^* \equiv m_1 m_2/(m_1 + m_2)$ can be regarded as an ***equivalent mass*** of the system. During the collision process, therefore, the variation of the relative velocity v^* with time is governed by

$$v^*(t) = v^*_o - \frac{p(t)}{m^*} \tag{7.7}$$

where $v^*_o = v_{1o} - v_{2o}$ (> 0) is the initial relative velocity.

At the end of the compression phase, i.e. at $t = t_c$, the two bodies possess the same velocity v_c, so that $v^*(t_c) = 0$, and Eq. [7.7] gives

$$p_c \equiv p(t_c) = m^* v^*_o = \frac{m_1 m_2}{m_1 + m_2}(v_{1o} - v_{2o}) \tag{7.8}$$

where p_c denotes the impulse imparted during the compression phase, or called the ***compression impulse***.

Integrating Eq. [7.4] over the compression phase ($t_o \le t \le t_c$), we have

$$-p_c = m_1(v_c - v_{1o}) \tag{7.9}$$

$$p_c = m_2(v_c - v_{2o}) \tag{7.10}$$

for bodies B_1 and B_2, respectively. By using Eq. [7.8], Eqs [7.9] and [7.10] lead to

$$v_c = v_{1o} - \frac{m^* v^*_o}{m_1} = v_{1o} + \frac{m^* v^*_o}{m_2} \tag{7.11}$$

The initial kinetic energy, K_o, of the system at $t = t_o$ is

$$K_o = \frac{1}{2}m_1 v_{1o}^2 + \frac{1}{2}m_2 v_{2o}^2 \tag{7.12}$$

and at the end of the compression phase ($t \le t_c$), the kinetic energy of the system reduces to

$$K_c = \frac{1}{2}m_1 v_c^2 + \frac{1}{2}m_2 v_c^2 = K_o - \frac{1}{2}m*(v*_o)^2 = K_o - \frac{p_c^2}{2m*} \qquad [7.13]$$

which indicates that the loss in the kinetic energy is proportional to the square of the compression impulse p_c and inversely proportional to the equivalent mass of the system.

From the energy point of view, the second term on the right hand side of Eq. [7.13] is exactly equal to the work done by the contact force $F(t)$ in the compression phase; and this work has been transformed to the deformation energy of the two bodies. That is

$$E_c^{ep} = W(p_c) = \frac{1}{2}m*(v*_o)^2 = \frac{p_c^2}{2m*} \qquad [7.14]$$

where W denotes the work done by impulse p, E^{ep} denotes the total elastic-plastic deformation energy of the system and the subscript c pertains to the compression phase.

Analysis of the restitution phase

Phenomenologically it is known that, as a result of the compression phase of the collision, part of the kinetic energy of the system could be lost due to a variety mechanisms, including plastic deformation of contact surfaces and generation of heat and sound. Consequently, the impulse they impart to each other during the restitution phase, p_r, would in general be smaller than the impulse they impart to each other in the compression phase, p_c. The ratio of these impulses is called the ***kinetic coefficient of restitution***

$$e \equiv \frac{p_r}{p_c} \equiv \frac{\int_{t_c}^{t_f} F(t)\mathrm{d}t}{\int_{t_o}^{t_c} F(t)\mathrm{d}t} \qquad [7.15]$$

In general, the value of e depends on the properties of the two bodies as well as their velocities and orientations when they collide, so it can be determined only by experiment or by a detailed analysis of the deformation of the two bodies during collision (see Section 7.1.2).

Figure 7.2(a) sketches the variation of the interactive force (i.e. the contact force) F with the total compression (i.e. the indentation) displacement δ, which is the sum of the compressions (indentations) of the two bodies. Figure 7.2(b) shows the variation of this force with time t; the area under the curve represents the impulse imparted to the bodies. Hence, the variation of imparted impulse $p(t)$ with time is shown in Fig. 7.2(c).

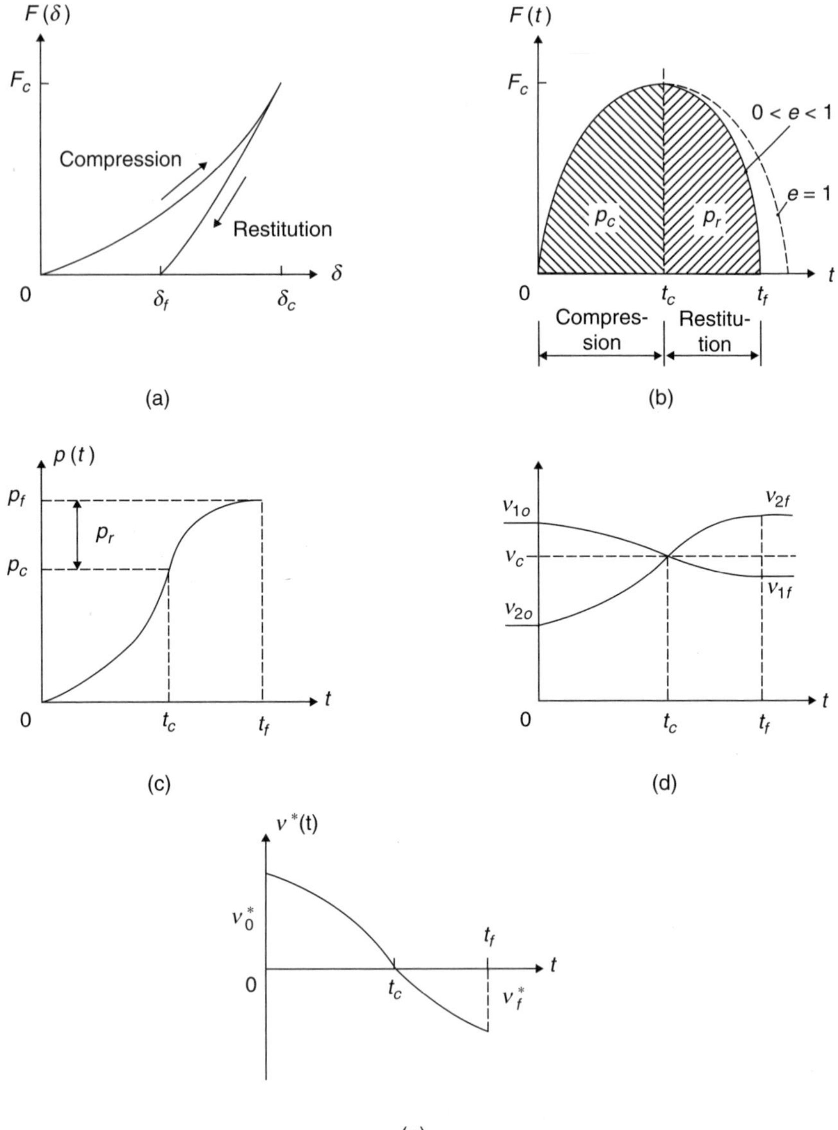

7.2 (a) Contact force varying with local deformation; (b) contact force varying with time; (c) impulse varying with time; (d) velocities of two bodies varying with time; (e) relative velocity varying with time.

Clearly, the kinetic coefficient of restitution defined by Eq. [7.15] can be found as the ratio of the two different shaded areas shown in Fig. 7.2(b) if the $F(t)$ curve is known. The total area under the $F(t)$ curve represents the total impulse

$$p_f \equiv p_c + p_r \tag{7.16}$$

Applying the theorem of impulse and momentum to the bodies B_1 and B_2, respectively, in the restitution phase, we obtain

$$-p_r \equiv -p_f + p_c = m_1(v_{1f} - v_c) \tag{7.17}$$

$$p_r \equiv p_f - p_c = m_2(v_{2f} - v_c) \tag{7.18}$$

It is known from Eq. [7.7] that the relative velocity will continue to reduce during the restitution phase:

$$v*(t) = -\frac{p(t) - p_c}{m*} \qquad t_c \leq t \leq t_f \tag{7.19}$$

and the final relative velocity is given by

$$v*_f \equiv v*(t_f) = v_{1f} - v_{2f} = -\frac{p_f - p_c}{m*} = -\frac{p_r}{m*} \tag{7.20}$$

The variations of velocities $v_1(t)$ and $v_2(t)$ with time t are sketched in Fig. 7.2(d), whilst that of the relative velocity $v*(t)$ is shown in Fig. 7.2(e). These velocity diagrams were proposed and discussed by Calladine (1990).

Comparing Eq. [7.20] with Eq. [7.8] leads to

$$\frac{v*_f}{v*_o} \equiv \frac{v_{1f} - v_{2f}}{v_{1o} - v_{2o}} = -\frac{p_r}{p_c} = -e \tag{7.21}$$

Therefore, we have another expression of e as

$$e \equiv \frac{v_{2f} - v_{1f}}{v_{1o} - v_{2o}} \tag{7.22}$$

Thus, the coefficient of restitution is related in a simple way to the relative velocities of the two bodies before and after the collision (in this definition, it is called ***kinematic coefficient of restitution***). If e is known, one can use Eq. [7.22] together with Eq. [7.1] (i.e. the conservation of linear momentum) to determine the final velocities of the two bodies after collision, v_{1f} and v_{2f}.

From the energy point of view, during the restitution phase only the elastic strain energy of the system will be released and transformed back to the kinetic energy of the two bodies. This elastic strain energy, $E^e{}_r$, is equal to the negative value of the work done by the imparted impulse p_r, $-W(p_r)$, in the restitution phase of the collision. In fact

$$W(p_r) = \int_{t_c}^{t_f} [p(t) - p_c] \mathrm{d}v* = -\frac{p_r^2}{2m*} \tag{7.23}$$

for which Eq. [7.20] is used. Therefore, similarly to Eq. [7.14], the elastic strain energy released in the restitution phase, E^e_r, is proportional to the square of the restitution impulse, that is

$$E_r^e = -W(p_r) = \frac{p_r^2}{2m*} \qquad [7.24]$$

Note that in the restitution phase a negative work is done by the contact force $F(t)$ (i.e. by the restitution impulse).

Combining Eqs [7.14], [7.24] and [7.15], it is found that

$$e = \sqrt{\frac{E_r^e}{E_c^{ep}}} = \sqrt{\frac{-W(p_r)}{W(p_c)}} = \sqrt{-\frac{W(p_f) - W(p_c)}{W(p_c)}} \qquad [7.25]$$

where $W(p_f) = W(p_c) + W(p_r)$ ($<W(p_c)$) is the net work done by the contact force $F(t)$ in the whole process of collision.

According to Eq. [7.25], the square of e is defined as the ratio of the elastic energy released during restitution and the total deformation energy stored during compression. In this definition, e is called the ***energetic coefficient of restitution***.

Discussion of coefficient of restitution

As seen above, we may have three different definitions for the coefficient of restitution e. It is called the kinetic coefficient of restitution, the kinematic coefficient of restitution and the energetic coefficient of restitution, as defined by Eqs [7.15], [7.22] and [7.25], respectively. As pointed out by Stronge (2000), all these three definitions for the coefficient of restitution are equivalent unless the bodies are rough, the configuration is eccentric or the direction of slip varies during collision. This explains why Eqs [7.15], [7.22] and [7.25] can be derived from each other for the case of direct central collision analysed herein.

From the definitions of e, it is seen that $0 \le e \le 1$, so that two extreme cases can be given below.

(1) If $e = 1$, the collision is ***completely elastic***. In this case, the restitution impulse must be equal to the compression impulse, i.e. $p_r = p_c$, and $p_f = 2p_c$. Accordingly, for the relative velocity we have $v^*_f = -v^*_o$.

Since there is no local energy dissipation due to plastic deformation or other sources (e.g. vibration, sound and heat), it can be shown that the total kinetic energy of the system remains the same before and after the collision, that is

$$K_o = \frac{1}{2}m_1 v_{1o}^2 + \frac{1}{2}m_2 v_{2o}^2 = \frac{1}{2}m_1 v_{1f}^2 + \frac{1}{2}m_2 v_{2f}^2 = K_f \qquad [7.26]$$

(2) If $e = 0$, the collision is ***completely inelastic*** (i.e. completely plastic). In this case, Eq. [7.22] indicates that $v_{1f} = v_{2f}$, so the two bodies remain together after collision.

In the compression phase, a considerable part of the kinetic energy is dissipated by the local plastic deformation; afterwards, there is no restitution phase and no elastic recovery at all since $p_r = 0$. Therefore, the final kinetic energy of the system is the same as that at the end of the compression phase. Thus, from Eq. [7.13]

$$K_f = K_c = K_o - \frac{p_c^2}{2m*} \qquad [7.27]$$

so that the energy dissipation in the collision process is given by

$$D = K_{\text{loss}} \equiv K_o - K_f = \frac{p_c^2}{2m*} \qquad [7.28]$$

In the simplest case analysed in Section 2.4.2, in which body B_2 has no initial velocity before collision, the loss in the kinetic energy is expressed by Eq. [2.72], which obviously is a special case of Eq. [7.28].

In a generic case, where $0 \le e \le 1$, since the elastic strain energy released in the restitution phase, E^c_r, is given by Eq. [7.24], Eq. [7.28] should be replaced by

$$D = K_o - K_f = \frac{p_c^2 - p_r^2}{2m*} = \frac{p_c^2(1-e^2)}{2m*} = \frac{(1-e^2)}{2} m*(v*_o)^2 \qquad [7.29]$$

7.1.2 Indentation caused by contact force

The analysis of the collision between two deformable bodies given above indicates that the behaviour of the system in both the compression and the restitution phases of the collision is dominated by the relationship between the contact force and the local deformations (i.e. indentations) of the two bodies in contact (as sketched in Fig. 7.2(a)). Obviously, this relationship depends not only on the elastic-plastic properties of the deformable bodies, but also on the local geometry of the contact surfaces.

Normal contact of elastic bodies: Hertz theory

When two deformable solid bodies are brought into contact they touch initially at a single point or along a line. Under the action of a very light load, they deform in the vicinity of the point of first contact so that they touch over a finite area, which is small compared with the dimension of the two bodies.

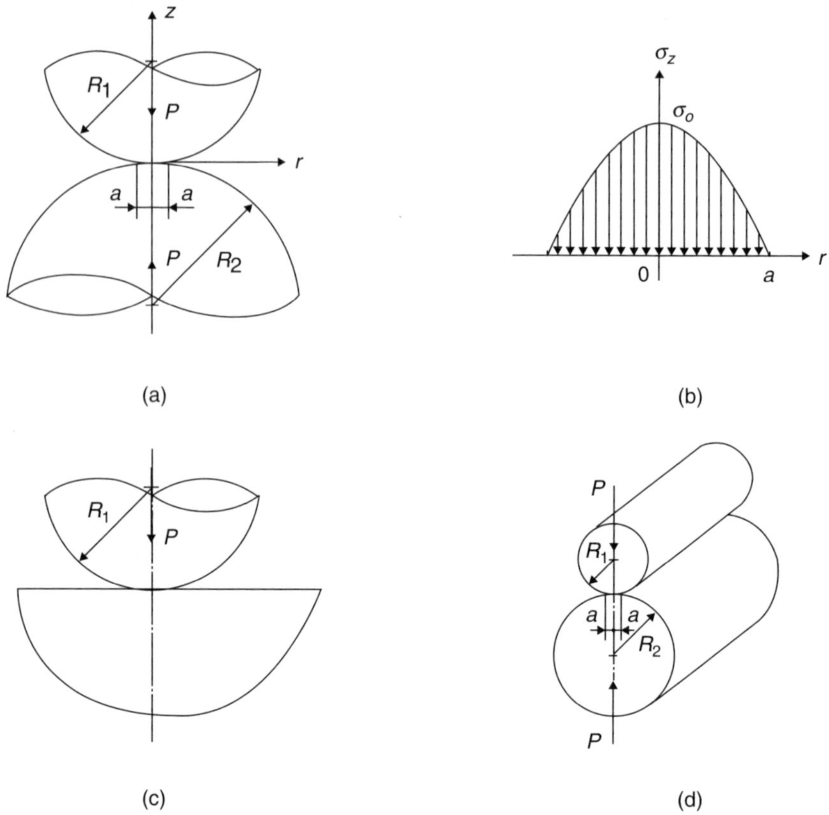

7.3 (a) Normal contact between two elastic spheres; (b) distribution of the normal pressure within the contact region; (c) contact between a sphere and an elastic half space; (d) contact between two cylinders.

In the following, it is assumed that the contact surfaces of the two bodies are smooth and the contact areas will develop axisymmetrically. Therefore, we can take the initial contact point as the origin of a cylindrical coordinate (r, θ, z) and let the z-axis be along the normal direction of the contact surfaces, so that (r, θ) forms the common tangential plane of the two bodies in contact, as shown in Fig. 7.3(a).

From the geometry, the profiles of the surfaces of the two bodies in contact can be approximately expressed by

$$z_1 = \frac{r^2}{2R_1} \qquad z_2 = \frac{r^2}{2R_2} \tag{7.30}$$

where R denotes the radius of curvature of the surface at the origin and subscripts 1 and 2 pertain to bodies 1 and 2, respectively.

In order to solve the distributions of the stress and displacement created by the contact force, the first step is the determination of the size and shape of the contact area as well as the distribution of normal pressure acting on it. In Hertz theory, the following assumptions are made:

- the contacting bodies are isotropic and elastic;
- the contact areas are essentially flat and small in comparison with the radii of curvature of the undeformed bodies in the vicinity of the interface;
- the contacting surfaces are perfectly smooth and frictionless, so only normal pressure needs to be taken into account.

The foregoing set of assumptions enable an elastic analysis to be conducted (refer to e.g. K.L. Johnson, 1985). Without going into the derivations, some of the major results of Hertz theory are summarised in what follows

(1) For two spherical surfaces in contact under force P (Fig. 7.3(a)), the contact pressure is distributed over a small circle of radius a given by

$$a = \left(\frac{3PR}{4E^*} \right)^{\frac{1}{3}} \quad [7.31]$$

Here the equivalent Young's modulus E^* and the equivalent radius R are defined by

$$E^* \equiv \left(\frac{1-\nu_1^2}{E_1} + \frac{1-\nu_2^2}{E_2} \right)^{-1} \qquad R \equiv \left(\frac{1}{R_1} + \frac{1}{R_2} \right)^{-1} \quad [7.32]$$

where E, ν and R with a subscript are the Young's modulus, Poisson's ratio and radius of the spheres, respectively; subscripts 1 and 2 pertain to spheres 1 and 2, respectively.

The maximum contact pressure is found to be

$$\sigma_o = \frac{3P}{2\pi a^2} = \left(\frac{6PE^{*2}}{\pi^3 R^2} \right)^{\frac{1}{3}} \quad [7.33]$$

which acts at the centre of the contact circle. The pressure distribution in the contact circle of radius a is given by

$$\sigma_z(r) = \sigma_o \left[1 - \left(\frac{r}{a} \right)^2 \right]^{\frac{1}{2}} \quad [7.34]$$

which is depicted in Fig. 7.3(b).

The total contact force P causes a relative displacement of the centres of the two elastic spheres, δ, owing to the local deformation, and they are related to each other by

$$\delta = \frac{a^2}{R} = \left(\frac{9P^2}{16E^{*2}R} \right)^{\frac{1}{3}} \qquad [7.35]$$

which can be rearranged as

$$P = k^* \delta^{\frac{3}{2}} \qquad [7.36]$$

with contact stiffness $k^* = 4E^*R^{\frac{1}{2}}/3$.

(2) For a spherical surface in contact with a flat surface under force P (Fig. 7.3(c)), we can take $R_2 = \infty$ (hence $R = R_1$) as a special case of (1). If we further assume that both bodies possess the same Young's modulus E and $\nu = 0.3$, then $E^* = 0.55E$ and

$$a = 1.089 \left(\frac{PR}{E} \right)^{\frac{1}{3}} \qquad \sigma_o = 0.388 \left(\frac{PE^2}{R^2} \right)^{\frac{1}{3}} \qquad \delta = 1.230 \left(\frac{P^2}{E^2 R} \right)^{\frac{1}{3}} \qquad [7.37]$$

(3) If a rigid punch (or projectile) is in contact with a flat elastic surface under force P, then as a special case of (1) we can take that $R_2 = \infty$ (hence $R = R_1$) with $E_1 = \infty$ (hence $E^* = 1.10E_2 \equiv 1.10E$ if $\nu = 0.3$) and obtain

$$a = 0.880 \left(\frac{PR}{E} \right)^{\frac{1}{3}} \qquad \sigma_o = 0.616 \left(\frac{PE^2}{R^2} \right)^{\frac{1}{3}} \qquad \delta = 0.775 \left(\frac{P^2}{E^2 R} \right)^{\frac{1}{3}} \qquad [7.38]$$

(4) In the case of line contact between two cylinders under load P (per unit length), see Fig. 7.3(d), the semi-contact-width is given by

$$a = \left(\frac{4PR}{\pi E^*} \right)^{\frac{1}{2}} \qquad [7.39]$$

where E^* and R are as defined in Eq. [7.32]. The maximum contact pressure is

$$\sigma_o = \frac{2P}{\pi a} = \frac{4}{\pi} \sigma_{zm} = \left(\frac{PE^*}{\pi R} \right)^{\frac{1}{2}} \qquad [7.40]$$

where σ_{zm} is the mean normal pressure in the contact region.

Normal contact of elastic bodies: Winkler foundation model

The difficulties of elastic contact stress theory arise because the displacement at any point on the contact surfaces depends on the distribution of

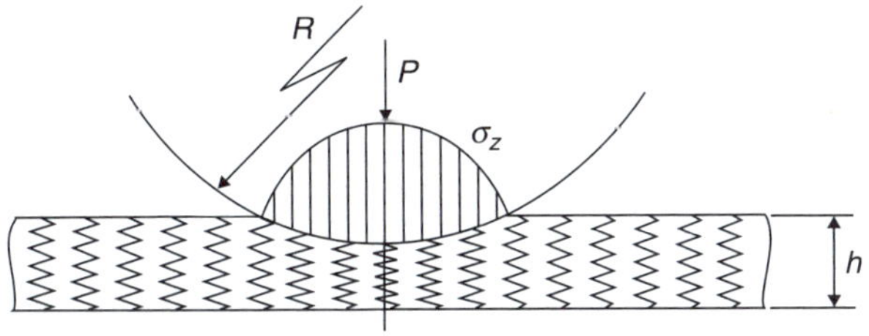

7.4 A rigid sphere is pressed on a Winkler foundation.

pressure throughout the whole contact. This difficulty can be avoided if the solids are modelled by a simple Winkler elastic foundation rather than an elastic half space. As shown in Fig. 7.4, an elastic foundation of depth h and elastic modulus k is supposed to rest on a rigid base and be compressed by an axisymmetric rigid indenter. The profile of the indenter is taken as the sum of the profiles of the two bodies (with radius of curvature R_1 and R_2, respectively) being modelled. By recalling Eqs [7.30] and [7.32], the profile of the indenter is given as

$$z(r) = z_1 + z_2 = \frac{r^2}{2}\left(\frac{1}{R_1} + \frac{1}{R_2}\right) = \frac{r^2}{2R} \qquad [7.41]$$

For the axisymmetric case, under compression by force P, the contact will be developed into a circular area of radius a, and it can be proven (see K.L. Johnson, 1985) that

$$P = \frac{\pi}{4}\left(\frac{ka}{h}\right)\frac{a^3}{R} \qquad \delta = \frac{a^2}{2R} \qquad [7.42]$$

while, for the two-dimensional contact of a long cylinder on a Winkler foundation

$$P = \frac{2}{3}\left(\frac{ka}{h}\right)\frac{a^2}{R} \qquad \delta = \frac{a^2}{2R} \qquad [7.43]$$

Equations [7.42] and [7.43] provide the relationships between the applied load and the size of the contact region. Comparing them with those obtained in Hertz theory (e.g. Eq. [7.35]), agreement can be reached if we choose $k/h = 1.70E^*/a$ for the axisymmetric case and $k/h = 1.18E^*/a$ for the two-dimensional case. If the depth of the foundation, h, is fixed, then we have to make the elastic modulus k inversely proportional to a, which increases with the indentation. In other words, the elastic modulus of the Winkler foundation in this model has to be reduced with the increase in P or in a.

Onset of plastic yield in normal contact

From detailed stress analysis based on Hertz theory, it is known that the maximum shear stress occurs at positions underneath rather than on the contact surfaces. Thus, in the case of two-dimensional contact of cylinders, by applying the Tresca yield criterion it is found (refer to, e.g. Hill, 1950 or Johnson and Mellor, 1983) that yield begins at a point $0.78a$ below the surface when the maximum contact pressure reaches the value

$$(\sigma_o)_y = 1.67Y \qquad [7.44]$$

where Y denotes the yield stress of the material in simple tension. Recalling Eq. [7.40], we find the load per unit length for initial yield as

$$P_y = \frac{\pi R}{E^*}(\sigma_o)_y^2 = 8.76\frac{Y^2 R}{E^*} \qquad [7.45]$$

If the von Mises yield criterion is adopted, the yield load is slightly higher with the coefficient in Eq. [7.45] being 10.1.

In the case of axisymmetric contact of spheres, by applying the Tresca yield criterion it is found that yield begins at a point $0.48a$ below the surface when the maximum contact pressure reaches the value

$$(\sigma_o)_y = 1.60Y \qquad [7.46]$$

and the corresponding load for initial yield is given by

$$P_y = \frac{\pi^2 R^2}{6E^{*2}}(\sigma_o)_y^3 = 6.74\frac{Y^3 R^2}{E^{*2}} \qquad [7.47]$$

Elastic-plastic indentation

When the yield point is first exceeded, the plastic zone is small and fully contained by material which remains elastic so that the plastic strains are of the same order of magnitude as the surrounding elastic strains, whilst the material displaced by the indenter is accommodated by an elastic expansion of the surrounding solid. As the indentation becomes more severe, an increasing pressure is required beneath the indenter to produce the necessary expansion. Eventually the plastic zone breaks out to the free surface and the displaced material is free to escape by plastic flow to the sides of the indenter. Whilst the rigid, perfectly plastic idealisation is adopted, this unconstrained mode of deformation can be analysed by the slip-line field method (see e.g. Hill, 1950; Johnson and Mellor, 1983), from which it is known that the unconstrained plastic flow (i.e. the fully plastic deformation) would occur if

$$\sigma_{zm} = cY \qquad [7.48]$$

where σ_{zm} is the mean normal pressure in the contact region and c has a value of about 3, depending on the geometry of the indenter and friction at the interface. In fact, it is known from Eq. [7.44] or Eq. [7.46] that at the onset of plastic yield, the mean normal pressure in the contact region $\sigma_{zm} \approx Y$, i.e. $c \approx 1$ in Eq. [7.48]. Therefore, the constrained plastic deformation takes place roughly in the range of $1 \le c \le 3$.

In last few decades, with the help of various approximate analytical models and finite element simulations, many useful results have been obtained from studies of the elastic-plastic stress and displacement fields caused by indentation (for more details, see K.L. Johnson, 1985).

In particular, if it is assumed that in the fully plastic deformation regime the edges of the impression neither pile up nor sink in, then

$$\frac{P}{P_y} = 5.3\frac{\delta}{\delta_y} \qquad [7.49]$$

where $P_y = 3.0Y$ and δ_y is related to P_y by Eq. [7.35]. It should be noted that the fully plastic deformation regime is reached only when P is very large ($P/P_y \approx 650$, i.e. $E^*a/RY \approx 40$).

By comparing Eq. [7.35] with Eq. [7.49] it is seen that in the elastic deformation stage P is proportional to $\delta^{3/2}$, but in the fully plastic deformation stage P is directly proportional to δ. Thus it is assessed that in the contained plastic deformation stage P is proportional to δ^q whilst q varies from 1.5 to 1.0.

Even in the case of large plastic deformation during loading, it is intuitive to expect the unloading process to be perfectly elastic. A simple check of this hypothesis was made by Tabor (1948). Consequently, the elastic recovery δ' is related to the maximum load P by

$$\frac{P}{P_y} = 0.38\left(\frac{\delta'}{\delta_y}\right)^2 \qquad [7.50]$$

By combining Eqs [7.49] and [7.50], the residual (permanent) indentation after the body is loaded to the fully plastic regime can be estimated as

$$\delta_f = \delta - \delta' = \left(0.182\frac{P}{P_y} - 1.62\sqrt{\frac{P}{P_y}}\right) \times \delta_y \qquad [7.51]$$

which is applicable only when P/P_y is large, say $P/P_y > 100$.

Indentation of a rigid sphere on a thin plate

Consider a thin plate of depth h resting on a rigid flat base and subjected to an indentation by a rigid sphere of radius R, as shown in Fig. 7.5(a). If the deformation remains in the elastic range, then the problem is similar to that shown in Fig. 7.4 for an indentation to a Winkler foundation, whilst the

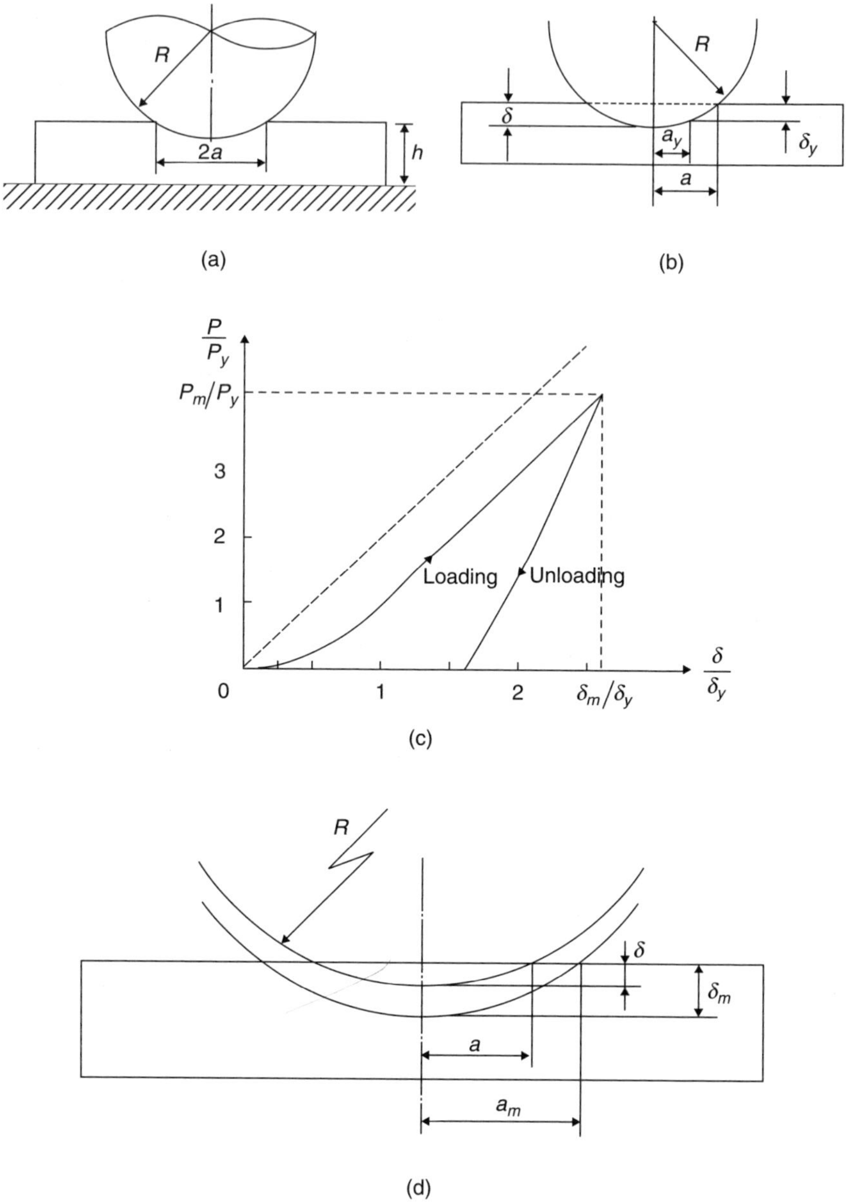

7.5 Indentation of a rigid sphere on a thin plate: (a) configuration; (b) elastic-plastic indentation, where a_y denotes the radius of plastic deformed region; (c) load–displacement curves; (d) configuration during unloading, where a_m denotes the maximum radius of the contact region.

elastic modulus of the foundation can be taken as that of the plate, i.e. $k = E$. Thus, by employing Eq. [7.42], the relation between load P and displacement δ is found to be

$$P = \frac{\pi ER}{h}\delta^2 \qquad [7.52]$$

The onset of yield will occur when the maximum compressive strain δ/h reaches the yield strain ε_y, i.e. when $\delta = \delta_y = h\varepsilon_y = Yh/E$. Hence, the yield load is

$$P_y = \frac{\pi Y^2 hR}{E} \qquad [7.53]$$

When the indentation proceeds further, $\delta > \delta_y$, an elastic-plastic stress distribution appears underneath the indenter. If the material is elastic, perfectly plastic, as shown in Fig. 7.5(b), the elastic-plastic boundary is at $r = a_y$, where the vertical displacement $u_z = \delta_y$. Since $u_z(r) = \delta[1 - (r/a)^2]$, the following relation must hold:

$$\left(\frac{a_y}{a}\right)^2 = 1 - \frac{\delta_y}{\delta} \quad \text{with} \quad \delta_y = \frac{Yh}{E} \qquad [7.54]$$

The stress distribution is expressed as

$$\sigma_z(r) = \begin{cases} Y & \text{plastic region } 0 \le r \le a_y \\ Eu_z/h = E\delta\left[1 - (r/a)^2\right]/h & \text{elastic region } a_y \le r \le a \end{cases} \qquad [7.55]$$

Thus, the contact force is calculated by

$$P = 2\pi\int_0^{a_y} Yr\mathrm{d}r + 2\pi\int_{a_y}^{a} E\delta\left[1 - (r/a)^2\right]r\mathrm{d}r = \pi YR(2\delta - \delta_y) \qquad [7.56]$$

which is valid for $\delta \ge \delta_y = Yh/E$. Combining Eqs [7.52], [7.53] and [7.56], we have

$$\frac{P}{P_y} = \begin{cases} \left(\dfrac{\delta}{\delta_y}\right)^2 & \delta \le \delta_y \\ 2\dfrac{\delta}{\delta_y} - 1 & \delta \ge \delta_y \end{cases} \qquad [7.57]$$

This relationship between the load and displacement is depicted as the solid line in Fig. 7.5(c). In fact, if a rigid, perfectly plastic material idealisation is adopted, then the relationship is simplified as $P/P_y = 2\delta/\delta_y$, as shown by the broken line in Fig. 7.5(c).

In the unloading process, only the elastic strains are recovered. If the maximum indentation displacement just before the occurrence of unloading is denoted by δ_m, then when the indenter is withdrawn to δ ($<\delta_m$, refer to Fig. 7.5(d)), the elastic stress released is

$$\sigma'_z = \begin{cases} E(\delta - \delta_m)/h & 0 \le r \le a \\ Eu_z/h = E\delta_m\left[1-(r/a_m)^2\right]/h & a \le r \le a_m \end{cases} \quad [7.58]$$

where a_m is the radius of the contact circle just before unloading, so that $a_m = 2R\delta_m$ and $a_m^{\;2} - a^2 = 2R(\delta_m - \delta)$. By deducting the integration of Eq. [7.58] from Eq. [7.57], we obtain the load–displacement relation in the unloading process as

$$\frac{P}{P_y} = \begin{cases} \left(\dfrac{\delta}{\delta_y}\right)^2 & \text{if } \delta_m \le \delta_y \\ \left(\dfrac{\delta}{\delta_y}\right)^2 - \left(\dfrac{\delta_m}{\delta_y} - 1\right)^2 & \text{if } \delta_m \ge \delta_y \end{cases} \quad [7.59]$$

This P–δ relation during unloading is also depicted in Fig. 7.5(c). When the load is entirely removed, i.e. $P = 0$, the final residual indentation is found to be

$$\delta_f = \begin{cases} 0 & \text{if } \delta_m \le \delta_y \\ \delta_m - \delta_y & \text{if } \delta_m \ge \delta_y \end{cases} \quad [7.60]$$

Energy dissipation due to indentation

If an indentation is carried on until the plastic deformation stage, i.e. until the indentation depth and the contact force reach δ_m ($>\delta_y$) and P_{m} respectively, then the work done by the contact force P in the loading process is found from an integration of Eq. [7.57] as

$$\begin{aligned} W(P)_{loading} &= P_y\delta_y\left[\frac{1}{3} - \frac{\delta_m}{\delta_y} + \left(\frac{\delta_m}{\delta_y}\right)^2\right] \\ &= P_y\delta_y \times \frac{1}{12}\left[1 + 3\left(\frac{P_m}{P_y}\right)^2\right] \end{aligned} \quad [7.61]$$

where P_{m} and δ_m follow the linear relation existing between P and δ as given by Eq. [7.57]

$$\begin{aligned} W(P)_{unloading} &= -P_y\delta_y\left[-\frac{\delta_m}{\delta_y} + \frac{7}{3}\left(\frac{\delta_m}{\delta_y}\right)^2 - \left(\frac{\delta_m}{\delta_y}\right)^3\right] \\ &= -P_y\delta_y \times \frac{1}{24}\left[-1 + 7\frac{P_m}{P_y} + 5\left(\frac{P_m}{P_y}\right)^2 - 3\left(\frac{P_m}{P_y}\right)^3\right] \end{aligned} \quad [7.62]$$

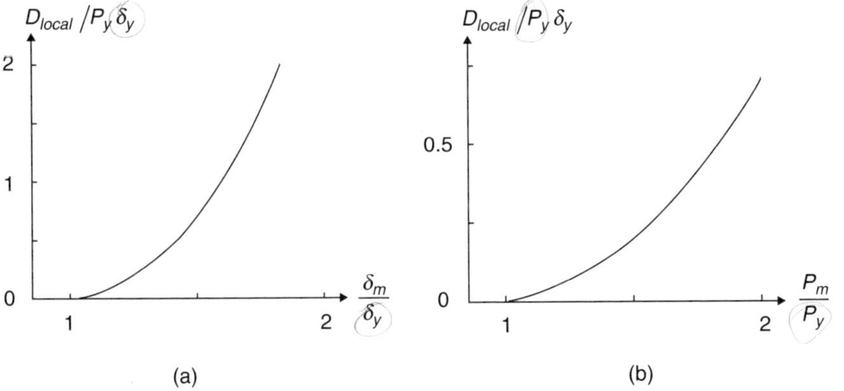

7.6 Non-dimension energy dissipation due to local indentation: (a) as a function of the maximum displacement; (b) as a function of maximum load.

Therefore, the net plastic energy dissipation after completing loading and unloading is

$$
\begin{aligned}
D_{local} &= W(P)_{loading} + W(P)_{unloading} \\
&= P_y\delta_y \times \frac{1}{3}\left[1-4\left(\frac{\delta_m}{\delta_y}\right)^2+3\left(\frac{\delta_m}{\delta_y}\right)^3\right] \\
&= P_y\delta_y \times \frac{1}{24}\left[3-7\left(\frac{P_m}{P_y}\right)+\left(\frac{P_m}{P_y}\right)^2+3\left(\frac{P_m}{P_y}\right)^3\right]
\end{aligned}
\qquad [7.63]
$$

where from Eq. [7.62], $P_y\delta_y = \pi Y^3h^2R/E^2$. The local energy dissipation is depicted in Figs 7.6(a)(b) as a function of δ_m/δ_y and as a function of P_m/P_y, respectively. It is seen that the energy dissipation increases quickly with δ_m/δ_y; for example, $D/(P_y\delta_y) = 0$ when $\delta_m/\delta_y = 1$ (purely elastic), $D/(P_y\delta_y) = 3$ when $\delta_m/\delta_y = 2$ (i.e. $P_m/P_y = 3$), but $D/(P_y\delta_y) = 43$ when $\delta_m/\delta_y = 4$ (i.e. $P_m/P_y = 7$).

When a structural component, e.g. a beam or a plate, is subjected to a transverse load via a rigid sphere, the local energy dissipation due to indentation can be estimated by Eq. [7.63] with P_m being taken as the collapse load of the component, P_s (refer to Section 2.2).

For example, for the beam shown in Fig. 2.12, the collapse load is (see Eq. [2.43]) $P_s = 4M_p/L = Ybh^2/L$, where b, h and L are the width, thickness and half-length of the beam, respectively. If $E/Y = 440$, $L/h = 20$ and $R = b$, then $P_m/P_y = P_s/P_y = bhE/\pi YRL = 7$, so that Eq. [7.57] gives $\delta_m/\delta_y = 4$, i.e. the maximum indentation $\delta_m = 4Yh/E = 0.009h$. Based on the local energy dissipation calculated from Eq. [7.63], in this particular case we have

$$
\frac{D_{local}}{D_{bending}} = \frac{43P_Y\delta_y}{4M_p\Delta/L} = 0.014\frac{h}{\Delta} \qquad [7.64]
$$

where Δ is the plastic deflection of the beam. This expression indicates that the local energy dissipation is of importance when and only when the global deflection of the beam is small in comparison with the thickness.

Together with a summary of the studies of indentation of beams and plates, the studies of indentation of laminates are reviewed by Abrate (1998).

7.1.3 Dynamic local deformation of structures under impact

Equivalent mass in a mass-spring model of a structure under impact

Usually, the structure's global deformation under quasi-static loads is relatively easy to determine by analysis, test or numerical simulation. This quasi-static load-deformation behaviour can then be lumped into the non-linear property of a spring in a mass-spring model (Fig. 7.7(a)).

As illustrated in Section 2.4.2, an impact of a projectile on a structure causes an immediate transformation of linear momentum, so that part of

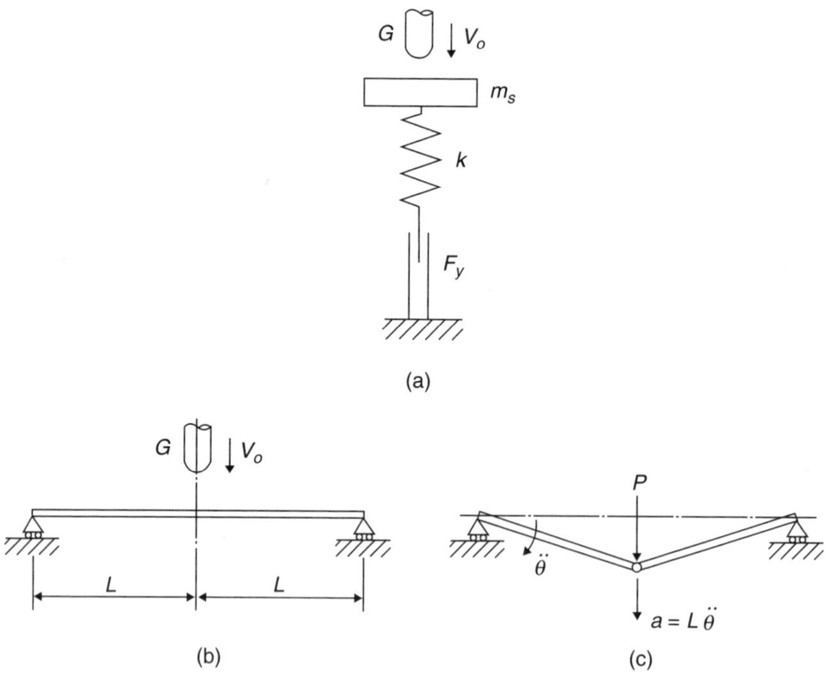

7.7 Impact on a structure: (a) the structure's global properties are represented by a mass-spring model; (b) impact on a beam; (c) collapse mechanism of the beam.

the initial kinetic energy of the projectile is instantaneously lost. It is known from Eq. [7.14] that the energy loss, i.e. the energy stored in the deformable particle (or contact spring) during the compression phase is

$$D = K_{\text{loss}} = K_o - K_c = \frac{1}{2}\left(\frac{1}{G} + \frac{1}{m_s}\right)^{-1} V_o^2 = \frac{m_s}{G + m_s} K_o \qquad [7.65]$$

where G is the mass of the projectile and $K_o = GV_o^2/2$ is the initial kinetic energy of the projectile with initial velocity V_o. Therefore, the difficulty in establishing the mass-spring model mainly lies in the determination of the equivalent mass of the structure, m_s, which is essential to the analysis of dynamic response of the structure under impact.

If the structure's dynamic response contains a large portion of elastic deformation, it has been suggested (Wu and Yu, 2001) that the equivalent mass in the mass-spring model, m_s, can be determined by equating the fundamental frequency of the mass-spring system to that of the structure's elastic vibration.

If the structure's global dynamic response is dominated by its rigid-plastic behaviour, then the equivalent mass can be calculated from its collapse mechanism. For example, consider a simply supported beam subjected to impact at its mid span, Fig. 7.7(b). In its rigid-plastic collapse mechanism two halves of the beam rotate about a plastic hinge located at the mid-span, Fig. 7.7(c). If the force applied at the mid-span is P, which produces acceleration a at that point, then the equation of motion for the rotation of a half of the beam about a support leads to

$$\frac{1}{2}PL - M_p = \frac{1}{3}\rho L^2 \ddot{\theta} = \frac{1}{3}\rho L^3 \frac{a}{L} \qquad [7.66]$$

which can be recast as

$$P - P_s = \frac{2}{3}\rho L a = m_s a \qquad [7.67]$$

where $P_s = 2M_p/L$ is the quasi-static collapse load of the beam. It is seen that the equivalent mass of the whole beam is $m_s = 2\rho L/3$ with ρ being the density of the beam's material. Note that this is just 1/3 of the total mass of the beam ($2\rho L$).

Modelling of indentation region

It is seen from Section 7.1.2 that in general the local behaviour of structures under indentation (or impact) is very complex. The load–displacement relationship depends not only on the elastic plastic properties of the indenter (or projectile) and the target structure, but also on the local geometry of the contact surfaces.

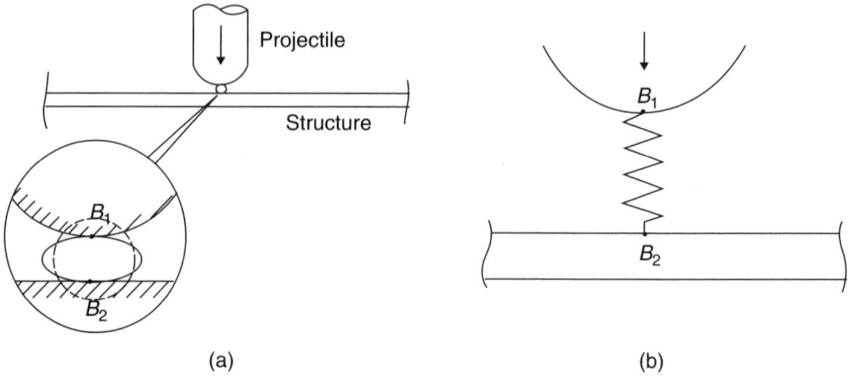

7.8 Model of indentation region: (a) an infinitesimal deformable particle; (b) a contact spring.

However, since the contact area is usually very small in comparison with the dimensions of the indenter (or projectile) and the target structure, the notably deformed material in the indentation (or impact) is contained only in a small volume. It would be appropriate, therefore, to adopt the assumption that all the local deformation of the projectile and the structure under collision can be lumped in an ***infinitesimal deformable particle*** between the contact points, as sketched in Fig. 7.8(a). Since the inertia of the infinitesimal deformable particle is negligible in a dynamic analysis, we may also model its behaviour as a ***contact element*** or a non-linear ***contact spring***, as shown in Fig. 7.8(b).

The deformable particle or the contact spring introduced serves as a cushion between the projectile and the structure's surface. If there is no such cushion, then points B_1 and B_2 in Fig. 7.8(a) or (b) will suffer a velocity discontinuity when two bodies first contact each other. With the assumed cushion, a short time interval will exist to decelerate the projectile and accelerate the structure whilst the cushion is experiencing a deformation. The loss in kinetic energy during the compression phase of the collision is indeed transformed into the deformation energy of that cushion.

The mechanical property of the infinitesimal deformable particle or the contact spring can be drawn from the indentation analyses given in Section 7.1.2. For instance, when analysing a beam or a plate subjected to impact of a rigid ball, Eqs [7.56] and [7.58] can be employed to quantify the properties of the infinitesimal deformable particle (or the contact spring) during its loading and unloading phase, respectively.

If the deformation of the structure after impact is negligible in the brief period of the collision process (i.e. within $0 \le t \le t_f$ with t_f denoting the time when collision is complete), as long as the indentation relationship is specified, the loss in the kinetic energy of the projectile during the compression phase of the collision can be reasonably attributed to the local energy dissipation due to indentation. For instance, if the impact is a rigid ball impinging on a thin plate, by equating Eq. [7.65] to Eq. [7.61], it is obtained that

$$D = \frac{m_s}{G+m_s} K_o = W(P)_{loading} = P_y \delta_y \times \frac{1}{12}\left[1+3\left(\frac{P_m}{P_y}\right)^2\right] \qquad [7.68]$$

Consequently, the maximum contact force in the impact process is found to be

$$P_m = P_y \sqrt{\frac{2m_s G V_o^2 E^2}{\pi(G+m_s)Y^3h^2R} - \frac{1}{3}} = \sqrt{\frac{2\pi Y R m_s G V_o^2}{G+m_s} - \frac{1}{3}\left(\frac{\pi Y^2 hR}{E}\right)^2} \qquad [7.69]$$

which depends on seven parameters: E, Y, h, R, G, V_o and m_s. The corresponding indentation depth is given by

$$\delta_m = \delta_y \times \frac{(P_m/P_y+1)}{2} = \frac{Yh}{2E}\left(\frac{P_m}{P_y}+1\right) \qquad [7.70]$$

This example shows that by combining the analysis of collision process and the analysis of quasi-static indentation, we are able to assess the local effects caused by impact on a structure, including the maximum contact force, the maximum local deformation (indentation) and the energy dissipated by this local deformation.

A simple model combining global and local behaviour of structures under impact

Wu and Yu (2001) have proposed a simple model to assess the dynamic response of elastic-plastic structures to impact. When the quasi-static structural and indentation behaviour of a structure is specified, by assuming that its dynamic deformation mode is broadly similar to the quasi-static one and that the material is strain-rate insensitive, the structure's response to impact can be simulated by a lumped mass-spring model, which consists of two masses and two non-linear springs, as sketched in Fig. 7.9(a).

In this model, G denotes the mass of the rigid projectile and m_s denotes the equivalent mass of the structure. Spring 1 represents the elastic-plastic behaviour of the structure itself, no matter whether the latter displays hardening, perfectly plastic or softening in its plastic range (Fig. 7.9(b)). Contact spring 2 represents the highly non-linear and inelastic mechanical property of the infinitesimal deformable particle between the projectile and the

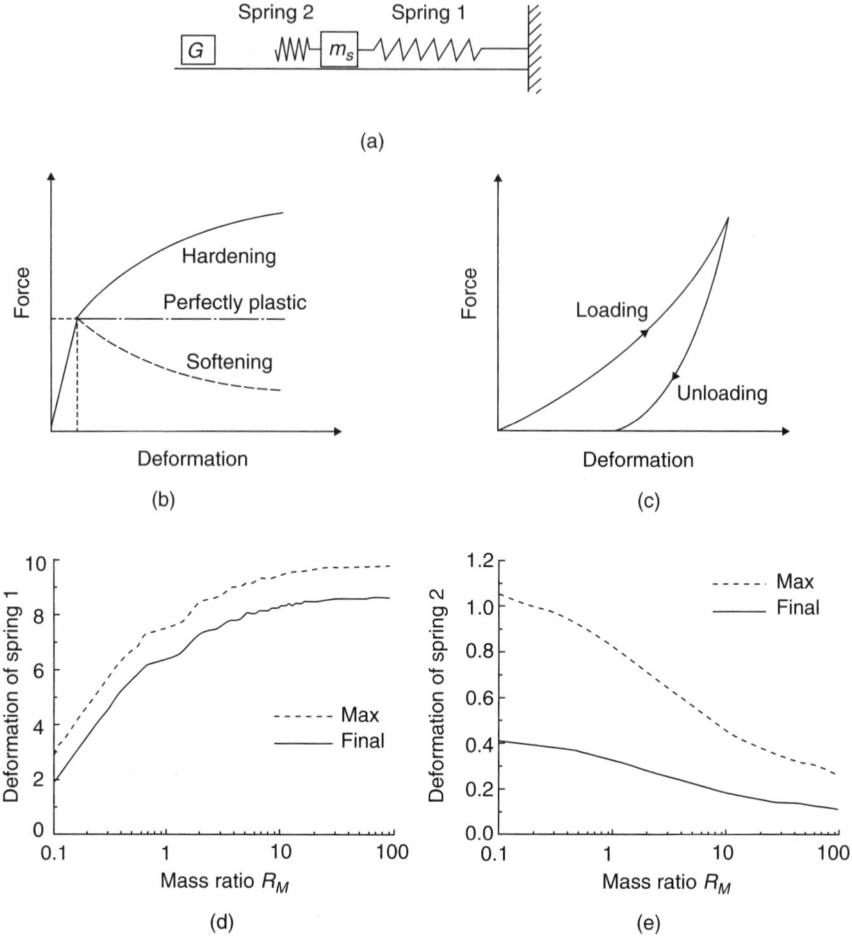

7.9 (a) A lumped mass-spring model; (b) property of spring 1; (c) property of spring 2 (contact spring); (d) displacement of spring 1, normalised by its maximum elastic displacement; (e) displacement of spring 2, also normalised by the maximum elastic displacement of spring 1. The results shown in (d) and (e) are based on $K_o/E^e_{max} = 20$, with K_o and E^e_{max} denoting the kinetic energy of mass G and the maximum elastic strain energy that can be stored in spring 1, respectively (reproduced with kind permission of Elsevier).

structure (Fig. 7.9(c)). For convenience, the load–displacement property of spring 2 can be approximated by

$$\frac{P}{P_c} = \begin{cases} A\left(\dfrac{\delta}{\delta_c}\right)^q & \text{loading} \quad d\delta > 0 \\ A'\left(\dfrac{\delta}{\delta_c}\right)^{q'} & \text{unloading} \; d\delta < 0 \end{cases} \qquad [7.71]$$

in which δ denotes the local deformation (indentation), coefficients A and A', and exponents q and q' are all drawn from a numerical simulation of indentation or determined by experiment. P_c and δ_c are the characteristic load and the characteristic indentation; for example, they can be taken as P_y and δ_y in case of an elastic-plastic thin plate being indented (or impinged) by a rigid sphere. Wu and Yu (2001) measured the indentation on an aluminium 6061 T6 beam by a hardened steel ball and fitted the experimental load–displacement curves using $q = 1.39$ and $q' = 1.8$.

Conducting numerical simulations based on this simple model, Wu and Yu (2001) demonstrated the effects of the mass ratio, structural stiffness/local rigidity and the hardening/softening factor on the maximum and final deformations of the model. For example, the effect of mass ratio $R_M \equiv G/m_s$ on the maximum and final deformations of spring 1 and spring 2 is as shown in Figs 7.9(d) and (e), respectively. The validity of the model for real structures is verified by impact tests on simply supported metal beams.

7.2 Inertia-sensitive energy-absorbing structures

7.2.1 Two types of energy-absorbing structures

To investigate the possibility of using small-scale dynamic tests together with a simple analytical formula in order to predict the behaviour of some full-scale prototype steel vehicle structures under dynamic conditions, Booth *et al.* (1983) conducted a series of 13 dynamic tests on thin-plated steel structures over a range of scales between 3 and 4. Their tests revealed a large and statistically significant deviation from linear scalability. Deformation and impact times at the larger scales were greater than expected and acceleration smaller. The deformations at full scale were 2.5 times larger than expected from the quarter scale model.

In attempting to explain the above experimental finding, Calladine (1983a) pointed out that the way in which metal structures absorb energy by gross distortion under impact conditions depends on the generic type of the structures. Typically, there are two types of energy-absorbing structures in terms of the shape of the overall static load–displacement curve: type I has a relatively 'flat-topped' curve (Fig. 7.10(a)), whilst type II has an initial peak load followed by a 'steeply falling' curve (Fig. 7.10(b)). The work by Booth *et al.* (1983) and Calladine (1983a) showed that the deformation of type II specimens is much more sensitive to the impact velocity than that of type I specimens; that is, when the total kinetic energy remains the same for all specimens, smaller final deformations result from higher impact velocities, and this phenomenon is much more significant for type II specimens than for type I specimens.

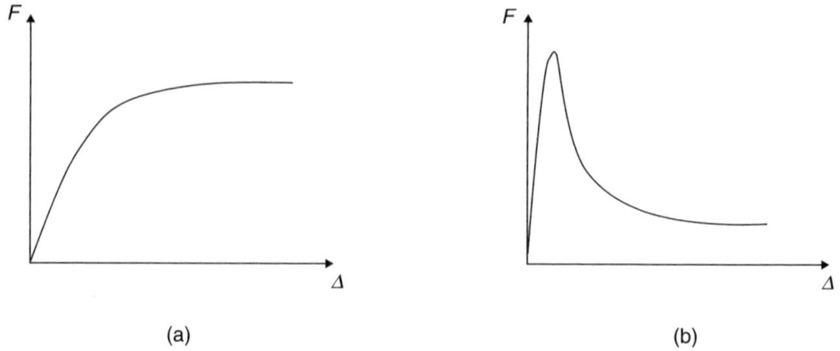

7.10 Two types of structures: (a) type I with a flat-topped load–displacement curve; (b) type II with an initial peak load followed by a 'steeply falling' curve.

The difference between type I and type II structures in their sensitivity to the impact velocity can be illustrated by Fig. 7.11. Suppose that the model's length scale is $1/\Lambda$ of the prototype. Owing to the strain-rate effect and inertia effect, which will be elaborated on later, the scaled dynamic collapse load of the small-scale model would be higher than its counterpart for the prototype, i.e. $(F_y\Lambda^2)_{\mathrm{model}} \equiv F''_y > F'_y \equiv (F_y)_{\mathrm{prototype}}$. When the impact energy remains scaled, the scaled final displacement of the model is smaller than its counterpart for the prototype, i.e. $(\Delta_f\Lambda)_{\mathrm{model}} \equiv \Delta''_f < \Delta'_f \equiv (\Delta_f)_{\mathrm{prototype}}$. When the collapse load is constant during the large deformation, typical for type I structures, the condition of 'equal scaled energy' requires that area A_1 = area A_2 as shown in Fig. 7.11(a), so that the difference between the model and the prototype in the final displacement is not significant. However, if the load–displacement curve is a 'steeply falling' one, typical for type II structures, then based on the same rule of 'area A_1 = area A_2', this difference will become very significant, as indicated by Fig. 7.11(b).

Clearly, the distinction between the two types of energy-absorbing structures and an understanding of the 'velocity sensitivity' of type II structures are of importance to the design of energy-absorbing structures, as well as to the scaling problem of model testing. The circular rings and tubes under lateral loads studied in Chapter 4 are in fact typical type I structures. In the following, our attention will be focused on the static and dynamic behaviour of type II structures.

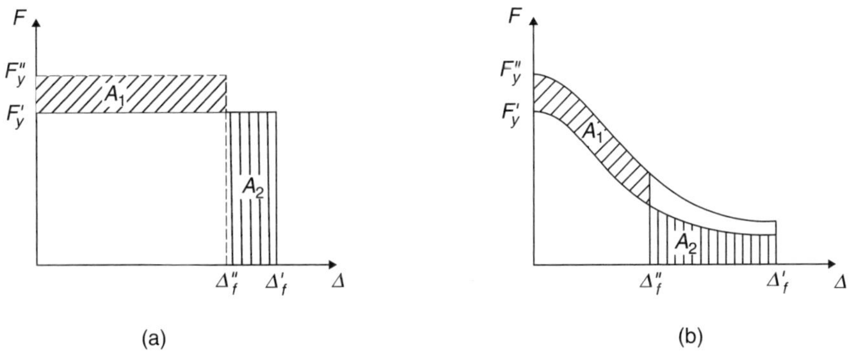

7.11 Sensitivity of structures to dynamic loading: (a) a type I structure; (b) a type II structure.

Calladine and English (1984) reported their tests on two sets of specimens. Their type I specimen was a circular tube resting on a flat base and their type II specimen consisted of two pre-bent plates, fastened by bolts near the top and clamped together between massive blocks at the bottom, as shown in Figs 7.12(a) and (b), respectively. The specimens were loaded by drop hammers having seven different weights and dropped from corresponding heights to give all the specimens the same input energy of K_o = 122 J. The velocity sensitivity which emerged from the tests on type II specimens was then explained by two relatively simple theories, which related the performance of specimens to the strain-rate sensitivity of the material and to the inertia effects.

7.2.2 Static behaviour of crooked plates

Since the pioneering work conducted by Calladine and English (1984), the pre-bent plate structure shown in Fig. 7.12(b) has served as a simple but typical type II structure for testing and analysis. Motivated by developing a way of controlling the peak load of the struts (or, in general, of thin-walled structures under axial loading) acting as energy absorbers, Grzebieta and Murray (1985, 1986) studied a strut with an initial kink at its mid-point, which is essentially similar to the pre-bent plate structures shown in Fig.

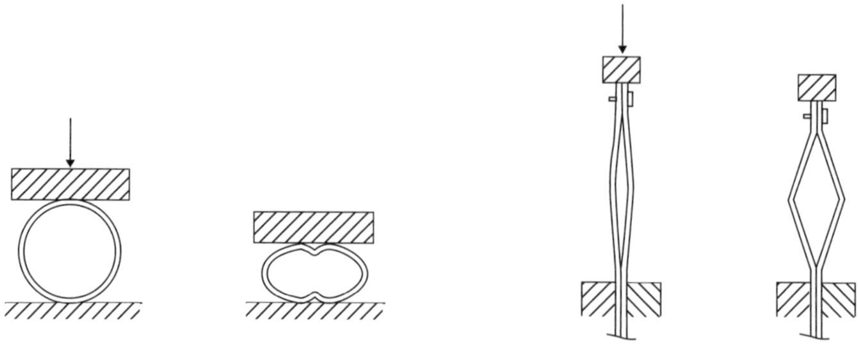

7.12 Typical specimens of two types of structure: (a) type I – a circular ring under compression; (b) type II – a pair of pre-bent plates.

7.12(b) except in the end supporting condition. In the following, we will call the pre-bent plate structures shown in Fig. 7.12(b) 'crooked plates' and analyse their behaviour.

Suppose the initial length of a half of a plate is L and both ends of the plate are simply supported. Figure 7.13(a) shows the initial configuration (solid line) and an elastically deformed configuration (broken line) of a quarter of the structure shown in Fig. 7.12(b). Let $y_o(x)$ and $y_1(x)$ represent the initial and current lateral coordinates, respectively, of points on the plate, then the elastically deformed profile of the plate is governed by the equation of static equilibrium which is

$$y_1'' + k^2(y_1 + y_o) = 0 \qquad [7.72]$$

where $k = (P/2EI)^{1/2}$ with $P/2$ being the axial force applied at the ends of one plate. By solving Eq. [7.72] with clamped end conditions, the elastic relationship between load P and the total lateral deflection at the middle point, $\delta = y(L) = y_1(L) + \delta_o$, and the relationship between load P and the vertical displacement Δ can be calculated.

For large inelastic deformation of the crooked plates, by considering both axial force and bending moment (refer to Eq. [2.22]) the load-carrying capacity of the crooked plates, P, is determined by the following equation

$$\frac{(P/2)\delta}{2M_p} + \left(\frac{P/2}{N_p}\right)^2 = \frac{(P/2)L\sin\theta}{2M_p} + \left(\frac{P/2}{N_p}\right)^2 = 1 \qquad [7.73]$$

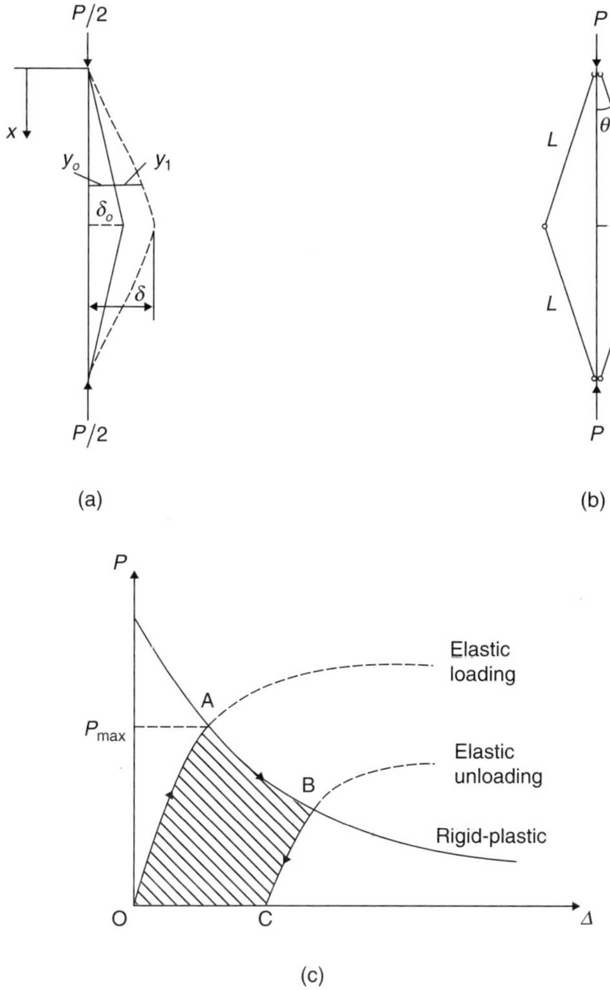

7.13 Deformation of crooked plates: (a) elastic deformation; (b) the large plastic deformation mode with four plastic hinges; (c) the axial load vs axial displacement relationship for the crooked plates.

where $M_p = Ybh^2/4$ and $N_p = Ybh$ are the fully plastic bending moment and fully plastic axial force, respectively, of a plate, with b and h being its width and thickness. $\delta = L\sin\theta$ is the total lateral deflection. The coefficient 2 in front of M_p comes from the rigid-plastic collapse mechanism (Fig. 7.13(b)) which contains plastic hinges at both ends of a half plate of length L.

Obviously, the initial collapse load P_s can be found from Eq. [7.72] if δ is taken as $\delta_o = L\sin\theta_o$. It follows that

$$P_s = 2N_p\left(\sqrt{1+\left(\frac{\delta_o}{h}\right)^2} - \frac{\delta_o}{h}\right) \qquad [7.74]$$

If the crookedness is very small, i.e. $\delta_o \ll h$, then Eq. [7.74] can be approximately recast as

$$P_s = 2N_p\left(\sqrt{1+\left(\frac{\delta_o}{h}\right)^2} - \frac{\delta_o}{h}\right) \approx 2N_p\left(1-\frac{\delta_o}{2h}\right) \qquad \frac{\delta_o}{h} \ll 1 \qquad [7.75]$$

On the other hand, if $\delta_o \gg h$, then Eq. [7.74] results in

$$P_s = 2N_p\left(\sqrt{1+\left(\frac{\delta_o}{h}\right)^2} - \frac{\delta_o}{h}\right) \approx N_p\frac{h}{\delta_o} = \frac{4M_p}{\delta_o} \qquad \frac{\delta_o}{h} \gg 1 \qquad [7.76]$$

When the deformation is large, the first term on the left hand side of Eq. [7.73] dominates, so that the load-carrying capacity of the crooked plates is approximated by

$$P = P(\Delta) \approx \frac{4M_p}{L\sin\theta} = \frac{4M_p}{\sqrt{L^2 - (L\cos\theta_o - \Delta)^2}} \qquad [7.77]$$

where $\Delta = L(\cos\theta_o - \cos\theta)$ denotes the vertical displacement at the top of the plates.

Combining the elastic behaviour and the rigid-plastic behaviour given by Eq. [7.77], Fig. 7.13(c) sketches the relationship between the axial load and the axial displacement for the crooked plates. The elastic-plastic response of the structure to an axial load will follow path O–A–B–C. Obviously, the elastic deformation of the structure makes the maximum axial load P_{max} much smaller than the rigid-plastic collapse load P_s predicted by Eq. [7.74] and the actual energy dissipation is the area surrounded by curve OABC.

7.2.3 Dynamic behaviour of crooked plates

General descriptions

Now consider the crooked plates shown in Fig. 7.12(b) under impact by a rigid striker, which has mass G and travels with initial velocity V_o before colliding with the top of the crooked plates. Experiments conducted by Tam and Calladine (1991) showed that the dynamic response of the crooked plates consists of two phases. In the first phase, whose duration is brief but finite, part of the initial kinetic energy of the striker is dissipated during collision and by axial deformation; and in the second phase, the dynamic response essentially follows the rigid-plastic deformation mechanism shown in Fig. 7.13(b).

It is clear that the analysis of the second phase is simple and straightforward; therefore, the emphasis of the studies is in understanding the energy dissipation mechanism in the first phase and identifying the transition from the first phase to the second phase.

Instantaneous energy loss due to inelastic collision

In an attempt to understand the notable difference between the dynamic behaviour of the crooked plates and their quasi-static behaviour, Zhang and Yu (1989) proposed a simple model which incorporates the energy loss based on the classical theory of inelastic collision between two bodies.

The system shown in Fig. 7.14(a) contains a rigid striker of mass G and a pair of rigid, perfectly plastic crooked plates. A half of each plate has length L and mass m. The initial crooked angle is θ_o and the crooked plates are supposed to be deformed into a four-hinge mechanism. When the impact velocity is not very high and the effect of stress waves is neglected, the crooked plates can be simplified as four rigid bars connected by hinges, where the bending moments M_1 and M_2 can be regarded as active ones.

This is a one degree-of freedom system. By taking angle θ as the generalised coordinate, the kinetic energy of the system at an arbitrary moment is

$$T = \frac{2}{3} mL^2\dot{\theta}^2 + 2L^2(m+G)\sin^2\theta\dot{\theta}^2 \qquad [7.78]$$

Using the Lagrange's equation of the second kind, we obtain the differential equation of motion of the system as

$$\ddot{\theta} + \frac{L^2(m+G)\sin\theta\cos\theta\dot{\theta}^2 + M_1 + M_2}{L^2[m/3+(m+G)\sin^2\theta]} = 0 \qquad [7.79]$$

The initial condition can be obtained by the Lagrange's equation in the case of impulsive loading; that is

$$\Delta\left(\frac{\partial T}{\partial \dot{\theta}}\right) = \hat{I} \qquad [7.80]$$

where $\partial T/\partial\dot{\theta}$ is the generalised momentum and $\Delta(\partial T/\partial\dot{\theta})$ is its instantaneous change due to the action of the generalised impulse $\hat{I}$. Consequently, from Eq. [7.80] we obtain the initial angular velocity at $t = 0$ as

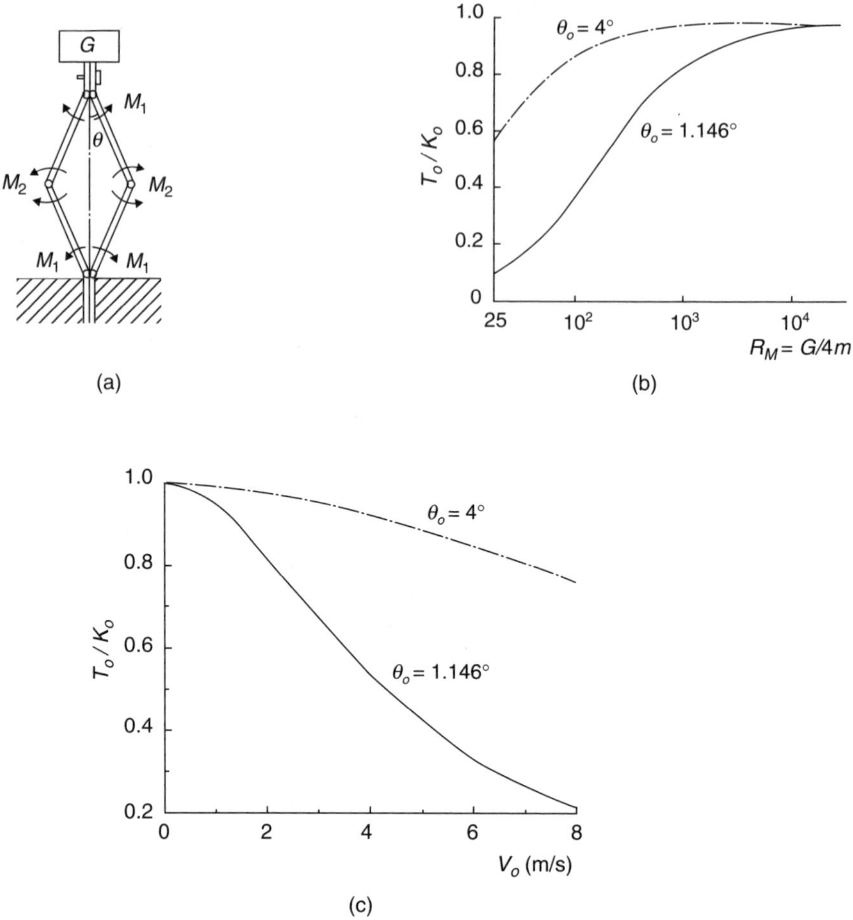

7.14 Zhang and Yu's (1989) model: (a) a rigid striker and a pair of rigid, perfectly plastic crooked plates; (b) T_o/K_o varies with mass ratio R_M; (c) T_o/K_o varies with the impact velocity when $K_o = 122$ J (reproduced with kind permission of Elsevier).

$$\dot{\theta}_o = \frac{GV_o \sin\theta_o}{2L[m/3+(m+G)\sin^2\theta_o]} \qquad [7.81]$$

Combining Eqs [7.78] and [7.81] leads to

$$\frac{T_o}{K_o} = \left[1+\frac{m}{G}\left(1+\frac{1}{3\sin^2\theta_o}\right)\right]^{-1} < 1 \qquad [7.82]$$

where $K_o = GV_o^2/2$ is the input energy carried by the striker and T_o is the kinetic energy of the system immediately after the impact. By noting $\theta_o \ll 1$ and introducing mass ratio $R_M \equiv G/4m$ = (mass of striker)/(mass of specimen), Eq. [7.82] is recast as

$$\frac{K_o}{T_o} = 1+\frac{1}{4R_M}\left(1+\frac{1}{3\theta_o^2}\right) \approx 1+\frac{1}{12R_M\theta_o^2} \qquad [7.83]$$

The dependence of T_o/K_o on $R_M \equiv G/4m$ is shown in Fig. 7.14(b) for the cases of $\theta_o = 1.146°$ and $\theta_o = 4°$, which were used in the tests by Calladine and English (1984).

Some interesting conclusions may be drawn from Eq. [7.82], Eq. [7.83] and Fig. 7.14(b).

(1) There is an instantaneous loss in kinetic energy ($T_o - K_o$) at the moment of impact.
(2) This energy loss depends only on the mass ratio $R_M \equiv G/4m$ and the initial crook angle θ_o, and neither on the impact velocity V_o itself nor on the mechanical properties of the material, except in that the collision must be 'perfectly inelastic', i.e. the two bodies must adhere to each other after impact.
(3) The behaviour of the system bears an analogy to the completely plastic central collision of two unequal masses; one of them, mass G, moves with velocity V_o before impact, while the other is initially at rest and has an effective mass

$$m_s = m+\frac{m}{3\sin^2\theta_o} \approx m\left(1+\frac{1}{3\theta_o^2}\right) \qquad [7.84]$$

the two terms of which come respectively from the 'longitudinal' and 'lateral' inertia of the crooked plates. In fact, using m_s defined by Eq. [7.84], and Eq. [7.82] results in

$$K_{\text{loss}} = K_o - T_o = \frac{m_s}{G+m_s}K_o \qquad [7.85]$$

which is exactly equal to the 'energy loss' calculated from the inelastic collision of two bodies, see Eq. [7.65].

(4) When varying the initial velocity of the striker V_o while keeping its initial kinetic energy $K_o = GV_o^2/2$ as constant, although V_o itself does not directly affect the energy loss (see item (2) above), the reduction in G accompanied by the increase in V_o will result in more energy loss, as verified by Eq. [7.82]. Figure 7.14(c) depicts the dependence of T_o/K_o on impact velocity V_o (via the variation of G) when $K_o = 122\,J = \text{const.}$

In the second phase of the dynamic response of the system, the rigid bars stably rotate about the four hinges, and the final rotation angle is proportional to T_o, which is the remaining energy of the system after the first (collision) phase. Since the final deformation of the crooked plates is mainly attributed to the hinge rotation, the fact that the value of T_o decreases rapidly with increasing V_o (i.e. with decreasing G) is the major source of the 'velocity sensitivity' of the crooked plates. In fact, the foregoing item (4) indicates that this 'velocity sensitivity' should be more correctly recognised as the 'inertia sensitivity' for type II structures.

The above analysis has also indicated that the 'inertia sensitivity' for type II structures is strongly influenced by the 'initial imperfection' of the structure. With an increase in the initial crook angle θ_o (or the initial crookedness δ_o) the inertia sensitivity will be severely weakened. In fact, Eq. [7.76] shows that the collapse load P_s decreases rapidly with the increase of δ_o. Therefore, with the 'initial peak load' being removed from its quasi-static load–displacement curve, the crooked plates as a type II structure (Fig. 7.10(b)) with a larger initial crook angle, θ_o, will no longer behave as a typical type II structure (see Fig. 7.10).

Effect of axial plastic deformation

Tam and Calladine (1991) conducted a more detailed study of the same problem, both experimentally and analytically. Their experiments were carried out in a drop hammer rig on a large number of specimens having the same initial geometry, but made in two different sizes and with two different materials chosen for their different strain-rate dependent characteristics in the plastic range. Details of the behaviour of these specimens during the dynamic deformation process were recorded.

They re-examined the applicability to this problem of the classical theory of collision and agreed that the results of Zhang and Yu (1989), e.g. Eq. [7.82], do provide a good first approximation. They pointed out that the deformation of crooked plates has two phases: the first involving only plastic compression of the specimen and the second involving rotation about the plastic hinges alone.

Tam and Calladine (1991) also modified the model proposed by Zhang and Yu (1989), so as to obtain a force-interaction of finite duration between the striker and the top of the crooked plates. To achieve this, the four rigid bars adopted in the collision model (Fig. 7.14(a)) are replaced by axially rigid, perfectly plastic rods. By allocating 1/3 of the total mass of the crooked plates on each side of the middle sections and adopting the (x, y) coordinates shown in Fig. 7.13(a), the equation of motion of the crooked plates in the lateral motion is found as

$$\frac{m}{3}\ddot{y} = \frac{N_p}{L}y \tag{7.86}$$

where $N_p = YA$ is the fully plastic axial force of a rod with Y and A being the yield stress and the cross-sectional area respectively. Solving Eq. [7.86] for y, we obtain the velocity at the top of the crooked plates as

$$v(t) = \frac{2}{L}y\dot{y} = \frac{\gamma\delta_o^2}{L}\sinh(2\gamma t) \tag{7.87}$$

where $\gamma \equiv (3N_p/Lm)^{1/2}$. On the other hand, the velocity of the striker is easily found as

$$V(t) = V_o - \frac{2N_p}{G}t \tag{7.88}$$

Figure 7.15(a) depicts a schematic plot of both V and v against time t. The velocity difference, $(V - v)$, represents the rate of shortening of the crooked plates when the rods are plastically compressed. When the V line intersects with the v curve at $t = t_1$, the first phase ceases and the second phase begins. The shaded area in the figure represents the total shortening produced in this phase. Since the deformation in the first phase occurs under the conditions of full plastic 'squashing', no bending moment is developed, so the energy dissipation in the first phase is associated directly with the shortening only. This phase ends when the lateral acceleration of the specimen is sufficient to enable the motion of the striker to be accommodated without further axial shortening of the specimen. During phase 1, energy is absorbed in much the same way as it is during the collision of two compact masses which adhere to each other after impact and the fraction of energy 'lost' depends strongly on the mass ratio of the striker and the specimen.

The kinetic energy T_o, which remains in the striker and the specimen at the end of phase 1, is absorbed in phase 2 by the rotation of the plastic hinges. The relevant yield stress in this process is that corresponding to the material of the specimen at the appropriate strain-rate. In the case where there is no strain-rate effect on the yield stress, the deformation in the second phase is essentially identical to what happens to the crooked plates under quasi-static loading. Therefore, Eq. [7.82] indeed provides a key to understanding the remarkable difference between impact performance and quasi-static performance of type II structures.

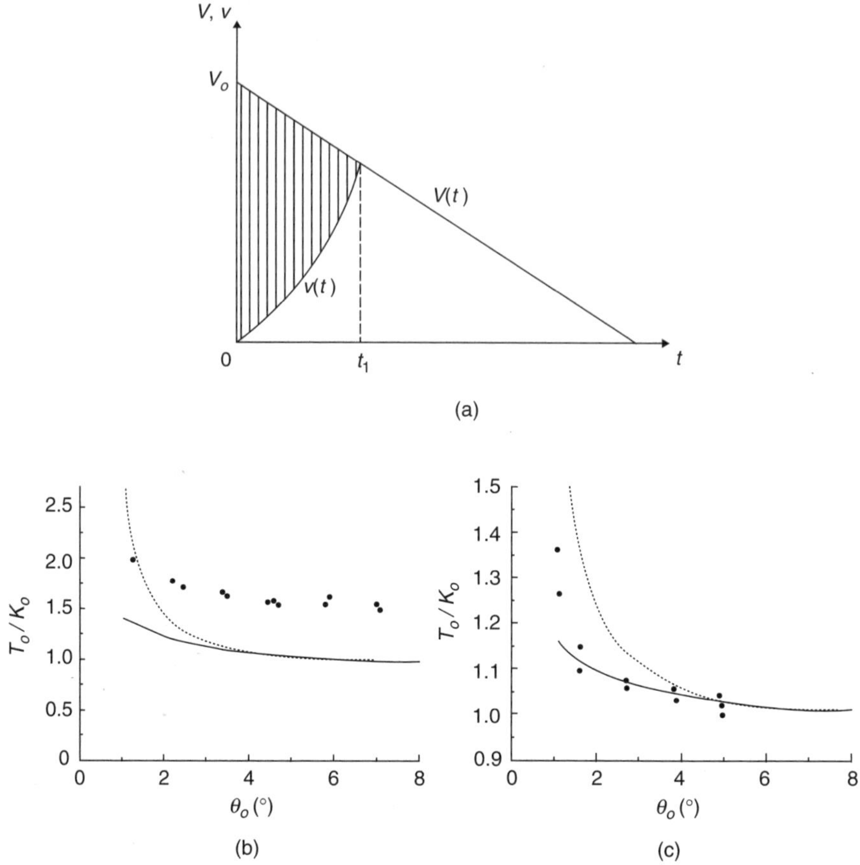

7.15 Tam and Calladine's (1991) model: (a) graphical representation of Eqs [7.87] and [7.88] by means of a plot of velocity against time; (b) T_o/K_o varies with θ_o for steel specimens; (c) T_o/K_o varies with θ_o for aluminium specimens. In (b) and (c), dots denote the experimental results, the solid and broken lines are predicted by Tam and Calladine (1991) and Zhang and Yu (1989), respectively (reproduced with kind permission of Elsevier).

The above remarks can be verified by Figs 7.15(b) and (c), in which the theoretical predictions are compared with experimental results. Aluminium specimens (Fig. 7.15(c)) displayed less rate dependency, so the results agree well with the rate-independent theories. On the other hand, mild steel is a rate-sensitive material, so the results for mild steel specimens are about 1.4 times the prediction by Tam and Calladine (1991). On the whole, it may be concluded that inertia is the dominant effect in the first phase and the behaviour of the second phase is more sensitive to strain-rate.

Effect of elastic deformation

It may be noted that neither Zhang and Yu (1989) nor Tam and Calladine (1991) took account of elastic deformation in their structural models. As a result, the resistance force of the structure would begin with a finite value ($2N_p = 2YA$) and the elastic deformation energy would be neglected. To remedy this, Su *et al.* (1995a) proposed a unified model that incorporates elastic, perfectly plastic constitutive relations and inertia effects into dynamic analysis. The model consists of four compressible elastic-plastic bars connected by four elastic-plastic 'hinges' of finite length.

By taking account of the complicated deformation history involving loading, unloading and reversed loading, the large deformation process is completely traced and the variation of the 'impact force' with time or with the vertical displacement at the top of the crooked plates is determined.

As a result of the involvement of elasticity, the dynamic response of the crooked plates can now be divided into four phases. In phase 1 the axial force increases rapidly to full yield of the bars. In phase 2 the load remains almost constant (at the peak load) for a short period. In phase 3 the load rapidly falls from the peak load. In phase 4 the load approaches another constant value which is much smaller than the peak load.

Figure 7.16(a) shows the variations of the axial force and bending moment with time, where $n \equiv N/N_p$ is the non-dimensional axial force, $m \equiv M/M_p$ is the non-dimensional bending moment, $\tau \equiv t/(mL/2N_p)^{1/2}$ is the non-dimensional time. It is seen that there is a quick transition in phase 3 from the axial force dominance to the bending moment dominance. Figure 7.16(b) further illustrates this transition by showing the stress loci in the (n, m) plane, where E_s, PI and PII denote the elastic regime, primary plastic regime and secondary plastic regime, respectively (refer to Yu and Johnson, 1982, or Yu and Zhang, 1996), and the outmost curve is the limit curve (see Fig. 2.8(c)). It is evident that, unlike in the static case, a plastic axial compression state is fully achieved in the structure after impact loading, and then it is quickly switched to a fully plastic bending state.

The partitioning of energy varying with time is depicted in Fig. 7.16(c), from which it is seen that the compression of bars dissipates a considerable portion (D_m/K_o) of the input energy in phases 1 and 2. It is particularly worth noting that with the incorporation of elastic-plastic axial compression in the model, the so-called 'initial energy loss' due to collision no longer exists. Although no local deformation at the collision interface is taken into

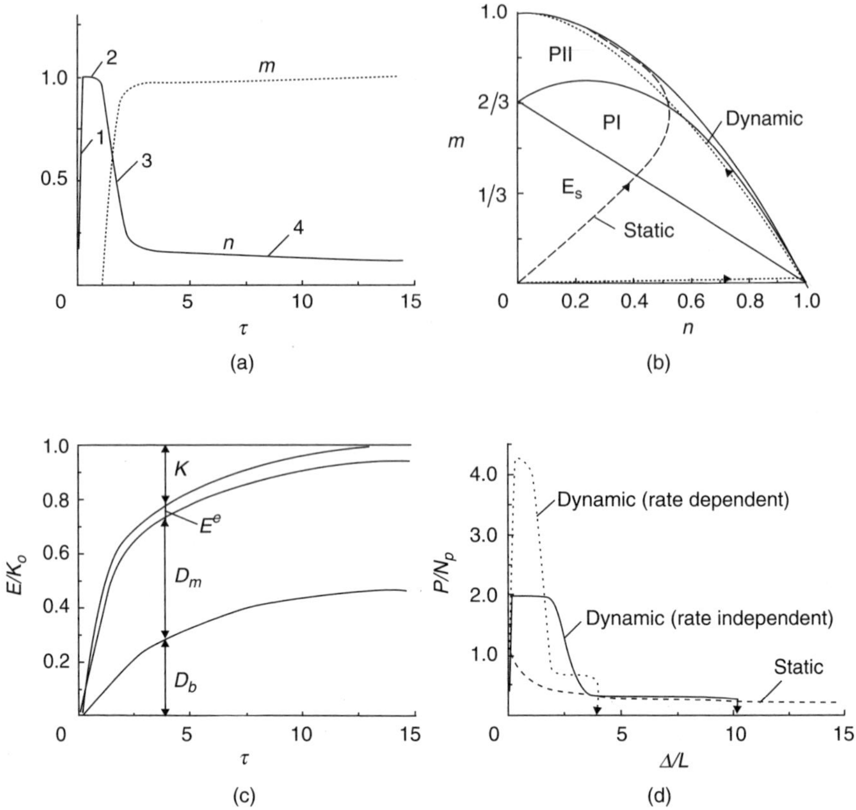

7.16 Su *et al.* (1995a) model: (a) the variations of the axial force and bending moment with time; (b) the stress loci in the (n, m) plane; (c) the partitioning of energy varying with time; (d) the load–displacement curves (reproduced with kind permission of Elsevier).

account, the 'lost' energy during collision has found its home in the elastic-plastic compression of the bars. In this sense, the compressive deformation of the bars plays the role of 'contact spring' that was suggested in Section 7.1.3 (e.g. spring 2 in the model shown in Fig. 7.9(a)).

Another important observation from the analysis of Su *et al.* (1995a) is that, even if the input energy is much larger than the elastic strain energy stored in the system (e.g. for the typical example illustrated, the elastic energy is about 4 % of the input energy), the inclusion of elasticity is essential because it enables one to determine the impact force in the early phase, in particular the peak load, which is of importance for the design of energy absorbers.

In a subsequent paper (Su *et al.* 1995b), the rate-dependent material properties of the Cowper-Symonds-type are incorporated into the model, whilst the impact force, especially the peak load, can still be predicted. Since the increase of the yield stress due to strain-rate sensitivity extends the range of elastic deformation, the inclusion of elasticity in a viscoplastic analysis becomes even more important. Figure 7.16(d) shows a typical case calculated. It is seen that the inertia has the effect of approximately doubling the peak load, while the strain-rate sensitivity of mild steel almost redoubles it.

The analysis given by Su *et al.* (1995a,b) indicates that the dynamic behaviour of crooked plates significantly differs from the quasi-static one of the same structure even when the effect of strain-rate on the material properties is excluded. The dynamic response and the final displacement are indeed dominated by the effective mass ratio G/m_s (for m_s see Eq. [7.84]) rather than the impact velocity V_o whilst the influence of the initial crookedness of the plates has also been considered. The rate dependency of material properties will further exaggerate the difference between the dynamic and static performances of type II structures.

Other work on crooked plates

Based on a discrete element model developed by Karagiozova and Jones (1995a) of the dynamic elastic-plastic buckling of plate structures, the same authors (1995b) studied the dynamic behaviour of crooked plates by taking into account inertia effects and the material rate sensitivity. The model (Fig. 7.17) consists of rigid bars with concentrated masses, two elastic-viscoplastic strain-hardening springs representing the axial flexibility and a non-linear spring representing the lateral flexibility. The equivalence between their model and the actual crooked plates was established by equating the corresponding static Euler buckling loads and the frequency of the elastic lateral vibration of the model with the first natural vibration frequency of the column. The predictions for the absorbed energy are in reasonable agreement with the experimental results reported by Tam and Calladine (1991).

Hönig and Stronge (2000) employed an elastic-viscoplastic material model of the Cowper-Symonds-type and finite element code ABAQUS to conduct a numerical simulation of crooked plates under impact. The material model included the measured effective-stress vs plastic strain curve. The

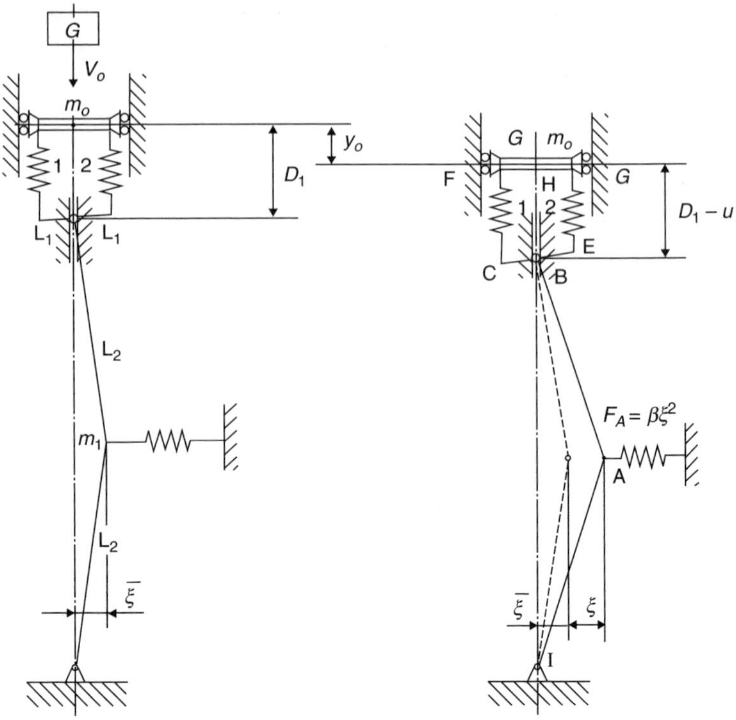

7.17 Karagiozova and Jones' (1995b) model (reproduced with kind permission of Elsevier).

numerical analysis used 2D beam and 3D shell elements. The striker was modelled by a point mass impacting a contact surface at the top of the plates. The simulation results compared well with experimental data reported by Tam and Calladine (1991), except that there were some differences in the peak load, which reveals that the boundary condition in the experiment was not very precisely defined. As shown in Fig. 7.18(a), which depicts the variation of the striker's velocity with time, only the viscoplastic model accurately predicts the measured dynamic response of mild steel plates. The numerical analysis also demonstrated that, for mild steel which

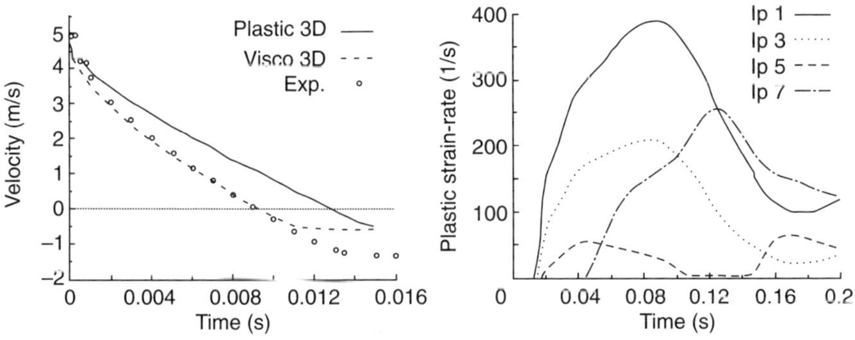

7.18 Numerical simulation by Hönig and Stronge (2000): (a) the velocity–time history; (b) variations with time of effective plastic strain-rate at the middle hinge calculated with a Dynamic Visco 3D model for impact speed $V_o = 4.8\,\text{m/s}$, where the number following 'Ip' pertains to the integration points 1 (inner concave surface), 3, 5 and 7 (outer surface), respectively (reproduced with kind permission of Elsevier).

has strain-hardening, the incorporation of strain-hardening is crucial for an accurate calculation of forces and energy dissipation at large deformation.

One of the interesting results from the numerical simulation is the prediction of the strain-rates, which cannot be obtained from previous theoretical models. When the impact speed is $V_o = 4.8\,\text{m/s}$, Fig. 7.18(b) depicts the variation of the strain-rates at the inner concave surface (solid line) and at the outer convex surface (chain line). The maximum strain rate in this case is about 400/s, but it will become 1400/s if the impact velocity increases to $V_o = 24.0\,\text{m/s}$.

7.2.4 Concluding remarks

After reviewing the various studies of crooked plates by different means, a deeper understanding of type II energy-absorbing structures has been gained. As pointed out by Tam and Calladine (1991), the thinking behind the distinction between two types of structures is straightforward. In type I structures (e.g. rings and beams) the absorbed energy increases linearly with deflection because the rotation of the plastic hinges is more or less directly proportional to deflection. But in type II structures the shape of the curve indicates that a disproportionately large amount of energy is absorbed in the first small increment of displacement, which is a direct consequence of the geometric effect whereby the endwise shortening of an initially straight rod containing a central hinge is proportional to the square of the angle of rotation of the hinge. These arguments are applicable not only to crooked plates, but also to many thin-walled structures under axial loading, e.g. struts, circular tubes and square tubes, etc.

When a type II structure is subjected to impact loading, the initial velocity suddenly applied in its axial direction has to be accommodated by its axial shortening and rapid rotation about the hinges. The latter implies not only a high strain-rate, but also a high lateral acceleration. Thus, the lateral inertia will significantly affect the dynamic behaviour of the structure. Equation [7.84] clearly indicates that the lateral inertia is dominant in this effect when the initial imperfection θ_o (crookedness) is small. As the crookedness increases, this effect will be quickly reduced. In fact, by adopting the concept of 'equivalent structure', a circular ring may be regarded as a crooked plate with $\theta_o = 45°$. Surely, it is a typical type I structure and its behaviour is entirely different from a crooked plate with small θ_o.

By conducting an elastic-plastic analysis either semi-analytically (Su *et al.* 1995a) or purely numerically (e.g. Hönig and Stronge, 2000), we are now able to paint a complete picture of the dynamic behaviour of type II structures. The energy lost during collision (Eq. [7.82] predicted by Zhang and Yu, 1989) can be accommodated by the elastic-plastic compression of the structure, so that the classical theory of collision between two bodies, as described in Section 7.1.1, is now consistent with a deformation analysis of the structure in its early response phase. It has also been demonstrated that elasticity, strain-hardening and strain-rate sensitivity all have negligible influence on the inertia-sensitive performance of type II structures.

8

Plastic deformation with ductile tearing

Structural failure as a result of excessive plastic bending/stretching is accompanied by ductile tearing. This brings about issues involving the assessment of ductile tearing energy under different loading conditions, the relative importance of this tearing and its possible interaction with the plastic deformation in regions far away from the tearing front. This chapter deals with these aspects of tearing, through several examples.

In all the problems discussed in previous chapters, the structures absorbed energy by plastic deformation and there was no separation of materials. The energy-absorbing mechanisms will be more complicated to analyse when tearing occurs. In this case, energy is absorbed by both ductile tearing and extensive plastic deformation in the region sufficiently far from the cracks. The issue hence is to understand the amount of energy dissipated by ductile tearing (Atkins, 1989) as well as that for plastic deformation in the far region. In this chapter, we present studies of measurement of ***ductile tearing energy*** for in-plane (mode I) and out-of-plane (mode III) tearing, splitting of metal tubes of both square and circular sections and cutting of a metal plate by a wedge. These represent possible approaches to this class of problems and have some interesting features contained within them.

8.1 Measurement of tearing energy

8.1.1 In-plane tearing

One convenient method of measuring the energy dissipated in mode I tearing is to use a deep edge notched tension (***DENT***) specimen, as shown in Fig. 8.1 (Cotterell and Reddel, 1977). In this specimen, symmetrical edge cracks are pre-cut. On monotonic loading, plastic flow occurs within a central zone before fracture. The total energy is composed of two parts: one involves plastic energy within this zone and the other is associated with the tearing along the ligament. Assume that this plastic zone is circular with the ligament as a diameter and that the tearing energy is proportional to

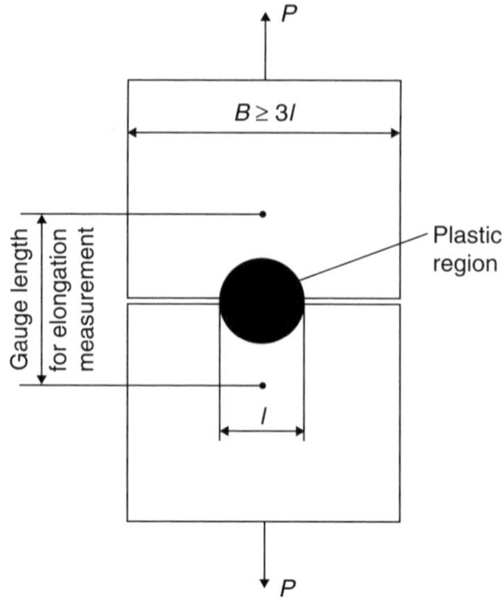

8.1 Sketch of a deep edge notched sheet specimen (DENT). The thickness of this sheet is *h*. The plastic flow is contained in a circular patch with ligament *l* as the diameter.

the new area created. The latter assumption is widely adopted in fracture mechanics. The term ***specific work***, or ***essential work***, is often used and here is denoted by w_t. It is simply equal to the tearing energy per unit of newly created area. The value of this specific work is regarded as being constant for a given material and sheet thickness, though we shall see that this is not always the case.

The plastic energy, D, for the total circular volume V is

$$D = w_p V \tag{8.1}$$

The plastic work per unit volume $w_p = \int \bar{\sigma} \mathrm{d}\bar{\varepsilon}$, where $\bar{\sigma}$ and $\bar{\varepsilon}$ are effective stress and strain, respectively. Consequently, the work done by the external load $W = \int P\mathrm{d}u$ (where P is the steady state force and u is the corresponding displacement) is dissipated by the tearing and plastic deformation in the confined central zone. That is, for a specimen of thickness h and initial ligament length l

$$\int P\mathrm{d}u = w_p(\pi l^2 h/4) + w_t hl$$

i.e.

$$\int P\mathrm{d}u/lh = w_p(\pi l/4) + w_t \tag{8.2}$$

For the assumed circular plastic zone, the representative effective strain and stress are similar for specimens of various initial ligament lengths. Equation

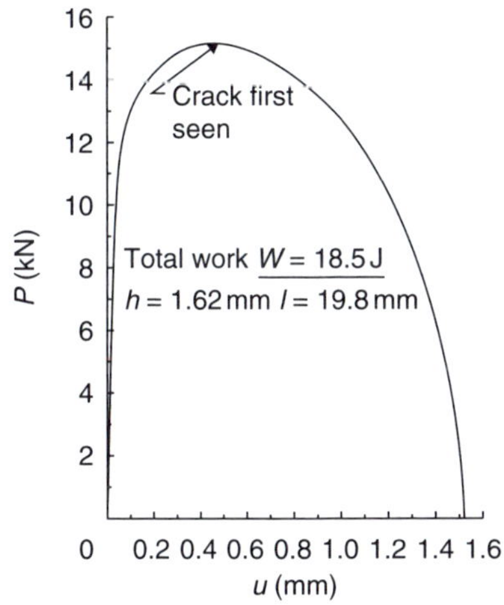

8.2 Load–displacement curve for a typical DENT specimen.

[8.2] immediately suggests that when the left hand side is plotted against the ligament length l, a straight line should be obtained; the ordinate intercept should be w_t and the slope should be indicative of the plastic work.

A typical DENT experimental load–displacement curve is shown in Fig. 8.2 (Cotterell and Reddel, 1977). This specimen is a cold-rolled low alloy steel, 1.62 mm thick. The central plastic zone in this case is circular with the diameter being the ligament length of 60 mm. Figure 8.3 shows a plot in accordance with Eq. [8.2]. The specific tearing energy obtained for these specimens of the same material and thickness is 240 kJ/m^2.

8.1.2 Out-of-plane tearing

The above idea of separating the plastic energy from the energy involved in the tearing region can be applied to the out-of-plane tearing case (mode III). One test method is to use a ***trousers type setup***, as shown in Fig. 8.4 (Mai and Cotterell, 1984). Since this is a steady-state process with a constant force, the tearing and plastic deformation can be better understood by visualising the material, such as AB, being pushed at end A through a fixed profile A–B–H. The material undergoes slight bending and tearing in region BC. Afterwards, bending proceeds until approximately point G, when the bending moment changes sign and unbending occurs until the legs become flat again at H. The total external work is dissipated in two parts of the metal sheet: one is associated with tearing (including the small initial bending) in region BC; and the other is the bending and unbending of the

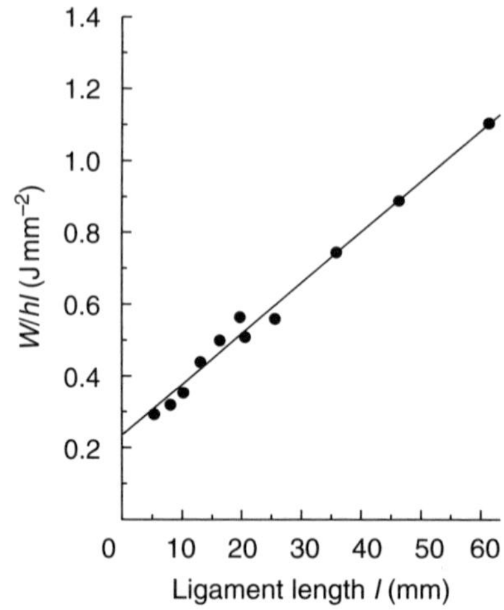

8.3 A plot of *W*/*lh* vs ligament length *l*. The ordinate intercept gives the tearing energy w_t.

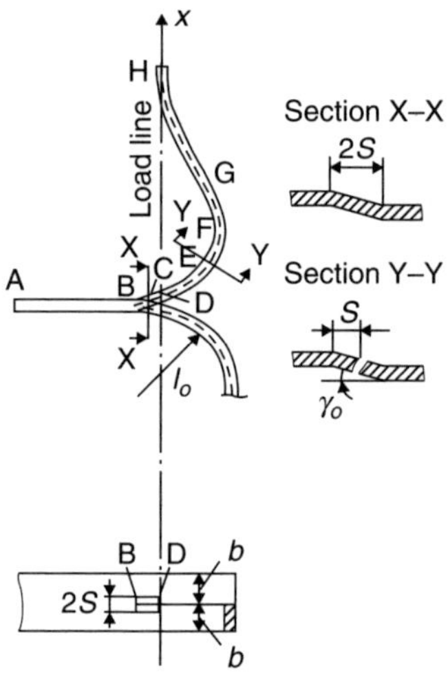

8.4 Sketch of an out-of-plane test (Mai and Cotterell, 1984).

two legs. For a rigid, perfectly plastic material, the energy of plastic bending to a radius R and its subsequent unbending back to its original flat sheet is similar to the travelling hinge described in Chapter 6. This energy is $2 \times 1/R \times M_o \times$ area bent, where M_o is (as before) the fully plastic bending moment per unit width. Denote the width of each of the two legs as b and the thickness as h. We can therefore write down the following equation of energy balance for an incremental length du

$$2P\mathrm{d}u = \left(\frac{2M_o}{R}\right)2b\mathrm{d}u + w_t h\mathrm{d}u$$

or

$$\frac{P}{h} = \left(\frac{4M_o}{Rh}\right)2b + w_t/2 \qquad [8.3]$$

The above equation is similar to Eq. [8.2] except that now it is written in terms of the steady state force P. Experiments suggest that the mean radius of curvature R is independent of the leg width b. For tests with the same material but with different values of leg width b, the value of tearing energy w_t is obtained when force P is plotted according to Eq. [8.3].

Figure 8.5 shows such a plot done by Mai and Cotterel (1984) for 1.6 mm thick low carbon steel and 2 mm thick 5251 aluminium alloy sheets. The values of tearing energy thus obtained are 1040 kJ/m² for steel and 600 kJ/m² for aluminium alloy.

In this arrangement, the radius of bending is not a known value and hence the bending energy is hard to determine accurately. Yu *et al.* (1988) devised a setup by bending a metal strip onto rollers of a known radius, and at the same time tearing the sheets. In this way, the bending energy can be assessed more accurately and the energy balance gives the tearing energy. Empirical equations were given for tearing energy in terms of sheet thickness.

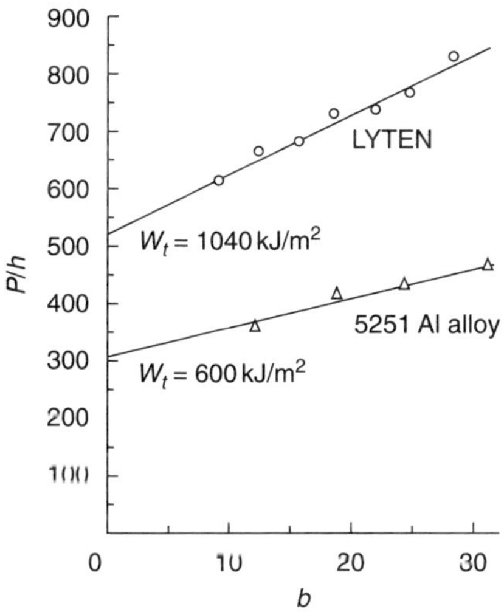

8.5 Plot of *P/h* against b. Again, the ordinate intercept gives the value of half tearing energy w_t: 1040 kJ/m² for steel and 600 kJ/m² for aluminium alloy.

8.1.3 Tearing energy in splitting square metal tubes

Splitting metal tubes has been investigated for its energy-absorbing capacity; see Section 8.1.4 (Stronge *et al.*, 1983 and 1984; Huang *et al.*, 2002a). The value of tearing energy needs to be known in order to predict theoretically the total energy dissipation. Lu *et al.* (1994) devised an experiment to determine the energy required for tearing the four corners of a square tube, see Fig. 8.6. In this setup, four corners are initially cut to a short length and the sidewalls of a square tube are attached to four rollers, which are fixed to a base plate. The four rollers are driven simultaneously by pulling up the four wire ropes attached. This motion causes bending of the sidewalls onto the rollers of a radius R and at the same time requires tearing along the four corners. One feature of this arrangement is that the two diagonal corners can be pre-cut to a given length. Hence, the process starts with bending of the sidewalls only. Once this pre-cut length is exhausted, additional force is required to tear the two corners. Typical load–displacement traces are shown in Fig. 8.7(a), while Fig. 8.7(b) shows one specimen after the test.

For each steady state during a test, the work balance is

$$P_n v_r = \frac{1}{1-\mu}[4D_b + nw_t h]v_s \qquad [8.4]$$

where n is the number of corners being torn, P_n is the corresponding external load, $D_b = M_p/R$ is the bending energy of each of the four sidewalls per unit increment of feeding length, h is the thickness, v_s is the linear velocity

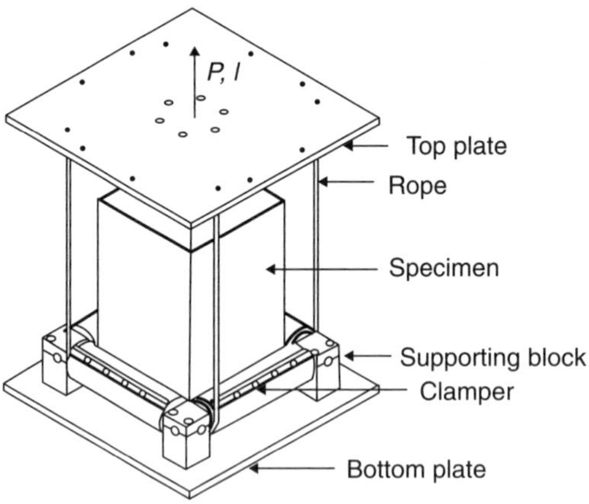

8.6 Schematic of the experimental setup. The bottom and top plates are to be attached to the base and crosshead of the testing machine. Details of the ropes and top plate are not shown (Lu *et al.*, 1994) (reproduced with kind permission of Elsevier).

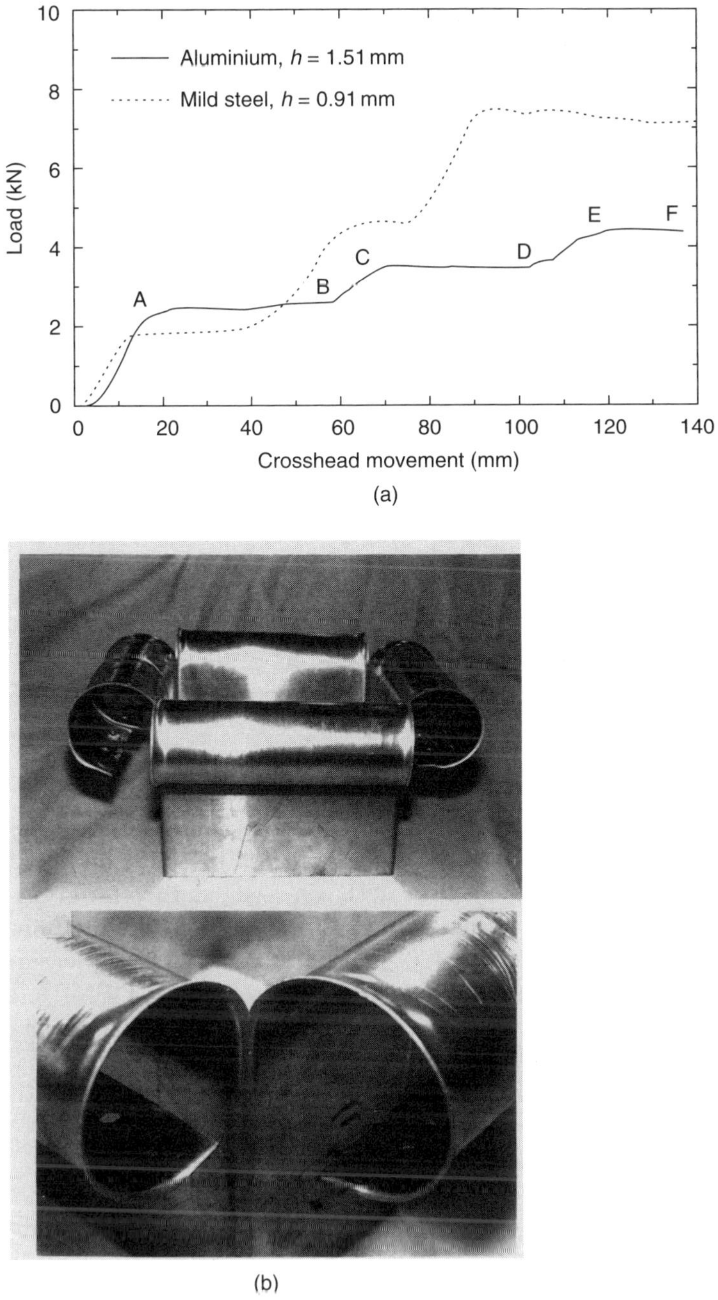

8.7 (a) Typical force–displacement traces. Steady state AB, CD and EF are three plateaus corresponding to, respectively, bending only, bending with tearing two corners and bending with tearing four corners (Lu *et al.*, 1994). (b) Photographs of a specimen after test.

of the rollers and v_r is the velocity of the ropes (equal to that of the crosshead). μ reflects the frictional effect of the roller bearings; its value was determined experimentally to be about 0.1.

Based on the above equation and experiments, the value of w_t can easily be obtained. Mild steel square tubes of thicknesses 0.7, 0.91 and 1.67 mm and aluminium ones of thicknesses from 0.47 to 1.51 mm were tested. It was found that

$$w_t = \begin{cases} 1360h^{0.61} & \text{for mild steel} \\ 211h^{0.43} & \text{for aluminium} \end{cases} \qquad [8.5]$$

where h is thickness in mm and w_t is specific work in J/m. Note that this tearing energy is dependent upon the thickness of the metal sheet and is not solely a material property. Recognising that the tearing energy is dissipated mainly by plastic deformation around the crack front, it is reasonable to relate it to the ductility (fracture strain ε_f) and ultimate stress of material (σ_u). Hence, empirically

$$w_t = \begin{cases} 8.8\sigma_u\, \varepsilon_f & \text{for mild steel} \\ 37.2\sigma_u\, \varepsilon_f & \text{for aluminium} \end{cases} \qquad [8.6]$$

Note that the constants 8.8 and 37.2 have units of mm and w_t is in $J/m^2 \times 10^{-3}$, when σ_u is in N/m^2. This equation will be used in Section 8.3.

8.1.4 Comments on the values of tearing energy

We have presented three testing methods for determining tearing energy in different loading situations. Lu *et al.* (1998) devised another method, which involves multiple tensile tests on a standard tensile test coupon (MTT), but of various gauge lengths. This method was successfully used by Mohammadi and Mahmudi (2001). The idea of separating tearing energy and plastic energy in the far field is similar to that in a DENT test. Nevertheless, the value of tearing energy is in general higher than from a DENT test for nominally identical materials. Table 8.1 shows typical values of tearing energy for mild steel specimens. It is clear that they can be of dif-

Table 8.1 Values of tearing enery for mild steel from various tests

Test method	MTT (Lu *et al.*, 1998)	DENT (Cotterell and Reddell, 1977)	Trousers test (Mai and Cotterell, 1984)	Splitting tubes (Lu *et al.*, 1994)
Tearing energy (kJ/m^2)	1520	240	1040	1826

ferent orders of magnitude depending on the loading conditions. A parameter describing mode mixture for plane stress cases has been proposed to elucidate this large variation (Fan *et al.*, 2002).

8.2 Axial splitting of circular metal tubes

Tube splitting is a problem involving plastic bending/stretching and tearing. When tubes are axially split, they are efficient in absorbing energy and can sustain long stroke (up to 90 % of tube length) with an almost constant load, which is desirable for energy absorbers.

One arrangement is to put one end of a circular tube against a die while applying a compressive force at the top. A curl stopper plate may be used to prevent curl formation and to enhance the axial load. Typical load–compression curves and deformed specimens are shown in Fig. 8.8 for mild steel tubes of diameter 50.8 mm and wall thickness 1.6 mm, but with dies of different values of radius R_{di} (Reddy and Reid, 1986). Note that pre-cuts may be made in order to initiate the splitting process.

Simplified analysis is presented here for the steady state load case, see Fig. 8.9 for a sketch without a curl stopper plate. When the tube of initial radius R_o and thickness h is pushed downwards at a rate v against a die of radius R_{di}, the tube wall bends in the axial (meridian) direction. This is accompanied by circumferential stretching, which is uniform before necking, and subsequent cracking occurs at a radius R_f. R_f is the corresponding radius at fracture (Fig. 8.9). There are five energy dissipation mechanisms: stretching ahead of the crack tips, plastic bending in the axial direction, plastic bending in the circumferential direction, crack propagation and friction. The circumferential strain increment is $d\varepsilon_\theta = dR/R$, leading to a total accumulated circumferential strain $\varepsilon_\theta = \ln(R_f/R_o)$. Assume that the tube wall yields with a fully plastic bending moment in both axial and circumferential directions and a fully plastic membrane force, $N_o = Yh$ (where Y is the yield stress), in the circumferential direction. The total stretching energy rate, $\dot{W}_s$, is therefore given by

$$\dot{W}_s = N_o \ln(R_f/R_o) 2\pi R_o v \qquad [8.7]$$

Energy rate for plastic bending in the axial direction with a die radius R_{di} is given by

$$\dot{W}_{b1} = 2\pi R_o M_o v / R_{di} \qquad [8.8]$$

The curvature in the circumferential direction before bending is $1/R_o$. Assuming that the strips are flat in the circumferential direction after splitting, the energy rate for plastic bending in the circumferential bending is

$$\dot{W}_{b2} = 2\pi R_o M_o v / R_o \qquad [8.9]$$

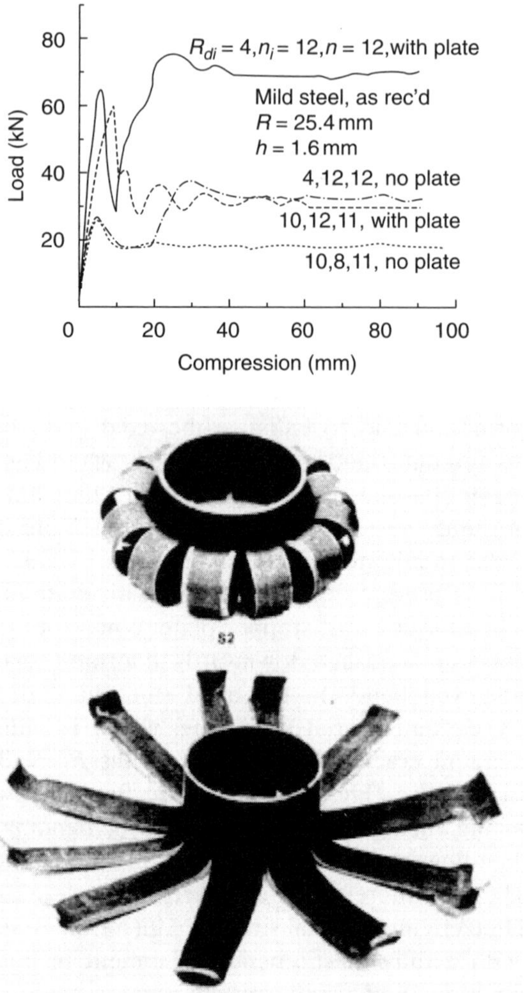

8.8 Load–compression curves of mild steel tubes and photographs of deformed specimens (curling and curls prevented, respectively) (Reddy and Reid, 1986) (reproduced with kind permission of Elsevier).

The energy rate dissipated by the propagation of n cracks is

$$\dot{W}_t = nw_t hv \tag{8.10}$$

Finally the frictional effect is considered. Assume that the resultant normal force acting on each strip is 45° to the horizontal. The normal force acting on each strip is, from equilibrium in the vertical direction, $\sqrt{2}P/n$. The frictional energy is

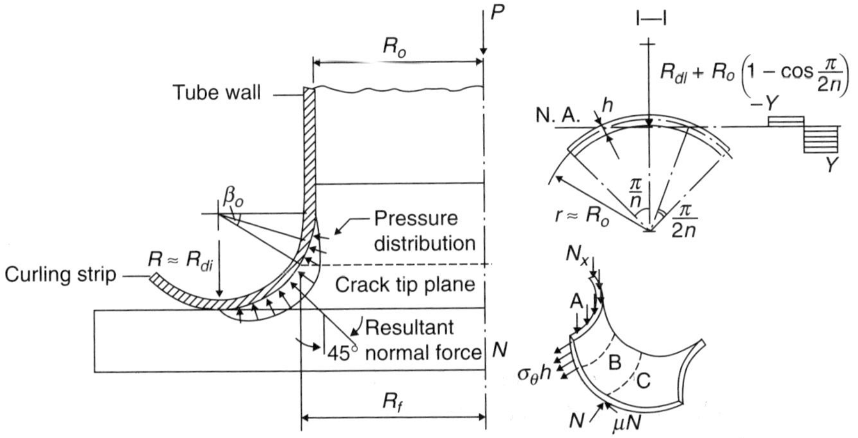

8.9 Sketch of a circular tube under splitting.

$$\dot{W}_{fri} = n\mu(\sqrt{2}P/n)v = \sqrt{2}\mu Pv \qquad [8.11]$$

where μ is the coefficient of friction.

The work done by the external load is Pv and this is dissipated by the sum of the above five energy components (Eqs [8.7]–[8.11]). Therefore, the external load is

$$P = \frac{1}{1-\sqrt{2}\mu}\left[2\pi YhR_o \ln\frac{R_f}{R_o} + \frac{\pi Yh^2}{2}\frac{R_o}{R_{di}} + \frac{\pi Yh^2}{2} + nw_t h\right] \qquad [8.12]$$

In the above equation, the values of R_f and w_t are not known. R_f could be estimated from the specimen after the test. Alternatively, by letting $\varepsilon_\theta = \ln(R_f/R_o)$ be equal to the strain corresponding to the onset of necking in a uniaxial tensile test, its value can be estimated. The value of tearing energy, w_t, needs to be determined independently, using appropriate methods described in Section 8.1. The number of cracks is observed from tests. It should be noted that this is not necessarily equal to the initial number of pre-cuts. Each tube of a given material and dimensions appears to have a characteristic number of cracks, regardless of the number of the pre-cuts. Atkins attempted to provide an explanation for this (Atkins, 1987). Another analysis by Huang *et al.* (2002b) for the number of cracks is given below.

The above analysis is largely due to Reddy and Reid (1986). In their comparison with experiments, the value of w_t was taken to be around 40 kJ/m^2 only, much smaller than the value obtained in splitting square tubes. The agreement between the above analysis and the experiments is close. A more realistic analysis would require a method of determining the value of tearing energy first and then incorporating suitable yield behaviour of the tube.

When curling is prevented, two more terms will be present: one corresponding to axial unbending, which has the same form as Eq. [8.8]; the other corresponding to the friction between the stopper and curls. The normal contact force can be determined from the condition of plastic unbending of the curls. It should be pointed out that a large proportion of energy is dissipated by plastic bending in the radial direction and friction, which raises the overall energy by a factor of as high as 2.3 for $\mu = 0.4$. Tearing energy accounts for about 2 % only in the analysis by Reddy and Reid (1986). This is the case because a separate evaluation of stretching energy before necking is made, which could be regarded as part of tearing energy in some cases.

This process of splitting circular tubes absorbs slightly more energy compared with other forms of axial deformation such as progressive buckling and inversion. One large advantage of this arrangement is the constant force after the initial peak and the long stroke. Also, the level of force can be adjusted by changing the radius of the die. Further modifications have been reported to reduce/eliminate the undesirable initial peak force by chamfering the end of the tube which is to be split.

Theoretical analysis by Huang, Lu and Yu (2002b)

Huang *et al.* (2002b) further investigated the problem of splitting circular tubes. They proposed an alternative theoretical model in an attempt to explain the characteristic number of cracks and the associated force. Their model invokes the idea of critical crack opening displacement in assessing the tearing energy. A non-dimensional ***critical separation*** $\gamma = \delta/h$ is used as a parameter and $\gamma = 1.0$ is taken unless specified otherwise. This in effect 'lumps' the circumferential stretching energy ahead of cracks into the tearing energy. Thus, β_o in Fig. 8.9 can be uniquely determined from the geometry of the problem and the value of γ, as

$$\beta_o = \cos^{-1}\left(1 - \frac{n\gamma h}{2\pi R_{di}}\right) \qquad [8.13]$$

where n is the number of cracks (or strips).

The applied force can be calculated from an energy balance. When the tube is moving downward (whilst cracks are propagating) with a speed v, the energy balance leads to

$$Pv = \dot{W}_p + \dot{W}_t + \dot{W}_{fri} \qquad [8.14]$$

where $\dot{W}_p$, $\dot{W}_t$ and $\dot{W}_{fri}$ denote the rate of energy dissipation in plastic bending, tearing and friction, respectively.

In this model, all plastic bending and stretching is confined to the curved strips (of radius R_o), bending with radius R. The rate of plastic bending becomes

$$\dot{W}_p = \frac{2\pi R_o m_p}{R_{di} + R_o[1 - \cos(\pi/2n)]} v \qquad [8.15]$$

where $R_{di} + R_o[1 - \cos(\pi/2n)]$ is the bending radius of the cross-sectional neutral axis and the fully plastic bending moment for one strip normalised with respect to the strip arc width is

$$m_p = \frac{nYR_oh^2}{2\pi}\left[\int_{\frac{\pi}{2n}}^{\frac{\pi}{n}}\left(\cos\frac{\pi}{2n} - \cos\theta\right)\mathrm{d}\theta + \int_0^{\frac{\pi}{2n}}\left(\cos\theta - \cos\frac{\pi}{2n}\right)\mathrm{d}\theta\right]$$

$$= \frac{nYR_oh}{\pi}\left(2\sin\frac{\pi}{2n} - \sin\frac{\pi}{n}\right) \qquad [8.16]$$

Hence, the rate of plastic bending dissipation is

$$\dot{W}_p = 2nYR_o^2h\frac{\left(2\sin\frac{\pi}{2n} - \sin\frac{\pi}{n}\right)}{R_{di} + R_o\left(1 - \cos\frac{\pi}{2n}\right)} v \qquad [8.17]$$

The tearing energy involves plastic work within the near-tip zone in the form of circumferential stretching ahead of the crack tip. Hence, the rate of tearing energy dissipation is

$$\dot{W}_t = \int \sigma_\theta \dot{\varepsilon}_\theta \mathrm{d}V \quad or \quad \dot{W}_t = Yn\delta_t h^2 v \qquad [8.18]$$

where V is volume. The rate of energy dissipation by friction is

$$\dot{W}_{fri} = 2\pi R_o \mu N v \qquad [8.19]$$

where N is the normal force per unit length. For a radius die, we assume that the resultant normal contact force is at 45° to the horizontal. N is related to the applied force P by

$$N = \frac{P}{\sqrt{2}\pi R_o(1+\mu)} \qquad [8.20]$$

By substituting Eqs [8.17]–[8.20] into Eq. [8.14] and introducing three non-dimensional parameters $f = P/2\pi R_o hY$, $\phi = 2R_o/h$ and $\eta = 2R_{di}/h$, the non-dimensional force is found as

$$f = \frac{1+\mu}{1-(\sqrt{2}-1)\mu}\left[\frac{\pi^2\phi}{8n^2\eta} + \frac{n\gamma}{\pi\phi}\right] \qquad [8.21]$$

Thus, the applied force is a function of both the non-dimensional curl radius (η) and the crack number (n). For the case of a radius die, we may assume that the curl radius is the same as that of the die. Assume that the number of cracks is such that the total force is a minimum, i.e. $\mathrm{d}f/\mathrm{d}n = 0$, then it leads to

$$n = \left(\frac{\phi^2}{4\gamma\eta}\right)^{\frac{1}{3}} \pi$$

Substituting this expression into Eq. [8.21] and taking the frictional coefficient $\mu = 0.2$, the force is

$$f = 1.23\left(\frac{\gamma^2}{\phi\eta}\right)^{\frac{1}{3}}$$

The coefficient would be 1.58 when $\mu = 0.4$.

The predicated crack number and force are plotted against the non-dimensional die (or curl) radius as shown in Figs 8.10(a) and (b), together with the test results from Reddy and Reid (1986). The comparison of n and f would suggest that annealing enhanced the ductility of both aluminium and mild steel tubes resulting in a larger value of γ: about 2 for annealed aluminium tubes and 0.5 ~ 1.0 for the rest. In these cases, tearing energy dominates the total dissipated energy and is twice as much as the plastic bending energy.

Splitting circular tubes using a conical die

Circular tubes can be arranged to split using other dies, for example conical ones (Huang *et al.*, 2002b). In this case, the magnitude of loads can be adjusted by varying the semi-angle of the cone-shaped die, instead of the radius R_{di} as for a curved die. Previous theoretical analysis applies in a similar manner. Nevertheless, both the curl radius and crack number are unknowns. An additional equation is obtained from equilibrium considerations. No closed form solution is available and numerical results are shown in Fig. 8.11(a) for the observed crack number. The test values seem undistinguishable for dies of different semi-angles, but they are within the theoretical curves for the semi-angles considered.

Figure 8.11(b) shows the variations of non-dimensional curl radius versus the die semi-angle for a typical tube with $\phi = 31$. The force is plotted against the ratio of the tube diameter to the thickness in Fig. 8.11(c) for dies with a semi-angle $\alpha = 45°$. There is an almost linear increase of force with die semi-angle for α in the range 30–90° (Fig. 8.11(d) for a tube with $\phi = 31$). General agreement between the theoretical prediction and experiment is obtained for all cases.

8.3 Axial splitting of square metal tubes

As another example of combined plastic deformation and tearing, we next discuss the problem of splitting square tubes. The tubes can be arranged to

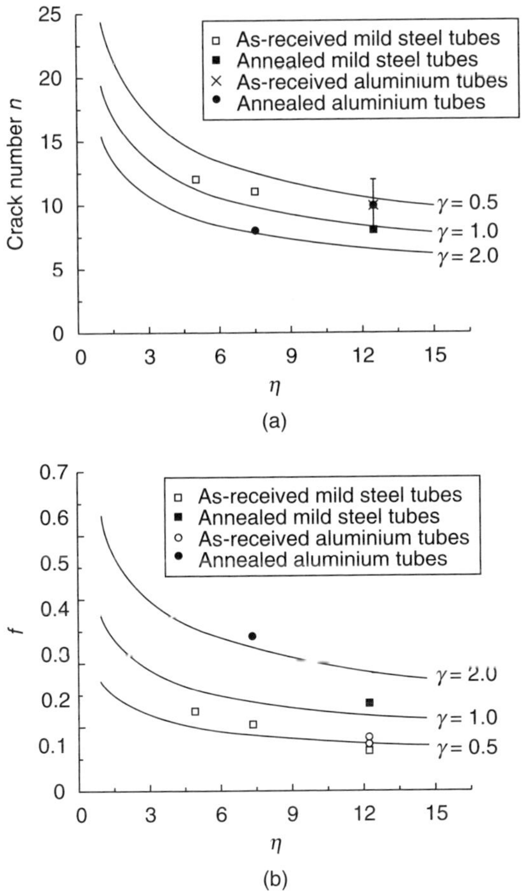

8.10 Theoretical and experimental values for number of cracks (a) and force (b) against dimensionless tube diameter (Huang *et al.*, 2002b).

be pressed against a flat surface or a die which is curved (Stronge *et al.*, 1983 and 1984) or pyramidal (Huang *et al.*, 2002a), see Fig. 8.12. Typical load–displacement curves are shown in Fig. 8.13 and specimens after testing in Fig. 8.14. In most cases, after an initial displacement of $10h$, a reasonably steady state is obtained with an almost constant load.

The total external work is dissipated by three major mechanisms: plastic bending of the four sidewalls, tearing of the four corners, and friction between the tube and the die. For the case of a curved die of radius R_{di} and without tube wall unbending (Fig. 8.12a), the bending energy rate is simply

$$\dot{W}_b = 4M_o b \frac{1}{R_{di}} v \qquad [8.22]$$

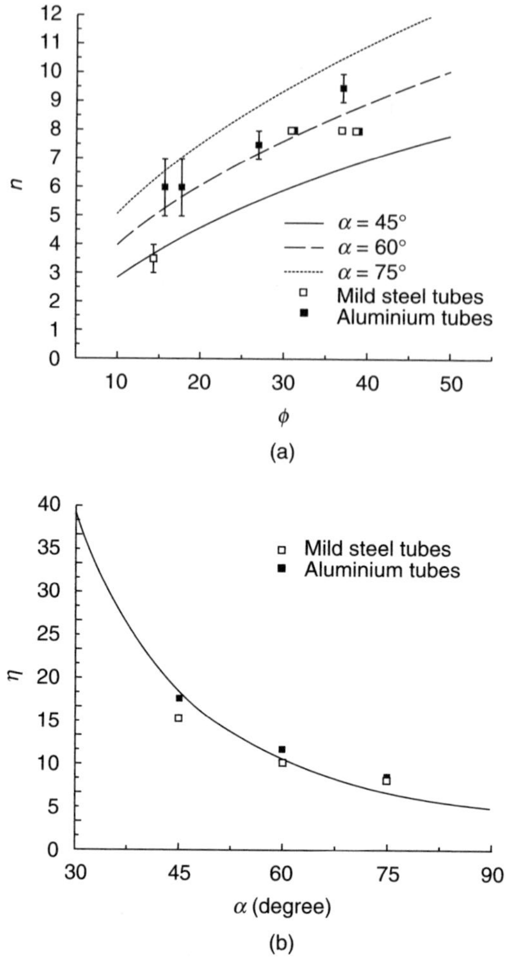

8.11 Tube splitting using a conical die: (a) number of cracks vs dimensionless tube diameter; (b) dimensionless curl radius vs die semi-angle; (c) dimensionless force vs tube diameter; (d) dimensionless force vs die semi-angle (Huang *et al.*, 2002b).

where b is the side length of the square tube. The rate of tearing energy for the four corners is

$$\dot{W}_t = 4w_t hv \qquad [8.23]$$

and the energy rate due to friction is

$$\dot{W}_{fri} = P\mu v \qquad [8.24]$$

Here it is assumed that the resultant contact force between the tube and the die is normal to the base. Hence, the external force P is given, from the energy balance, as

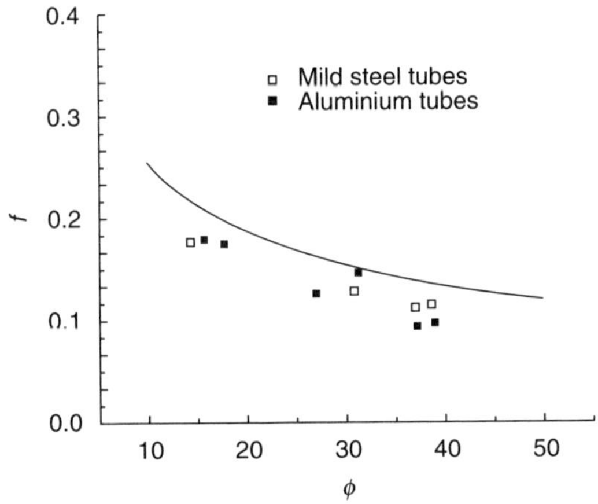

(c)

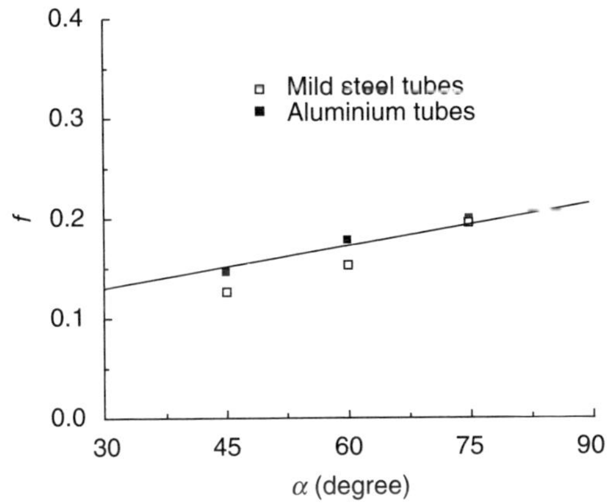

(d)

8.11 Continued

$$P = \frac{4}{1-\mu}(M_o b/R_{di} + w_t h)$$ [8.25]

The value of w_t should be obtained independently, based on the methods described in Section 8.1. Values of the other parameters can be determined also.

When a pyramidal die is used, plastic bending of the tube walls occurs with a natural unconfined curl radius R, the value of which will be discussed later. The tearing energy has the same expression as in Eq. [8.23]. The energy rate due to friction is now

$$\dot{W}_{fri} = 4\mu N v.$$ [8.26]

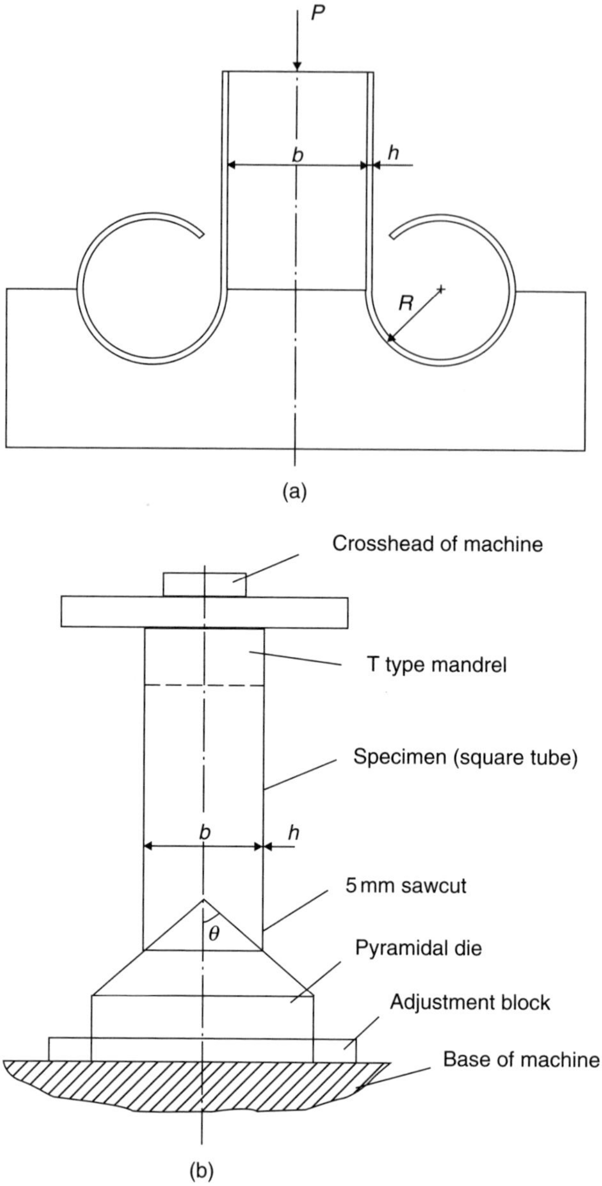

8.12 Sketch of splitting square tubes using a curved die (a) and a pyramidal die (b).

Here $N = P/4(\sin\theta + \mu\cos\theta)$ is the normal force for each side and θ is the die semi-angle.

Again, from the energy balance, it can be shown that the force required is

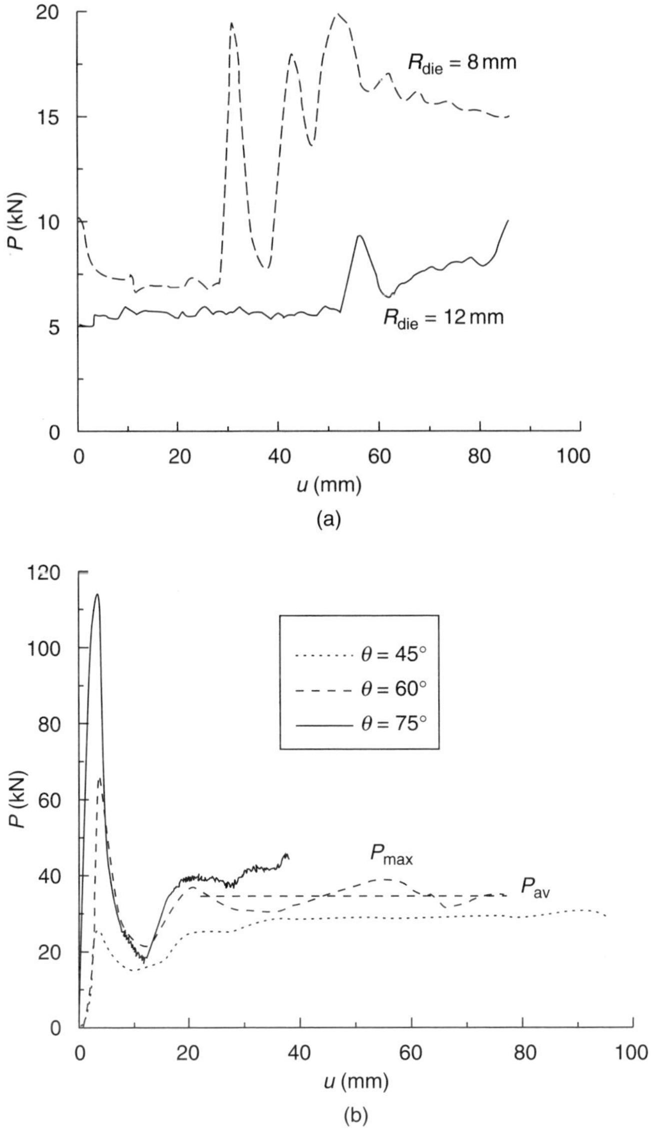

8.13 Typical force–displacement curves for splitting square tubes: (a) curved die; (b) pyramidal die (Huang *et al.*, 2002a)

$$P = \frac{4(M_o b/R + w_t h)}{1 - \mu/(\sin\theta + \mu\cos\theta)} \tag{8.27}$$

From experiments, it is found that the curl radius mainly depends on the semi-angle of the die and is little affected by the thickness and material of the tube, see Figs 8.15(a) and (b). Here the average radius is 20.4 mm for

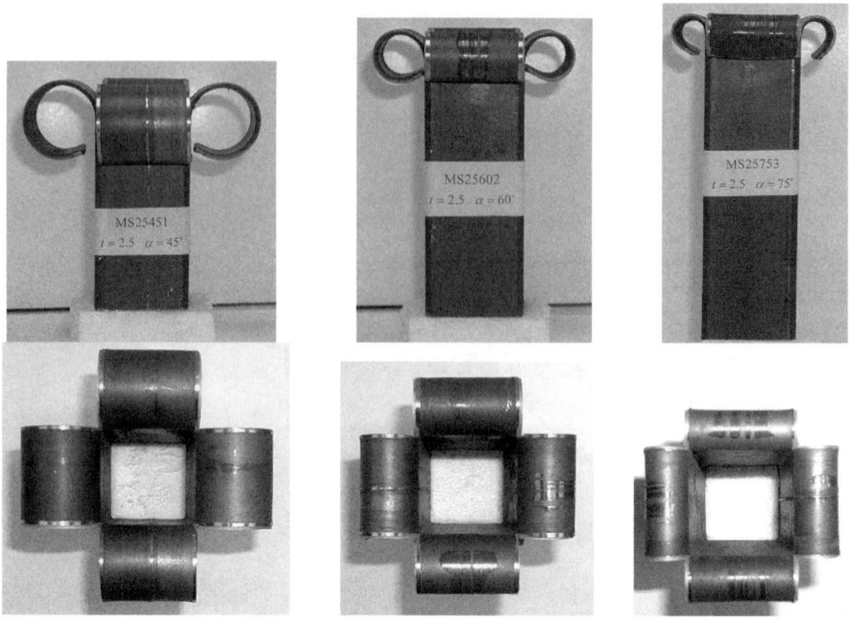

(a) (b) (c)

8.14 Photograph of specimens after test. All mild steel specimens have a side length of 50 mm and a thickness of 2.5 mm. The die semi-angle is 45° (a), 60° (b) and 75° (c), respectively (Huang *et al.*, 2002a) (reproduced with kind permission of Elsevier).

$\theta = 45°$, 12.1 mm for $\theta = 60°$ and 7.8 mm for $\theta = 75°$. The following empirical formula may be obtained for the radius of curls

$$R = \frac{32.7}{\sin\theta} - 25.7 \qquad [8.28]$$

where the two coefficients have units of mm. This formula is applicable only for square tubes with external dimensions 50 mm × 50 mm; there is not enough evidence to say whether or not curl radius is related to the overall dimensions. The value of tearing energy is estimated using Eq. [8.6]. Substituting Eq. [8.28] into Eq. [8.27], the applied axial load P can be calculated directly.

Figure 8.16 shows the calculated axial load against the die semi-angle for the mild steel tubes. Good agreement with experiment is obtained, in general, but the experimental value is higher than the theoretical one for the mild steel tube with $h = 3.0$ mm and $\theta = 45°$, and it is lower with $\theta = 75°$. A larger die semi-angle leads to a larger change in curvature of the curl and hence increases the operating load. Therefore, a given tube, especially of aluminium, can be made to split at different load levels by changing the die

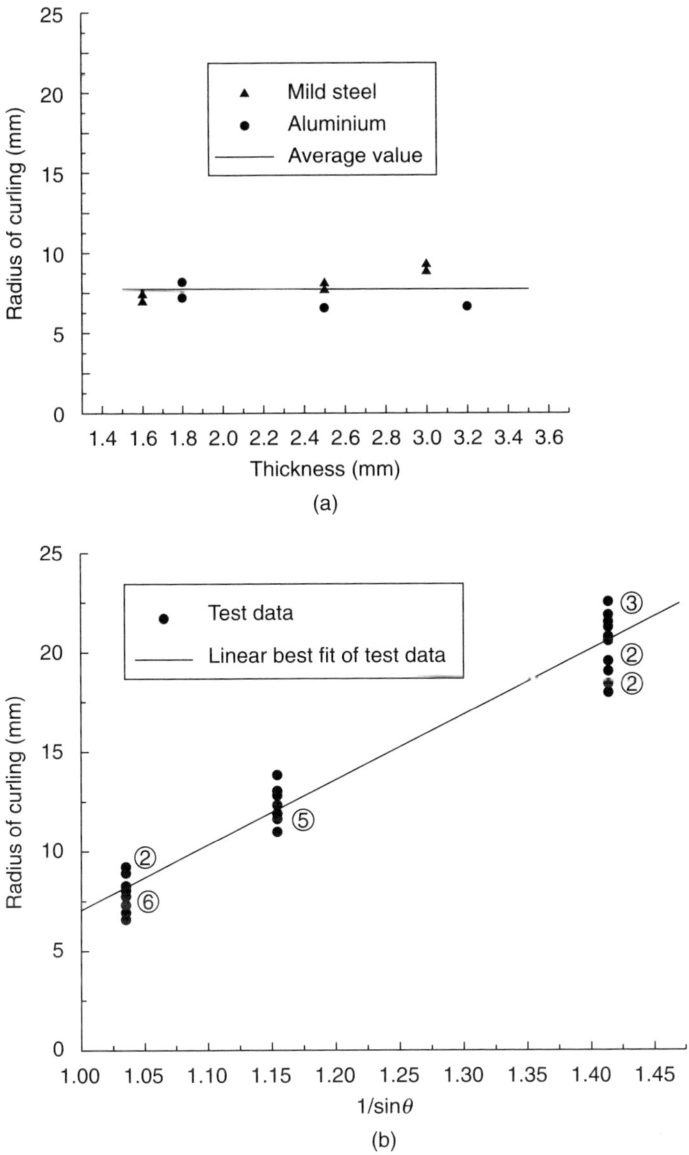

8.15 (a) Curl radius versus tube thickness; (b) curl radius as a function of the die semi-angle (Huang *et al.*, 2002a).

angle. For example, the load for an aluminium tube with $h = 1.8$ mm against a die of $\theta = 75°$ is about twice that for $\theta = 45°$, and it is three times that for tubes with $h = 3.2$ mm as reported by Huang *et al.* (2002a).

An assessment of the contribution from each of the energy dissipation mechanisms can be made. For example, taking a mild steel tube with

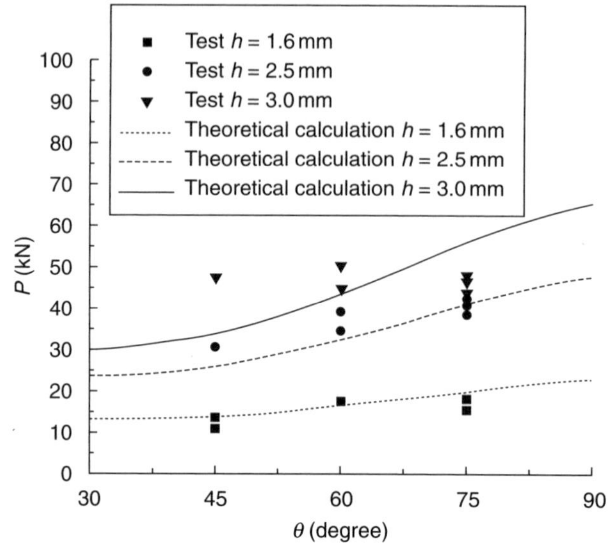

8.16 Theoretical and experimental values for force against the die semi-angle (Huang *et al.*, 2002a).

h = 2.5 mm against a die with $\theta = 45°$, the percentage contributions from plastic bending, fracture and friction are, respectively, in the percentage ratio 43:28:29; and for a similar aluminium tube, they are in the ratio 47:17:36.

The above analysis can be slightly modified when plastic unbending occurs as a result of the tube edge impinging on the tube wall after one complete roll. Dynamic tests indicate that the force may not differ significantly from the static value. It is possible that fracture strain is reduced in dynamic loading, leading to a decrease in tearing energy. Further, friction may dissipate less energy as a result of reduction in frictional coefficient. These two factors may compensate for the increase of flow stress caused by the strain-rate effect. Similar experimental observations and arguments are made later in the case of cutting a plate by a wedge (Section 8.5).

8.4 Piercing of metal tubes

8.4.1 Experiments

Metal tubes can be pierced, laterally, using various indenters. Lu and Wang (2002) studied the energy absorption of square tubes pierced with six different pyramidal punches of square or circular cross-section with 12.7 mm sides or diameter. The square tubes have a side length a = 40 mm and a thickness h = 1.6 or 2.5 mm. The tube length L varies from 40 (= a)mm to

340 (= 8.5a)mm. For h = 1.6 mm, the measured yield stress is Y = 350 MPa, ultimate stress σ_u = 370 MPa, fracture strain ε_f = 0.2; and for h = 2.5 mm, the yield stress is Y = 420 MPa, ultimate stress σ_u = 450 MPa and fracture strain ε_f = 0.2.

Figure 8.17 shows the test results for a specimen with h = 1.6 mm, L = 2.5a and a punch of semi-angle θ = 30°. The post-test specimen is shown in Fig. 8.17(a) and the corresponding load–punch displacement curve is shown in Fig. 8.17(b). Here, penetration and associated petal formation of the specimen, together with friction are the three forms of energy absorption. Plastic deformation is localised within the central area; hence this piercing mode is termed the ***local penetrating mode***.

In the load–displacement curve, five distinctive stages may be observed (Fig. 8.17(b)). The load initially increases in stage I with elastic-plastic deformation of the top surface of the specimen. The sharp head of the indenter

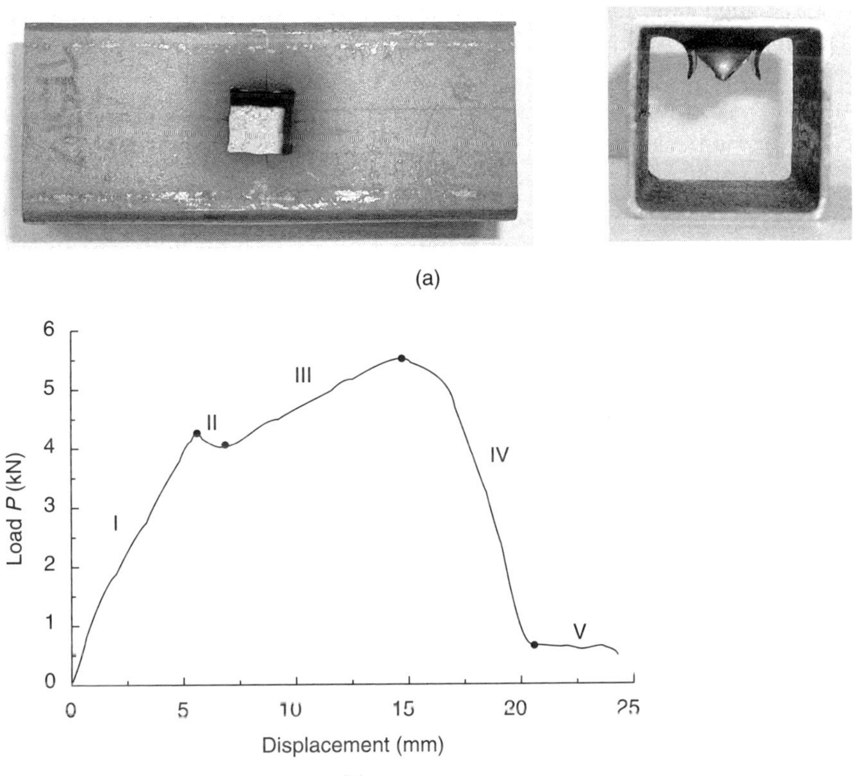

8.17 Typical results of tube piercing: (a) a specimen; (b) load–displacement curve (Lu and Wang, 2002) (reproduced with kind permission of Elsevier).

then squeezes a little into the top wall. The load plateaus in stage II when the punch is breaking through the top wall as a result of excessive stretching around the bottom face of the top wall. In stage III, the load steadily increases again. This corresponds to four cracks propagating from the corners of the punch; petals on the four sidewalls begin to occur between these cracks. As the petals expand, the contact area enlarges, increasing the friction forces acting on the punch. Also, the instantaneous plastic bending area becomes larger, contributing to an increasing punch force. The maximum load is attained when the shank of the punch enters the deformation zone. The load then rapidly decreases in stage IV as there is a gradual reduction of contact area. Finally in stage V, there is contact only between the sides of the shank and the sidewall petals. The load plateaus again at a much lower level, being the load due to friction between the sides of the shank and the petals.

Figure 8.18 is the result for a test with a tube of the same thickness (h = 1.6 mm) but of a much shorter tube ($L = a$ = 40 mm). Structural collapse of the whole tube takes place in this case, see Fig. 8.18(a). The corresponding load–punch displacement curve shown in Fig. 8.18(b) exhibits different features from the previous case of long tubes. Here there are four stages evident (A, B, C and D). This deformation behaviour is called a ***global collapse mode***.

The load–displacement curves for tubes of different length are plotted in Fig. 8.19, other conditions being the same. From the series of tests, it is found that the failure mode is mainly governed by the tube length. For both tubes with h = 1.6 mm, L = 100 and 75 mm, the mode of deformation is by penetration with localised plastic deformation; the two curves are almost identical. For the 58 mm long tube, the interaction between the penetration process and global structural collapse is complex. It appears that after the punch tip has penetrated the tube top wall, structural deformation occurs over the whole tube. Hence, the failure mode varies from initial penetration to structural collapse, presumably whichever path requires the least load. When the tube length is reduced to 40 mm, the tube fails by global collapse. Similar observations can be made for 2.5 mm thick specimens.

The effect of punch semi-angle is presented in Fig. 8.20. It shows that, as the semi-angle of the pyramidal punch increases, the maximum load also increases. This is largely due to the rate of plastic deformation required for petal formation, such that a punch with a greater semi-angle requires a larger maximum load to pierce the tube. But the punch displacement corresponding to the maximum load decreases with the semi-angle of the punch. This is because of the smaller travel distance for the punch to pass completely through the material (in order to exhaust the shank length). Even though punches with a small semi-angle do not attain as high a load

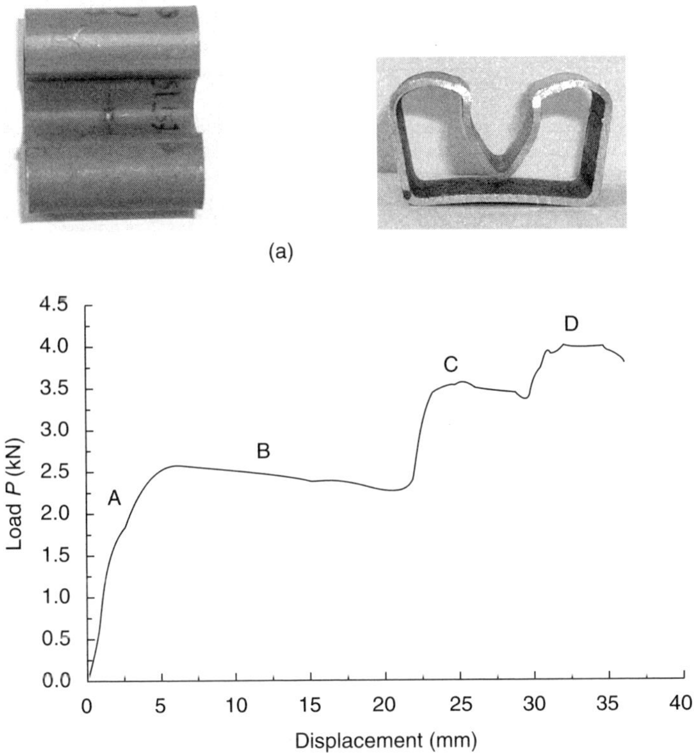

(b)

8.18 Global bending of short tubes: (a) a specimen after test; (b) load–displacement curve (Lu and Wang, 2002) (reproduced with kind permission of Elsevier).

as those with larger semi-angles, the load is maintained for a longer period during penetration and the total energy absorbed can be similar.

The geometry of a punch, whether pyramidal or conical, does not greatly affect the load–displacement curve, provided the semi-angle is the same. Also, when the pyramidal punch is rotated about the punch axis by 45°, the result does not vary much. The overall energy-absorption behaviour seems similar.

8.4.2 Theoretical analysis

Incremental plastic analysis would need to be employed if we want to calculate theoretically the detailed variation of load with punch displacement for stages I to IV (Fig. 8.17(b)), based on an idealised deformation mechanism. However, here we present only a simple analysis for the

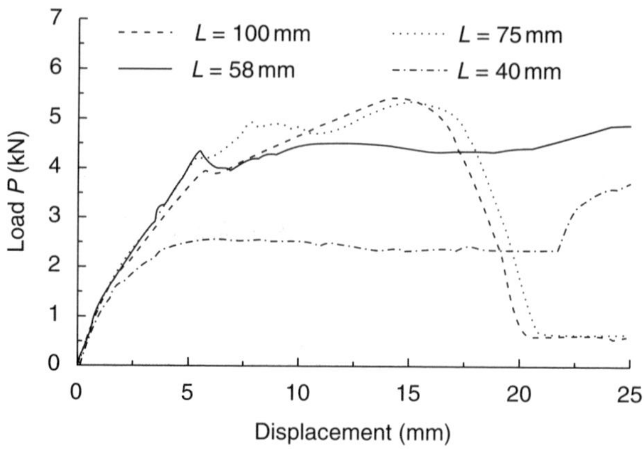

8.19 Load–displacement curves for tubes of four different lengths.

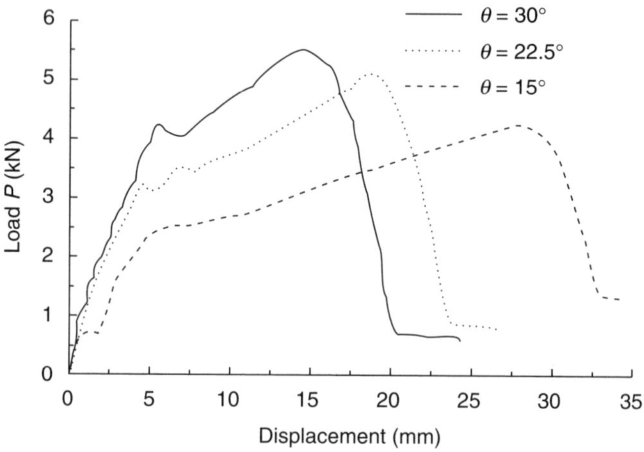

8.20 Load–displacement curves for indenters of three different semi-angles (Lu and Wang, 2002).

average force in the penetrating phase, i.e. stage III and the crack initiation force corresponding to stage II (Fig. 8.21). Let u be the total penetration of the pyramidal punch. Total work done by the external load $W_e = Pu$ is balanced with internal energy dissipation. Assume that the length of the four cracks is proportional to u and is $\sqrt{2}u\tan\theta$. The total tearing energy is written as

$$W_t = 4w_t h\sqrt{2}u\tan\theta \quad [8.29]$$

w_t is given by Eq. [8.6]. Plastic bending occurs mainly in the form of travelling plastic hinges. Assuming that the mean radius of curvature is R, the total plastic bending energy for the four sides is

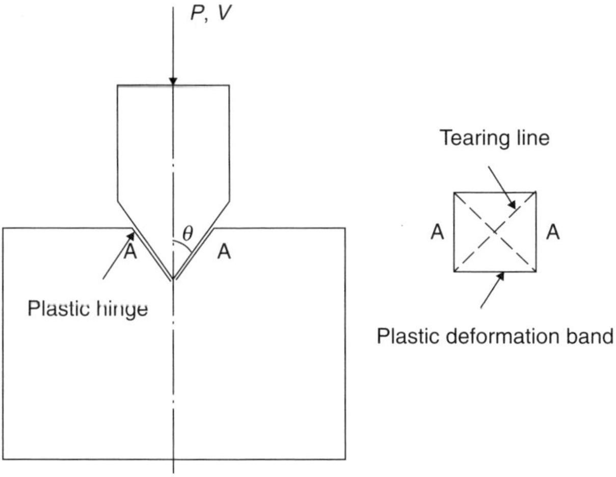

8.21 Sketch of plastic bending and tearing for total energy calculation.

$$W_b = 4M_o \frac{1}{2} u \tan\theta u \frac{1}{R} = 2M_o u \tan\theta\beta \qquad [8.30]$$

where β represents the total angle through which the plastic hinges rotate.

The energy dissipation due to friction is

$$W_{fri} = 4N\mu u / \cos\theta \qquad [8.31]$$

where $N = P/[4(\sin\theta + \mu\cos\theta)]$ is the normal force for each contacting side. μ is the frictional coefficient, whose mean value is assumed to be 0.2. $u/\cos\theta$ represents the total sliding distance.

From the energy balance together with Eqs [8.29]–[8.31], the load required to penetrate the surface of a specimen by a pyramidal punch of square section with a semi-angle θ is

$$P = [0.05\sigma_u \varepsilon_f h + 0.5Yh^2\beta] \frac{\sin\theta + \mu\cos\theta}{\cos\theta - \mu\sin\theta} \qquad [8.32]$$

If we discard the terms for plastic bending energy of the tube walls and the friction, the analysis should lead to an estimate for a force corresponding to the onset of crack propagation, i.e. the critical tearing force. From Eq. [8.32], this critical tearing force is

$$P_f = 0.05\sigma_u \varepsilon_f h \tan\theta \qquad [8.33]$$

By assuming that the total plastic bending angle β has the same value as the punch semi-angle and taking $\mu = 0.2$, Eqs [8.32] and [8.33] provide a reasonable estimate of the average load compared with the experiments.

In principle, the plateau stresses for global collapse mode without tearing in Fig. 8.18(b) can be worked out, as the initial deformation is bending only. Also, an equation can be established for the critical tube length when the mode changes between the local and global deformation. (This is performed on the assumption that a tube adopts a particular mode when the corresponding energy is a minimum compared with other possible modes.) For 1.6 mm thick specimens, Lu and Wang (2002) obtained a critical length of 44 mm ($\approx 1.1a$). It is interesting to note that when the actual tube length is around the critical value, the collapse mode can start as local penetration with petals, followed by a global structural collapse, as observed in the experiment for a square tube with $h = 1.6$ mm and $L = 58$ mm.

Static piercing experiments on circular tubes were performed earlier by Johnson *et al.* (1979). Behaviour of the circular tubes was very similar to the square ones discussed above, with two distinctive modes. Further, the two ends of a circular tube may undergo reverse ovalisation for tubes of certain lengths.

8.5 Cutting of metal plates by a wedge

Our last example involves cutting of a metal plate by a wedge (Lu and Calladine, 1990). This problem is a simplified version of penetration of the bow of a ship into the side deck of another ship or a ship grounding (Jones and Birch, 1987). A typical experimental setup is shown in Fig. 8.22. Typical specimens and load–displacement curves are shown in Fig. 8.23. In the initial cutting stage, both curves show a rapid increase of force to an initial high peak force. This corresponds to cutting motion of a wedge into a plate, which remains flat with no bending. This process lasts until the cutting length L is about $3h$. Depending on the inclination of plates with respect to the (vertical) direction of wedge movement, two distinct modes of deformation are observed. When the plate is vertical ($\beta = 0°$), cutting is accompanied by forward and backward bending of flaps and the force-cutting length curve shows a corresponding periodic fluctuation. When the plate is inclined at, say, $\beta = 10°$, the two flaps bend in one direction continuously and the force increases gradually, after the initial cutting stage.

Energy is dissipated by plastic bending of the two flaps, plastic deformation and tearing in the tip vicinity, and friction. Nevertheless, the cutting action at the tip is not exactly the same as the ductile tearing we have encountered before. Indeed, through the thickness, the plate is partially cut – a process similar to metal indentation – and partially torn due to excessive tensile strain. Therefore analytically it is a much more complex problem than the previous ones and we present only empirical results here, based on a dimensional analysis.

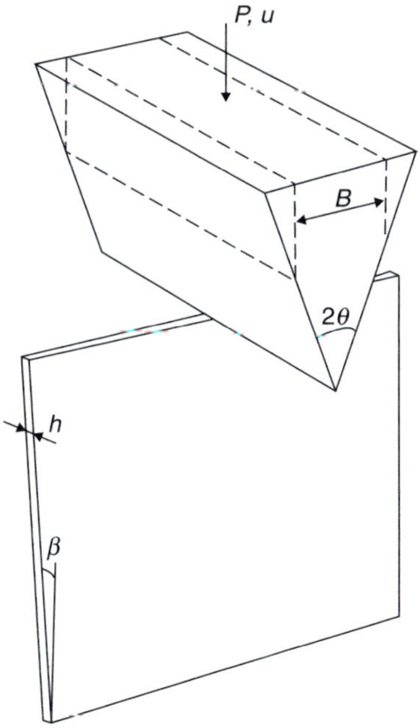

8.22 Sketch of the experimental setup for cutting a plate by a wedge (Lu and Calladine, 1990).

Bending and stretching dissipate energy by plastic deformation. The ductile cutting/tearing here is mainly a plastic straining process too. Furthermore, the contact force between the wedge tip and the tearing front is very high, suggesting that the friction is a result of plastic deformation at the micro-level. Hence, the only significant material property is the yield stress Y. For a given inclination angle β and a given wedge angle 2θ, the external work dissipated W is only related to Y, plate thickness h and the cutting length L. The size of plates is not relevant as the deformation is localised without reaching the clamping frame. Also, elastic energy is very small and hence the elastic modulus is immaterial.

From the dimensional analysis presented in Section 3.1, there are four physical variables and two primary (fundamental) dimensions: force and length (the dimensions of time and mass are not involved in this static problem). From the Buckingham theorem, there are only two (four minus two) independent dimensionless groups. Here we choose groups W/Yh^3 and L/h. When all the test results are plotted accordingly, as shown in Fig. 8.24 for a wedge with an included angle of $2\theta = 40°$, all the curves collapse into a single curve. (Note that the value of Y is obtained by dividing the measured ***Vicker's hardness*** number by three.) This suggests that the above dimensional analysis is successful. Specifically, it appears that using the yield stress as the only material property, as initially assumed, is sufficient in this

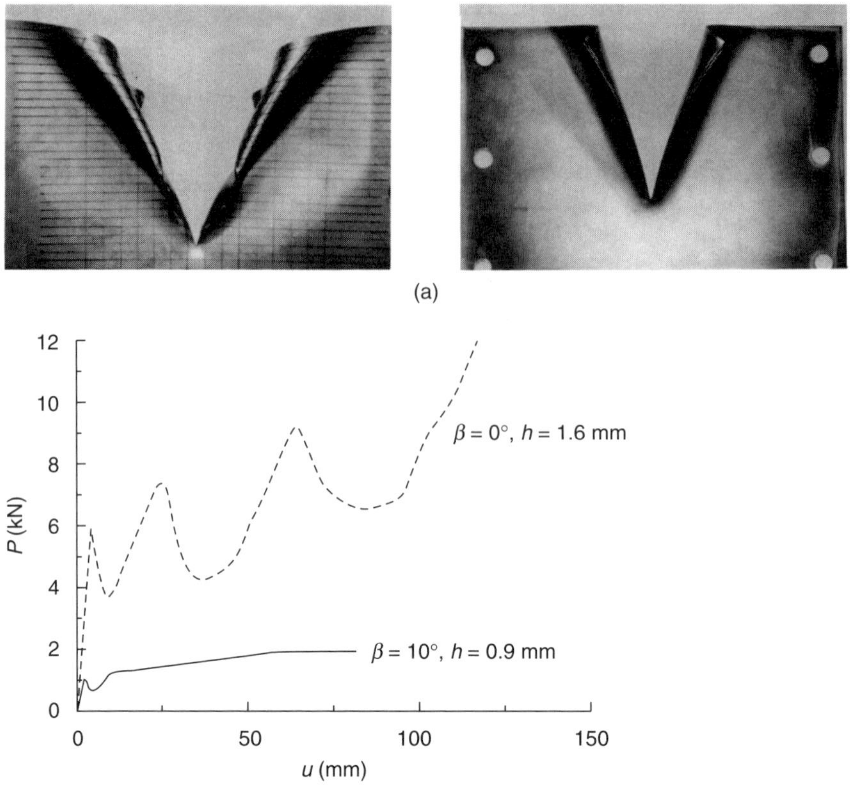

(b)

8.23 Typical experimental results: (a) specimens with $\beta = 0°$ and $\beta = 10°$; (b) load vs cutting displacement (Lu and Calladine, 1990) (reproduced with kind permission of Elsevier).

problem; there is no real need to single out explicitly another material parameter such as the tearing energy or the critical crack opening displacement δ.

Note that Fig. 8.24 is double logarithmic and that for $L/h \geq 5$, all the curves may be approximated as a single straight line with a slope of 1.3. This immediately suggests the following empirical equation for energy

$$\frac{W}{Yh^3} = C\left(\frac{L}{h}\right)^n$$

$$= C_{1.3}\left(\frac{L}{h}\right)^{1.3} \qquad [8.34]$$

or

$$W = C_{1.3} Y L^{1.3} h^{1.7} \qquad [8.35]$$

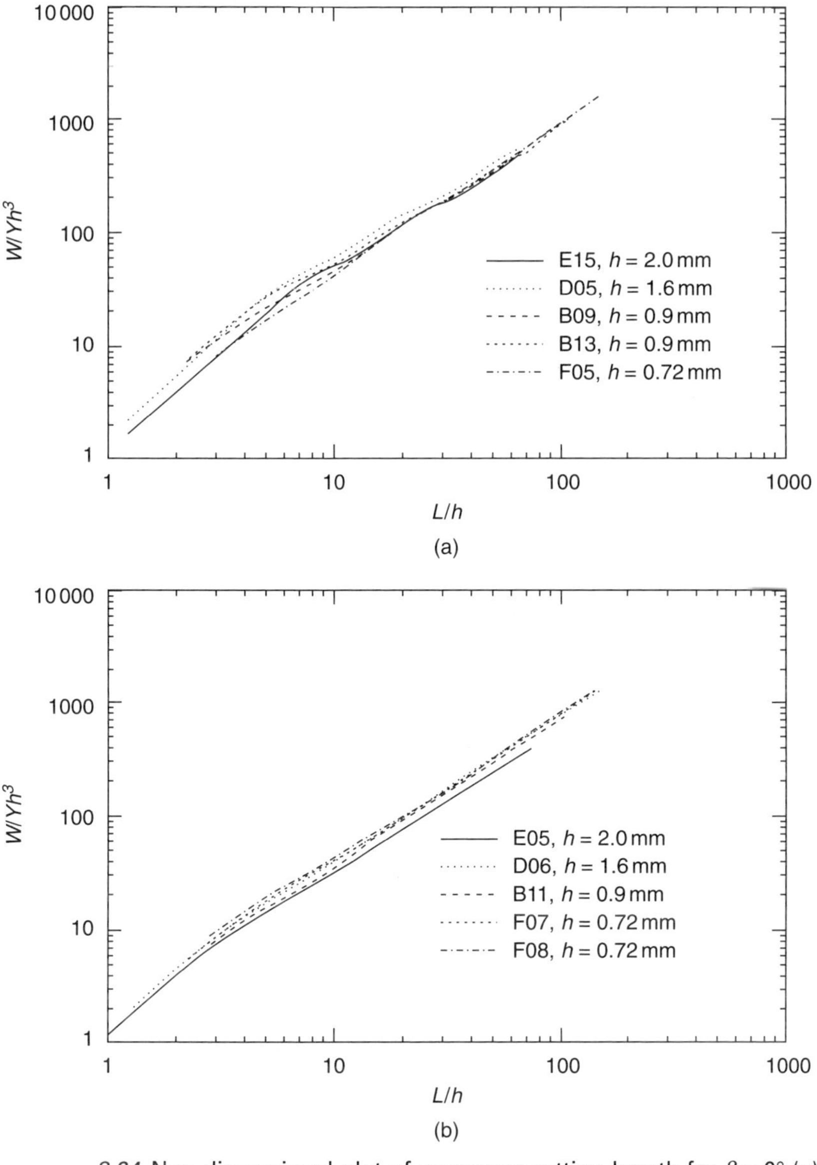

8.21 Non-dimensional plot of energy vs cutting length for $\beta = 0°$ (a) and $\beta = 10°$ (b) (Lu and Calladine, 1990).

where C, n and $C_{1.3}$ are constants obtained by best fitting a straight line to the data in Fig. 8.24. Their values are listed in Table 8.2 for specimens with different materials and testing conditions. Some tests were conducted with a truncated wedge of width B (Fig. 8.22) and results are also given in this table.

Table 8.2 Summary of test results for plate cutting (Lu and Calladine, 1990)

Workers	Material	β (°)	2θ (°)	B (mm)	Static (S) or dynamic (D)	No. of specimens	C	n	$C_{1,3}$
Lu and Calladine (1990)	Mild steel	0	40		S	2	3.5	1.2	2.4
		10	40		S	6	1.9	1.3	2.0
		0	20		S	5	3.1	1.2	2.3
		10	20		S	5	1.9	1.3	1.9
		0	20	20	S	2	2.8	1.2	2.2
		0	20	10	S	4	2.5	1.2	1.9
		10	20	10	S	8	2.1	1.2	1.9
		20	20	10	S	3	0.9	1.4	1.1
	Aluminium	10	20		S	1	1.0	1.5	2.2
	Brass	10	20		S	1	1.4	1.3	1.4
	Copper	10	20		S	1	1.4	1.4	2.2
	Dural	10	20		S	1	1.6	1.2	1.2
Goldfinch (1986) and Prentice (1986)	Mild steel	10	20		D	11	0.9	1.5	2.0
	Aluminium	10	20		D	13	1.9	1.2	1.5
	Brass	10	20		D	13	0.6	1.4	1.0
	Copper	10	20		D	7	2.2	1.2	1.4
	Dural	10	40		D	3	2.8	1.0	0.8
Jones and Birch (1987)	Mild steel	0	15		D	5	3.8	1.4	5.4
		0	30		D	25	3.9	1.3	4.6
		0	45		D	29	4.8	1.3	4.5
		0	60		D	27	4.3	1.3	4.6

Source: reproduced with kind permission of Elsevier.

By differentiating this energy W with respect to the cutting length L, the cutting force P is obtained as

$$P = \frac{\partial W}{\partial L} = 1.3C_{1.3}Yh^{1.7}L^{0.3} \qquad [8.36]$$

For example, for mild steel plates with a wedge of $2\theta = 40°$, we have

$$W = \begin{cases} 2.0YL^{1.3}h^{1.7} & \text{for } \beta = 10° \\ 2.4YL^{1.3}h^{1.7} & \text{for } \beta = 0° \end{cases} \qquad [8.37]$$

Dynamic tests using a drop hammer reveal that in general the energy absorbed is less than the static case, see Table 8.2 and Fig. 8.25. (An exception to this is mild steel plates, where the values of $C_{1.3}$ for both dynamic and static tests are close.) Typically, the energy absorbed in dynamic tests is about 75 % that for static tests. This is unexpected, as both strain-rate and inertia effects, which are present in dynamic loading, tend to enhance the energy-absorbing capacity. The main reason in this case is that friction plays an important role in this problem.

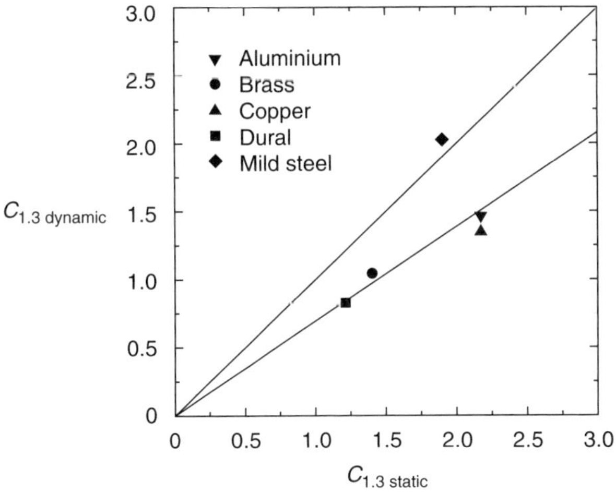

8.25 Comparison of dynamic test results with the static ones (Lu and Calladine, 1990).

It is well known that the frictional effect is reduced in dynamic cases and this decreases the energy dissipation. The resultant dynamic effect in terms of the energy dissipation, whether enhancing or decreasing it, is therefore dependent upon the relative degrees of contribution. When the strain rate and inertia effects are larger than the reduction due to friction, the overall energy absorbed does not change very much, as in the case of mild steel, which is strongly rate sensitive. On the other hand, the energy reduction due to friction can be more than the enhancement as a result of strain-rate and inertia effects, and the overall energy in a dynamic case can be smaller than its static counterpart. This seems to explain the overall energy reduction for most materials presented in Fig. 8.25. (These materials such as aluminium are relatively rate insensitive.) Also, for plates with $\beta = 0°$ the reduction in $C_{1.3}$ in dynamic loading is less pronounced. This indicates a stronger inertia effect, similar to the type II structures presented in Section 7.2.

Theoretical analysis for the static loading case has been conducted by, for example Wierzbicki and Thomas (1993), by adopting as a fracture parameter critical opening displacement (COD). The theoretical model leads to the above empirical formulae; however, some aspects of detailed experimental observations remain to be explained (Calladine, 1993).

8.6 Concluding remarks

This chapter presents issues and several problems involving plastic deformation and ductile tearing. The work presented is largely experimental, because theoretical analysis is challenging owing to the complexity of the problems. For example, one major difficulty is the assessment of tearing energy and an understanding of its possible interaction with other forms of plastic deformation. Except for one analysis of circular tube splitting, we have not been able to establish theories as current knowledge is very limited in this respect. Another relevant work is by Wierzbicki and Thomas (1993) where the tearing energy appears to affect plastic deformation in the far region through a process of minimising total energy.

In the examples presented, it turns out that the tearing energy sometimes contributes only a small percentage of the total energy. However, the tearing process is important in ensuring the desired energy dissipation mechanisms, for example in ensuring a force plateau over a long stroke. On the other hand, friction can be a major energy dissipation mechanism, which leads to the following idea. Contrary to what one might expect, energy-absorption performance in the dynamic case is not necessarily better than it is in the static counterpart. On the contrary, it can be worse. This is a direct result of the large contribution which friction makes towards the total energy dissipation; in the dynamic situation friction is smaller than in the static case. This point needs to be noted in the design and assessment of energy-absorbing systems.

9
Cylindrical and spherical shells

This chapter discusses four more cases involving analysing the plastic response of shell structures: inversion of tubes, tube nosing, spherical shells under point loading and propagating collapse of a submarine pipeline. Apart from tube nosing, all these cases involve an almost steady-state, propagating plastic zone. In the latter three cases, an equilibrium approach is employed. This gives results which are in better agreement with experiment than those from the energy method, which has been used extensively in the previous chapters.

9.1 Tube inversion

Energy is dissipated when a circular tube is inverted either externally or internally. This process can be adopted in designing a collapsible steering column or other energy-absorbing devices. One of the key advantages of such a device is that a steady-state with a constant force is achieved, which is ideal for energy absorption. Tube inversion can be realised with or without a die (the latter is called ***free inversion***), Fig. 9.1. Figure 9.2 shows photographs of externally and internally inverted tubes on their respective dies (Al-Hassani *et al.*, 1972). Typical load–displacement curves are given in Figs 9.3(a) and (b), for external inversion and internal inversion, respectively. In both cases, the aluminium tube is 1.6 mm thick, has an outside diameter of 50.8 mm and is 89 mm long. The die radius is 4 mm. Both curves exhibit very similar characteristics; they have two initial peaks, followed by a steady-state process with a virtually constant force. The displacement at which the force becomes constant is approximately 0.5 the diameter of the tube. In the initial transient stage, the tube wall does not conform to the die face; rather, it acts more like an edge contact followed by plastic bending at some distance away from the edge, leading to the first peak force. The latter stage before the steady-state involves frictional contact between the tube wall and die, and plastic bending and stretching of the tube wall.

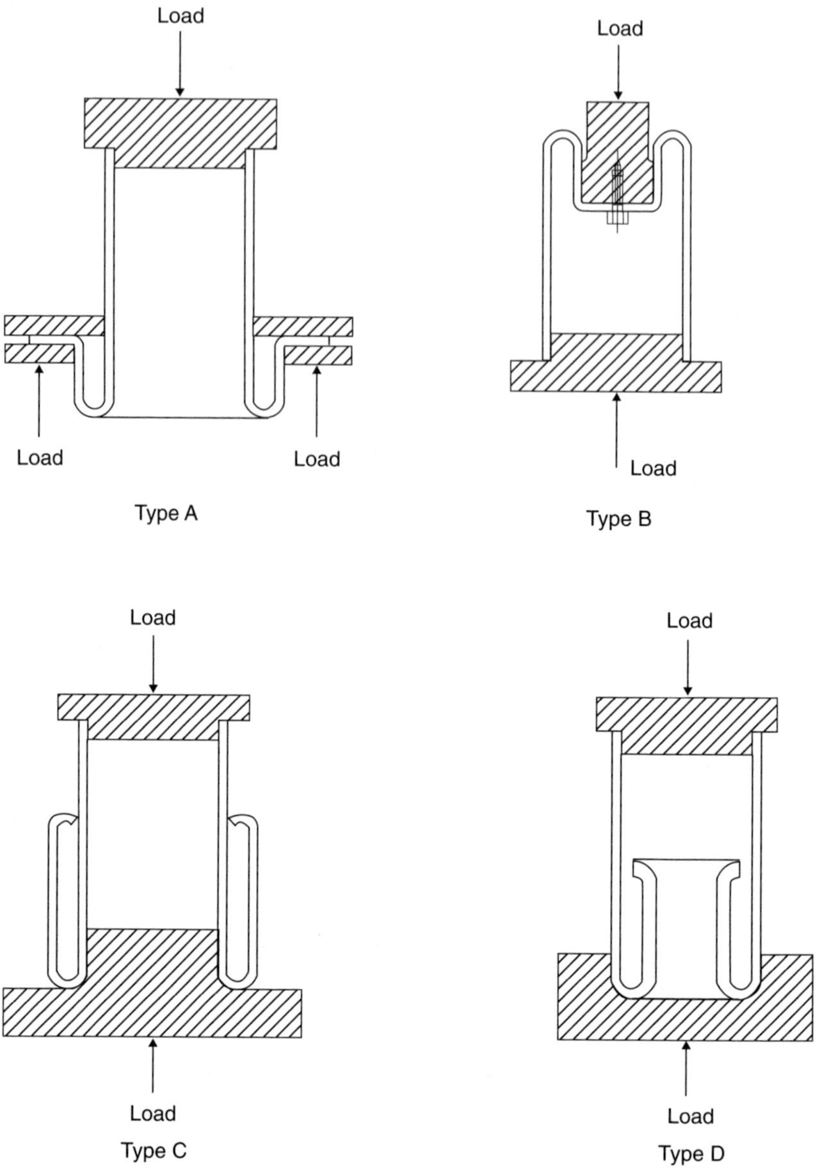

9.1 Sketch of four types of tube inversion (Al-Hassani *et al.*, 1972) (reproduced with kind permission of the Council of the Institution of Mechanical Engineers).

9.2 Photographs of sectioned inverted tubes on their respective dies (Al-Hassani *et al.*, 1972).

The bending radius is not necessarily equal to the die radius. In fact, it has been found that the radius is not constant. This process has been successfully reproduced using the finite element software ABAQUS (Reid and Harrigan, 1998). Figure 9.4 shows load–displacement curves of the experiment and ABAQUS for mild steel tube internal inversion. Axisymmetric solid continuum elements were used. The material model was elastic, linearly strain-hardening. A frictional coefficient of 0.15 was used in order to match the experimental results.

Simple analyses can be made to determine the steady-state force (Guist and Marble, 1966). Consider free external inversion of a circular tube as shown in Fig. 9.5. When the inner part of the tube is pushed downwards, the material first enters a toroidal region at point A, where bending occurs in the meridian direction. The free-forming radius of curvature in the meridian direction is assumed to be constant (= b). This is also known as the ***knuckle radius***. When the material exits this toroidal region at B, it is straightened back to zero curvature in the meridian direction, but the tube radius now becomes $R + 2b$. There is hence circumferential stretching with a strain of $2b/R$. Assuming that the material obeys Tresca's yield criterion, the strain increments in the meridian and circumferential directions are equal but of opposite sign from the associated normality rule. There is no change in thickness.

Let the inner tube move downwards by a unit distance. The area of surface passing through the toroidal region is then $(2\pi R)/2$. The stretching energy is

$$W_s = \pi Rh\frac{2b}{R}Y = 2\pi hbY \qquad [9.1]$$

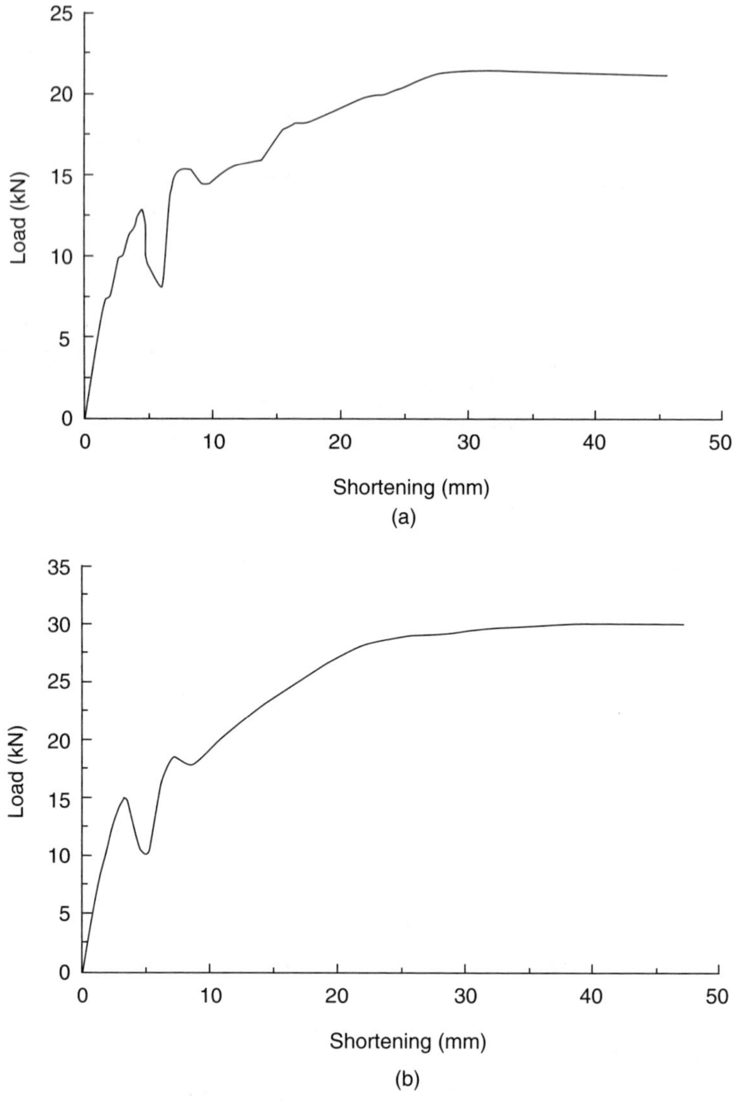

9.3 Typical load–displacement curves: (a) external inversion; (b) internal inversion (Al-Hassani *et al.*, 1972).

where Y is yield stress and h is thickness. The bending energy is

$$W_b = 2\pi R \frac{1}{b} M_o = \frac{\pi h^2 R}{2b} Y \qquad [9.2]$$

Noting that $M_o = Yh^2/4$ and equating internal energy with the work done by the external load P, we have

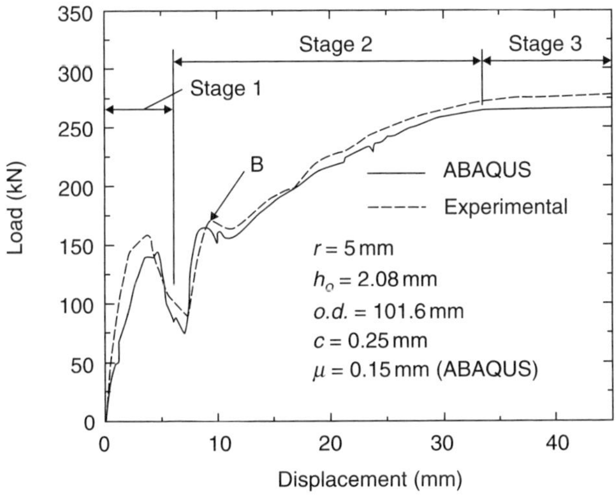

9.4 Experimental and ABAQUS results for quasi-static internal inversion of mild steel tubes (Reid and Harrigan, 1998) (reproduced with kind permission of Elsevier).

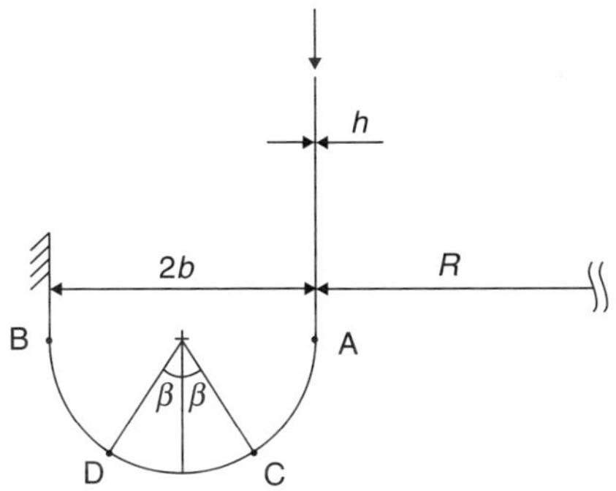

9.5 Sketch of a tube wall under free external inversion.

$$P \cdot 1 = W_s + W_h$$

or

$$P = 2\pi hbY + \frac{\pi h^2 R}{2b} Y \qquad [9.3]$$

Now we argue that the value of radius b must be such that force P is minimised. This indicates that bending and stretching dissipate the same amount of energy. Hence

$$b = \frac{1}{2}\sqrt{Rh} \qquad [9.4]$$

and by substituting into Eq. [9.3]

$$P = 2\pi Yh\sqrt{Rh} \qquad [9.5]$$

The knuckle radius b predicted by Eq. [9.4] is about twice that measured from experiment, whilst Eq. [9.5] underestimates the actual steady-state force. When the measured values of b are used, Eq. [9.3] produces forces about 15 % lower than the experimentally measured ones.

Calladine (1986) postulated that hinge circles may form at C and D instead of A and B, respectively (Fig. 9.5). The position of C and D varies with b. Subsequently, Reddy (1992) assumed a rigid, linear hardening material model and applied the work balance equation in the form of rate of energy. He obtained

$$P = 2\pi RYh\left\{\frac{h}{4b} + \frac{1}{\sqrt{3}}\ln\left(1+\frac{2b}{R}\right)\left[1+\frac{E_p}{2Y}\ln\left(1+\frac{2b}{R}\right)\right]\right\} \qquad [9.6]$$

where E_p is the modulus of linear strain-hardening. The value of radius b is obtained by letting $dP/db = 0$, which leads to

$$\left(\frac{b}{R}\right)^2 \frac{8}{\sqrt{3}}\left[1+\frac{E_p}{Y}\ln\left(1+\frac{2b}{R}\right)\right] - \frac{h}{R}\left(1+\frac{2b}{R}\right) = 0 \qquad [9.7]$$

Equation [9.7] can be solved numerically to find out the value of b which, when substituting back into Eq. [9.6], leads to the steady-state force. It was found that for $h/R = 0.02$ and 0.1, $E_p/Y = 3.5$ and 5.5, respectively and excellent agreements between experiment and theory are achieved.

The above approach for free inverting tubes can be applied to inverting tubes using a die. For example, Al-Hassani *et al.* (1972) considered external inversion of a tube using a die. They assumed that the material obeys a stress–strain relationship described by

$$\sigma = A(B+\varepsilon)^n \qquad [9.8]$$

where A, B and n are constants. The knuckle radius is obtained by minimising the load, which turns out to be the same as Eq. [9.4]. The corresponding force is

$$P = \frac{4\pi RhA}{n+1}\left[B + \frac{2}{\sqrt{3}}\ln\left(1+\sqrt{\frac{h}{R}}\right)\right]^{n+1} \qquad [9.9]$$

which agrees well with their experiments.

The force–displacement relationship in the transitional stage has been worked out by Lenard (1978) and Reddy (1989) for both external and internal inversions with a die. It was assumed that the tube wall conforms with the die completely and there is no friction between the two. The steady-state force can also be obtained. Interested readers may refer to the original papers. Alternatively, this relationship may be obtained by an equilibrium approach which will be employed in analysing tube nosing in Section 9.2. Dynamic effects in tube inversion have been studied (Colokoglu and Reddy, 1996; Miscow and Al-Qureshi, 1997; Reid and Harrigan, 1998).

9.2 Tube internal nosing

A process closely related to tube inversion is ***tube nosing*** (see Fig. 9.6 for internal nosing). The load increases with end displacement of the tube (Fig. 9.7). As the tube is pushed downwards, the tube wall bends, conforming to the die in the meridian direction. At the same time, circumferential compression occurs. Excluding the case of wrinkling as a result of circumferential compression, a lower bound and an upper bound analysis can be performed (Reid and Harrigan, 1998).

As in the previous section, assume that the material yields according to the Tresca criterion and that there is no interaction between the membrane force resultants and bending moments. The stress in the thickness direction is small and $\sigma_t \approx 0$. Both the meridian stress σ_x and circumferential stress σ_θ are compressive and $\sigma_\theta \geq Y$. Normality rule indicates that the meridian strain increment $\dot{\varepsilon}_x$ is zero. Because the volume is a constant during deformation (i.e. $\varepsilon_\theta + \varepsilon_t + \varepsilon_x = 0$), we have

$$\varepsilon_\theta = -\varepsilon_t \qquad [9.10]$$

or

$$Rh = R_o h_o \qquad [9.11]$$

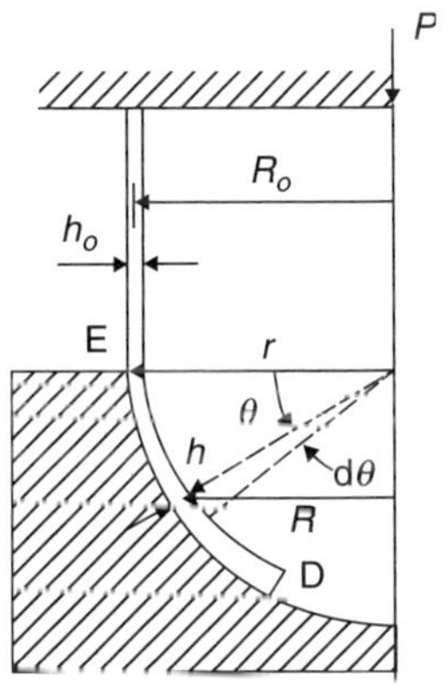

9.6 Sketch of a tube wall under internal inversion and nosing.

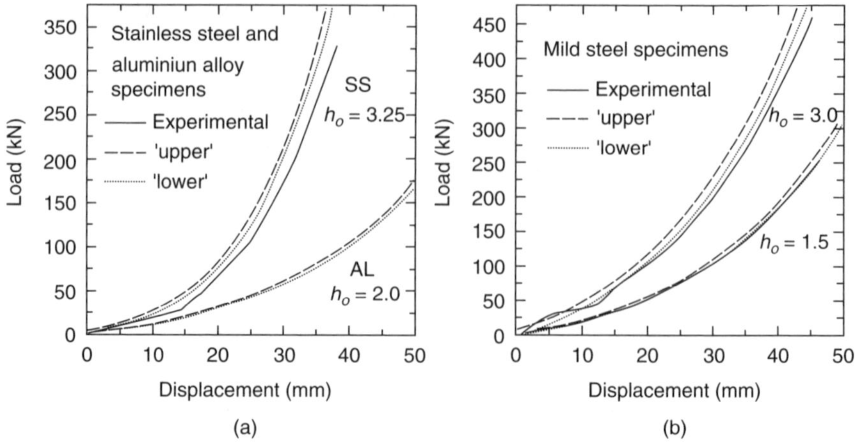

9.7 Experimental and analytical load–displacement curves for tube nosing (die radius: 50 mm): (a) stainless steel and aluminium alloy specimens; (b) mild steel specimens (Reid and Harrigan, 1998) (reproduced with kind permission of Elsevier).

where R and h are the current radius and thickness as shown in Fig. 9.6. A frictional coefficient μ equal to 0.15 is also assumed. The material is rigid, linear hardening. Hence, the hoop stress is given as

$$\sigma_\theta = Y + E_p\left(1 - \frac{R}{R_o}\right) \tag{9.12}$$

Lower bound analysis involves consideration of equilibrium and yield conditions. Consider the tube under internal nosing as sketched in Fig. 9.6 and let the contact pressure between the tube wall and die be p. The equilibrium equations in the thickness and meridian directions are

$$pr\left(R + \frac{h}{2}\cos\theta\right) = \sigma_x Rh + \sigma_\theta h\left(r - \frac{h}{2}\right)\cos\theta \tag{9.13}$$

and

$$-R_o h_o \frac{d\sigma_x}{d\theta} = upr\left(R + \frac{h}{2}\cos\theta\right) + \sigma_\theta h\left(r - \frac{h}{2}\right)\sin\theta \tag{9.14}$$

respectively. Eliminating p from these two equations leads to

$$-\frac{d\sigma_x}{d\theta} = \mu\sigma_x + \frac{\sigma_\theta}{R}\left(r - \frac{h}{2}\right)(\mu\cos\theta + \sin\theta) \tag{9.15}$$

In Eq. [9.15], the only variables are σ_x and θ, because R is related to θ via the die geometry and σ_θ is given in terms of R by Eq. [9.12]. The boundary condition is that at the leading edge of the tube $\theta = \theta_o$ and $\sigma_x = 0$. Hence Eq. [9.15] can be integrated numerically to obtain the meridian stress σ_x along line ED (Fig. 9.6). In particular, the stress at E, $(\sigma_x)_E$, gives the axial force as

$$P_L = 2\pi R_o h_o (\sigma_x)_E \tag{9.16}$$

An upper bound solution can be obtained from the conservation of energy. Hence, the rate of external work done must be equal to the rate of plastic dissipation, which is plastic bending in the meridian direction, circumferential compression and friction at the interface. If the rate of tube end movement is $\dot{s}$, the corresponding rate of work is $P\dot{s}$. Hence

$$P\dot{s} = \dot{W}_b + \dot{W}_c + \dot{W}_{fri} \tag{9.17}$$

The rate of plastic bending in the meridian direction is simply

$$\dot{W}_b = \frac{2\pi R_o h_o^2 Y \dot{s}}{4(r - h_o/2)} \tag{9.18}$$

When the tube moves downwards at a rate of $\dot{s}$, the deforming tube wall slides along the die surface at an angular rate of $\dot{s}/r$. Hence the circumferential strain increment is

$$\dot{\varepsilon}_\theta = \frac{\dot{R}}{R} = \frac{1}{R}\frac{\mathrm{d}R}{\mathrm{d}\theta}\frac{\dot{s}}{r} \tag{9.19}$$

The energy dissipation rate is the product of stress σ_θ and strain increment over the whole deforming volume

$$\dot{W}_c = 2\pi R_o h_o \dot{s} \int_{\theta_E}^{\theta_D} \frac{\sigma_\theta}{R}\left(\frac{\mathrm{d}R}{\mathrm{d}\theta}\right)\mathrm{d}\theta \tag{9.20}$$

The interface pressure is obtained from the previous equilibrium equations and then the rate of the frictional energy is

$$\dot{W}_{fri} = 2\pi R_o h_o \mu r \dot{s}\left(\int_{\theta_E}^{\theta_D} \frac{\sigma_x}{r - h/2}\mathrm{d}\theta + \int_{\theta_E}^{\theta_D} \frac{\sigma_\theta \cos\theta}{R}\mathrm{d}\theta\right) \tag{9.21}$$

Here σ_θ and σ_x are given by Eqs [9.12] and [9.15], respectively. Substituting Eqs [9.18], [9.20] and [9.21] into [9.17], an upper bound estimate of force can be made.

Numerical results from both the lower and upper bound analyses are given in Fig. 9.7, for specimens of stainless steel, aluminium alloy and mild steel. They agree well with the experiments. The analyses can be easily followed for other loading cases such as transient stages of tube inversion, provided that the tube wall conforms the die surface.

9.3 Inversion of a spherical shell

Plastic deformation of a spherical shell under a point load (Fig. 9.8) or crushing by a rigid plate involves a similar feature to that of tube inversion: the plastic deformation is concentrated within a small, moving zone. This problem has been studied by several researchers (Wasti, 1964; Leckie and Penny, 1968; Morris and Calladine, 1969; Updike, 1972; de Oliveira and Wierzbicki, 1982). Here we present an analysis by means of equilibrium consideration (Updike, 1972; Calladine, 1986).

Figure 9.8 shows a sketch of a spherical shell being inverted by a radial force P. Plastic deformation is concentrated within a small knuckle region whose size is independent of the central deflection w, as will be seen later. Other portions of the shell remain rigid. Thus, as deflection w increases, adjacent undeformed material enters into this knuckle of plastic deformation. As it passes the outer plastic-hinge circle, plastic bending occurs in the meridian direction and a curvature is being imparted. At the same time, some material within the knuckle moves towards the axis of symmetry and, as it passes the inner plastic hinge circle, a reversal curvature is being imparted, leading to an inverted central part. From experiment, the radius of curvature for the inverted portion is approximately the same as that before inversion, but in the opposite sense. Hence the inverted profile can be assumed to be the mirror image of its original shape.

The instantaneous plastic deformation can be regarded as rotation of the knuckle about the outer hinge. Hence, all the material within this region experiences circumferential compression, with a yield stress Y, as the material is assumed rigid, perfectly plastic. Note that the inner and outer plastic hinge circles must lie at the same horizontal level.

Referring to Fig. 9.8, let the current position of the knuckle be defined by radius r. The size of the knuckle is given by l. The slope at the hinge points is then r/R. Assume that the profile of the knuckle is parabolic. The dimension ξ is given from the geometry by

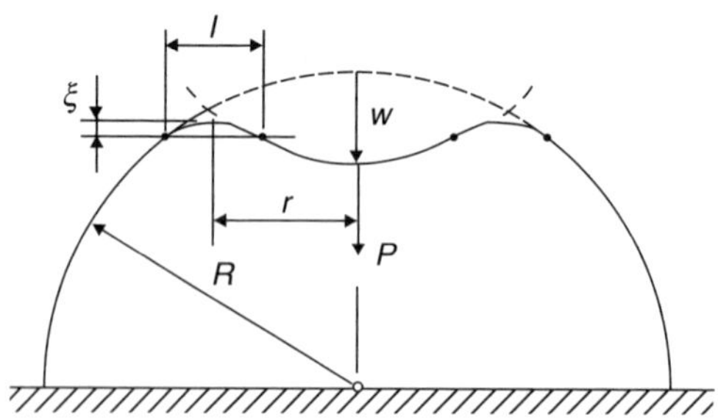

9.8 Sketch of a spherical shell being inverted by a radial force *P*.

$$\xi = \frac{rl}{4R} \qquad [9.22]$$

We now study the forces and bending moments acting on the knuckle. Figure 9.9(a) shows these for a knuckle of unit circumference, on the right hand side of the axis of symmetry. At each plastic hinge, there is a fully plastic bending moment per unit length M_o and a resultant membrane force tangential to the knuckle meridian, represented by vertical and horizontal force components V and H per unit length, respectively. Hence

$$\frac{H}{V} = \frac{R}{r} \qquad [9.23]$$

Figure 9.9(b) shows an element subtending a small angle θ, when viewed from above. The hoop stress is Y and the total hoop force over the knuckle is Ylh with h being the thickness. The radial equilibrium equation leads to

$$H = \frac{Yhl}{2r} \qquad [9.24]$$

Note that the knuckle shown in Fig. 9.9(a) has a unit circumference and the corresponding resultant of the two hoop forces is therefore Ylh/r. This force points radially outwards and acts at the centroid of the assumed parabola, which is $(2/3)\xi$ from the base.

Equilibrium in the tangential direction is automatically satisfied for this axisymmetric problem, while the second equilibrium equation is obtained from moment consideration. Taking moment about a point in the hinge circles leads to

$$Vl = 2M_o + \frac{2\xi Yhl}{3r} \qquad [9.25]$$

Substituting Eq. [9.22] into the above equation, we obtain

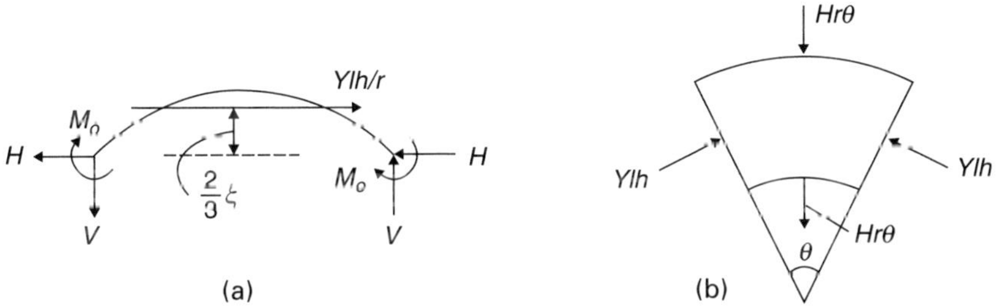

9.9 (a) Enlarged cross-sectional view of the right-hand toroidal region of Fig 9.8 with forces and couples; (b) axial view of a portion of the same toroid, showing horizontal forces (Calladine, 1993).

$$V = \frac{2M_o}{l} + \frac{Yhl}{6R} \quad [9.26]$$

Eliminating Hr from Eqs [9.23] and [9.24], we have

$$V = \frac{Yhl}{2R} \quad [9.27]$$

Solving Eqs [9.26] and [9.27] for l and V

$$l = 1.22R^{0.5}h^{0.5} \quad [9.28]$$

$$V = \frac{0.61Yh^{1.5}}{R^{0.5}} \quad [9.29]$$

It turns out that both l and V are independent of r. Hence the size of the knuckle does not change and the vertical force per unit length remains constant. The load P increases with deflection w merely as a result of the increase in total circumference of the hinge circles. From the overall equilibrium, the total axial load P is

$$P = 2\pi rV \quad [9.30]$$

But from the geometry of the shallow arc

$$r = w^{0.5}R^{0.5} \quad [9.31]$$

Consequently

$$P = \pi(1.5)^{0.5}\,Yh^{1.5}w^{0.5} \quad [9.32]$$

This equation is plotted in Fig. 9.10 as the solid line. It agrees well with a more detailed step by step upper bound analysis without assuming the knuckle shape (Morris and Calladine, 1969), if the present curve is shifted to the right by approximately one thickness. This is because of the local deformation near the apex in the early stage of deformation, which was considered by Morris and Calladine (1969), but not in the analysis presented above. Furthermore, when the load is first applied, the spherical shell acts like a flat plate with a clamped edge. The initial collapse load is therefore [Eq. 5.1] $P = 4\pi M_o$ and Eq. [9.32] is valid only when $P \geq 4\pi M_o$.

The above analysis is for a point load. Now if the spherical shell is crushed by a rigid flat plate, the displacement of the plate is approximately $w_p = 0.5w$.

de Oliveira and Wierzbicki (1982) used the same kinematics as that described above, but an energy method was then followed instead of the equilibrium consideration. The value of l is determined by minimising the value of force V. They obtained

$$l = 1.73R^{0.5}h^{0.5} \quad [9.33]$$

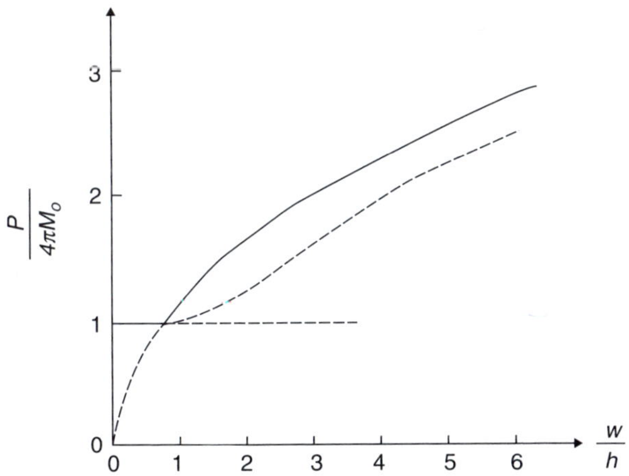

9.10 Non-dimensional force–deflection curves. Solid line: present analysis. Dotted line: analysis by Morris and Calladine (1969) (Calladine, 1986) (reproduced with kind permission of Springer-Verlag).

$$V = \frac{0.58Yh^{1.5}}{R^{0.5}} \qquad [9.34]$$

Equation [9.33] gives a value of l about 40 % higher than Eq. [9.28]. Hence, minimisation of the total energy may not lead to the same results which could be obtained from equilibrium consideration, although the total force may not be sensitive to the details of the deformation.

9.4 Propagating collapse of a submarine pipe

We finally present an analysis of plastic collapse of a submarine pipeline. A submarine pipe may undergo ***propagating collapse*** under external pressure after an initiation of buckling due to, for example, accidental denting. Figure 9.11 is a sketch of such a case. A transition, plastic deforming zone of length L propagates and the original circular section becomes dog-bone shaped on exiting the deformation zone. The question is to find out the minimum external pressure which causes this propagating collapse, for a given pipe material and dimensions.

Major plastic energy dissipation mechanisms are circumferential bending and longitudinal stretching/compression, as the length of the curved generators is obviously longer than their original length L. Most of the early analysis involves calculation of external pressure causing ring-type circumferential bending, and this is presented first. A simple ***beam-on-foundation model*** will be presented next, which takes into account the longitudinal stretching effect.

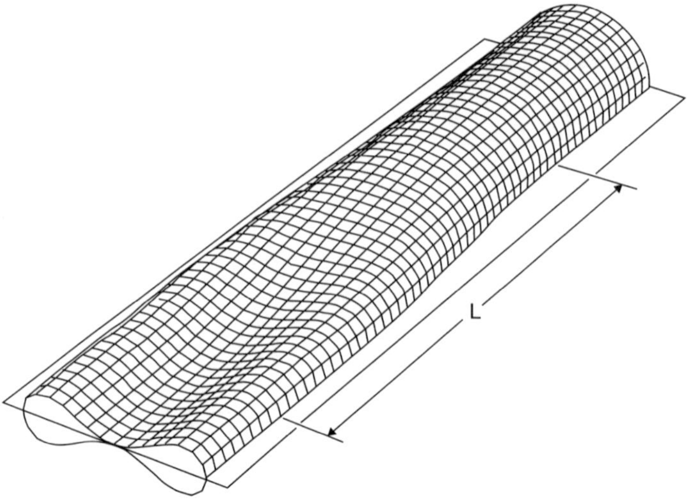

9.11 Sketch of a buckle propagating along a pipe under external pressure (Kamalarasa and Calladine, 1988) (reproduced with kind permission of Elsevier).

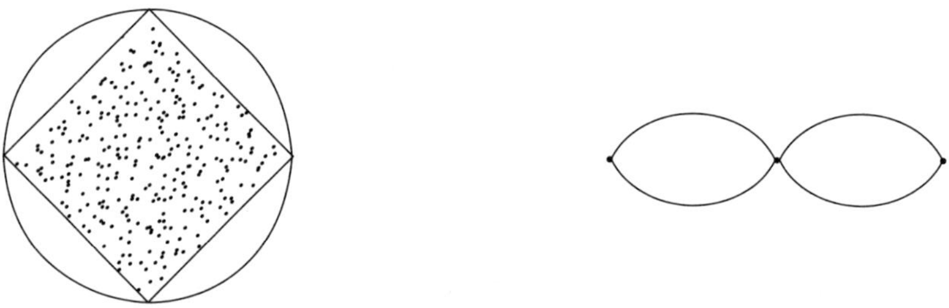

9.12 A simple ring model for buckle propagation of pipelines under external pressure (Palmer and Martin, 1975).

9.4.1 Ring model

Palmer and Martin (1975) made the first theoretical analysis of this problem, by assuming that the collapse pressure of this pipe is the same as that for a ring. A four-hinge plastic collapse mode can be assumed (Fig. 9.12), very much like the case of a ring collapsing under point loads discussed in Section 4.1. After collapse, the cross-sectional area reduces by $\Delta A = 2R^2$, here R is the initial pipe radius. The work done by the external pressure is therefore $p\Delta A = p2R^2$. As before, the total hinge rotation of the four hinges is $4 \times \pi/2 = 2\pi$, and the plastic energy is $2\pi M_o$, where $M_o = Yh^2/4$ is the fully plastic bending moment per unit length. Applying energy balance

$$p\,2R^2 = 2\pi M_o \tag{9.35}$$

From this, we immediately obtain the pressure as

$$p_o = \frac{\pi}{4} Y \left(\frac{h}{R}\right)^2 \tag{9.36}$$

This equation underestimates the actual pressure measured from experiment, especially for small values of R/h. Various researchers have attempted to study the collapse process and the effect of strain-hardening (Steel and Spence, 1983; Chater and Hutchinson, 1984; Kyriakides *et al.*, 1984; Croll, 1985; Wierzbicki and Bhat, 1986). In particular, Wierzbicki and Bhat (1986) obtained an expression for pressure as

$$\frac{p}{p_o} = 1 + 1.09 \left(\frac{E_p}{Y}\right)^{0.7} \left(\frac{h}{R}\right)^{0.7} \tag{9.37}$$

Similar expressions have been given to incorporate the strain-hardening effect. Nevertheless, the other important effect, stretching/compression, is still missing in the ring models and is dealt with next.

9.4.2 Plastic beam-on-plastic foundation model

A plastic beam-on-plastic foundation model can be used (Kamalarasa and Calladine, 1988) to take into account the stretching effect which is neglected in the ring model (Fig. 9.13). The beam represents the stretching effect and the plastic foundation represents the resistance owing to circumferential bending. The profile of the beam resembles the shape of the transition zone in Fig. 9.11. The plastic foundation is softening and its characteristic is assumed as in Fig. 9.13(b). Three plastic hinges (A, B, C) exist in the beam, each with a fully plastic bending moment M_p. The task is to find the minimum external pressure represented by force intensity q.

Assume that both AB and BC are parabolas. Because of continuity of slope at A, B and C, the coordinates of B are given as $(\beta L, \beta h_1)$. Here h_1 is the total height and β is a numerical fraction. The bending moment diagram, net vertical force distribution and the free body diagrams for segments AB and BC are shown in Figs 9.13(c), (d) and (e), respectively. Because of the linearity of the foundation, the shape of the net force is identical to that of the beam profile.

From the vertical equilibrium of segment AB, we have

$$\mu = \frac{\beta}{3} \tag{9.38}$$

Let $\Delta f = f_o - f_1$. Moment equilibrium for AB and BC (about C) gives

$$24 M_p = \beta^3 L^2 \Delta f \tag{9.39}$$

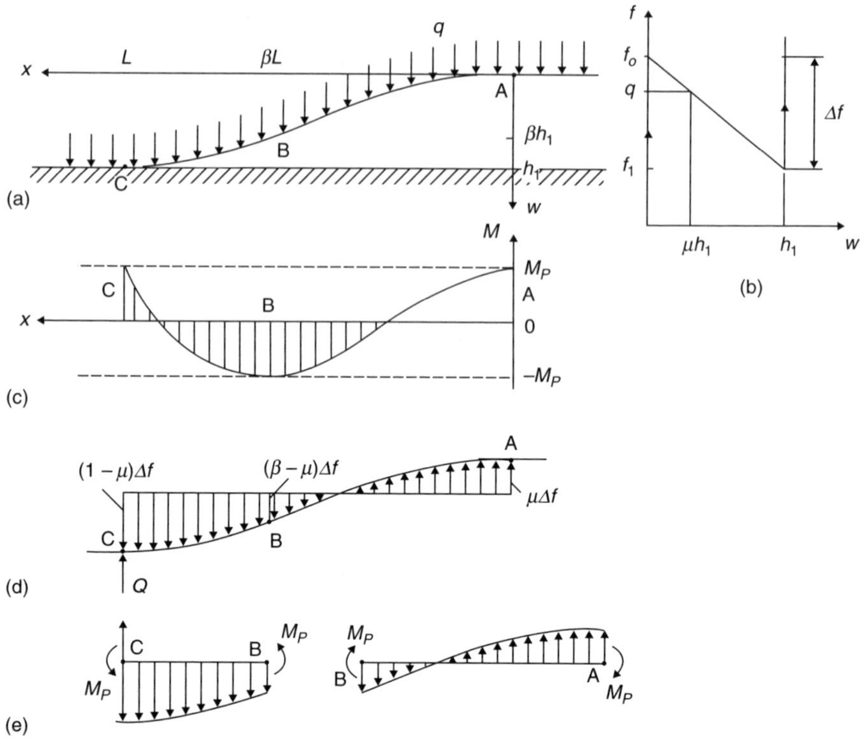

9.13 A beam-on-plastic foundation model proposed by Kamalarasa and Calladine (1988): (a) general layout and coordinate system; (b) constitutive relation for the foundation; (c) bending moment diagram; (d) net vertical force distribution; (e) free body diagrams for portions AB and BC (reproduced with kind permission of Elsevier).

and

$$24M_p = (1-\beta^2)(3+\beta)L^2\Delta f \qquad [9.40]$$

These two equations lead to a quadratic equation for β

$$\beta^2 - 5\beta + 3 = 0 \qquad [9.41]$$

Hence

$$\beta = 0.697 \qquad [9.42]$$

and

$$\mu = \frac{\beta}{3} = 0.232 \qquad [9.43]$$

Therefore, the external load intensity q is

$$q = 0.768 f_o + 0.232 f_1 \tag{9.44}$$

Also, substituting $\beta = 0.697$ into Eq. [9.39], we have

$$L = 8.4\left(\frac{M_p}{\Delta f}\right)^{\frac{1}{2}} \tag{9.45}$$

If we ignore the effect of beam bending, an energy balance suggests that the external work qh must be equal to the plastic work of the foundation, which is the area under the f–w curve shown in Fig. 9.13(b). This leads to

$$q' = 0.5(f_o + f_1) \tag{9.46}$$

which is lower than that given by Eq. [9.44]. This ratio of q/q' is

$$\frac{q}{q'} = 1 + 0.536\frac{\Delta f}{f_o + f_1} \tag{9.47}$$

The above equation suggests that the actual propagation pressure is a fixed multiple of the pressure causing circumferential bending of the cross-sectional rings. The latter has been obtained experimentally by squashing a ring with two curved indenters (Fig. 9.14) to produce a final cross-section approximately identical to that of a buckled pipe shown in Fig. 9.11. The change of cross-sectional area (ΔA) can be measured from the deformed specimen. Hence from the total energy absorbed per unit tube length u (the area underneath the load per unit tube length–displacement curve), the mean pressure causing ring collapse is obtained from

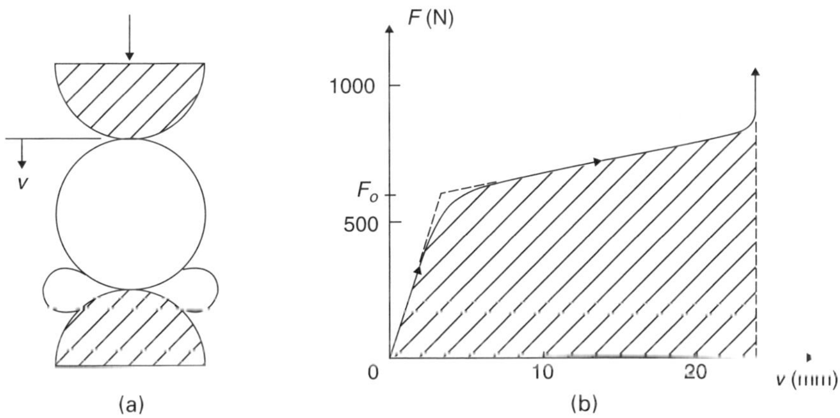

9.14 (a) Compression test of a ring between curved dies: initial and final configurations. (b) Typical force displacement plot (stainless steel specimen with R/h = 22.2) (reproduced with kind permission of Elsevier).

$$p' = \frac{u}{\Delta A} \tag{9.48}$$

This scheme takes into account the strain-hardening effect experimentally. Kamalarasa and Calladine (1988) demonstrated from their experiments that the propagating pressure is simply given by

$$p = 1.4p' \tag{9.49}$$

The length of the ***transition zone*** is given by Eq. [9.45]. Now recall that the beam bending actually represents the longitudinal stretching effect of the pipe and the foundation represents the circumferential bending. We expect

$$M_p \propto h \tag{9.50}$$

and

$$\Delta f \propto h^2 \tag{9.51}$$

Hence, from Eq. [9.45]

$$L \propto h^{-\frac{1}{2}} \tag{9.52}$$

The only other variable with length dimension is tube radius R and dimensional analysis suggests that

$$L = C\frac{R^{\frac{3}{2}}}{h^{\frac{1}{2}}} \tag{9.53}$$

Empirically, the value of the constant C is found to be 3.6 for both stainless steel and aluminium pipes.

9.5 Concluding remarks

The problems analysed in this chapter, with the exception of tube nosing, involve a steady-state plastic deformation. Both energy method and static equilibrium consideration may be employed in order to obtain analytical expressions; however, they do not always produce the same answer. In some cases, the size of the plastic zone is obtained by minimising the total energy, which does not give an accurate prediction for this size of plastic zone. In fact, the method may fail for some problems: the last example of buckle propagation of submarine pipelines is included in this chapter to demonstrate this point, although strictly speaking it is not a problem of energy-absorption in our context.

In some cases, the energy method does give reasonable estimates of crushing force, although the details of plastic deformation are not well predicted. This may indicate that in those cases the external force is not sensitive to the plastic deformation modes and size and hence that the energy method may be sufficient for the purpose of calculating energy absorption of structures. But an accurate of prediction of deformation details will need equilibrium considerations.

10
Cellular materials

Cellular materials have good energy-absorption characteristics. This chapter presents their stress–strain relations, the fundamental mechanics at the cell level and their impact response. Materials discussed include honeycombs, foams, woods and cellular textile materials.

Gibson and Ashby (1997) give a comprehensive treatment of various properties of cellular solids in their book. In this chapter, we will start our discussion by presenting the basic properties and mechanics for honeycombs, foams and woods, which are relevant to energy absorption.

10.1 Honeycombs

Honeycombs are a typical type of cellular material. They (and foams) are widely used as the core structure in sandwich panels, for example. Honeycombs can also be used alone as good energy-absorbing materials. Their structure is essentially two-dimensional and regular. Hence they are easier to analyse than foams, which have three-dimensional cell structures. In this section, we describe the crushing behaviour of honeycombs. Foams will be discussed in the next section.

10.1.1 Cell structure, relative density, stress–strain curves and densification strain

Most honeycomb cells are hexagonal in section (Fig. 10.1), but other shapes are also possible such as triangular, square, rhombic or circular (Chung and Waas, 2002a, b). The materials from which the cells are made are man-made polymers, metals, ceramics and paper. In this section, we will restrict our discussion to honeycombs with hexagonal cells.

As shown in Fig. 10.1, a typical honeycomb consists of series of hexagonal cells whose dimensions are defined by cell wall lengths, l and c, the angle between two cell walls θ and the cell wall thickness h. Deformation caused by loading in the global plane of the honeycomb, X_1X_2, is known as in-plane

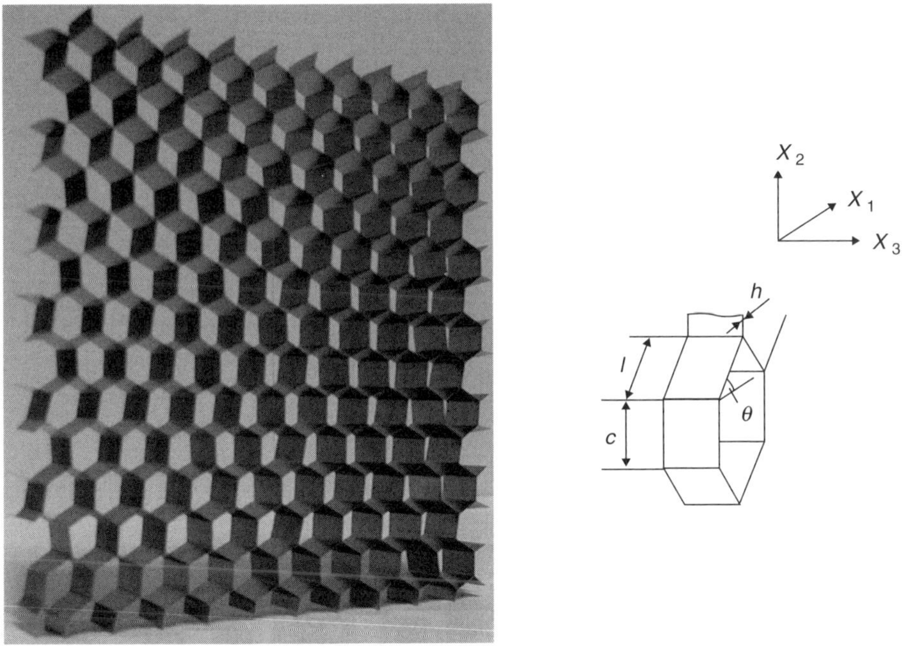

10.1 A honeycomb with definitions of parameters for a cell.

response, while that caused by loading in the X_3 direction is out-of-plane response. They will be discussed separately.

One of the most important parameters characterising cellular materials is ***relative density***, which is defined as ρ^*/ρ_s, where ρ^* is the overall density of the cellular material and ρ_s is the density of the solid of which the cellular material is made. The corresponding porosity, the fraction of pore volume in the cellular material, is therefore $1 - \rho^*/\rho_s$. For the honeycomb shown in Fig. 10.1, when $h \ll l$, then

$$\frac{\rho^*}{\rho_s} = C_1 \frac{h}{l} \qquad [10.1]$$

where C_1 is a numerical constant and is dependent upon the details of cell geometry. A more detailed analysis gives (Gibson and Ashby, 1997)

$$\frac{\rho^*}{\rho_s} = \frac{h/l(c/l+2)}{2\cos\theta(c+\sin\theta)} \qquad [10.2]$$

For regular cells, $l = c$ and $\theta = 30°$, Eq. [10.2] gives

$$\frac{\rho^*}{\rho_s} = \frac{2}{\sqrt{3}}\frac{h}{l} \qquad [10.3]$$

Some honeycombs are obtained by first gluing stamped sheets of the material along specific strips and then expanding the glued sheets. As a result, one-third of the cell walls (of length c) have double-wall thickness. The density of such honeycombs is

$$\frac{\rho^*}{\rho_s} = \frac{8}{3}\frac{h}{l} \qquad [10.4]$$

For $h = 0.094$ mm and $l = 9.53$ mm, Eq. [10.4] gives $\rho^*/\rho_s = 2.63\,\%$.

More exact expressions for the relative density could be obtained. However, Eqs [10.3] and [10.4] are simple to use and are accurate enough unless $h/l > 1/4$; hence they are adopted in most analyses.

Typical stress–strain curves for uniaxial compression either in the X_1 or X_2 direction are sketched in Fig. 10.2 (Gibson and Ashby, 1997). Each curve essentially consists of three stages. In the first stage, the response is linear-elastic. This stage terminates when a critical stress is reached and this critical stress level is maintained almost constant over a large range of strain (stage 2). Finally, the stress increases rapidly with strain, as a result of compaction of cells or densification.

The global external load is transferred to the cell walls at the cell level and they deform very much like structures. Their deflection is reflected as strain at the macro-level. Hence, the structural response of honeycomb cells

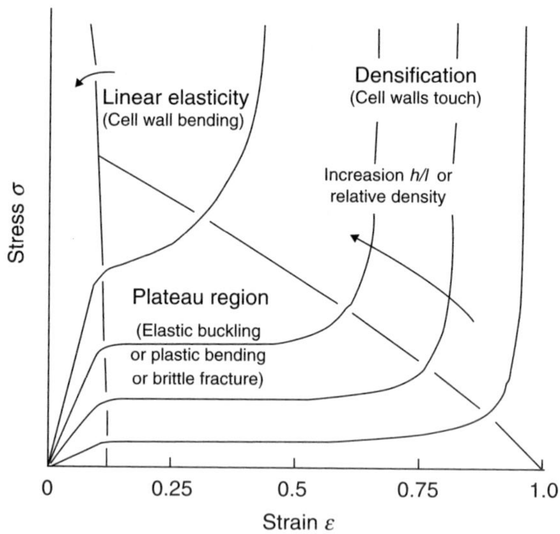

10.2 Sketch of typical stress–strain curves for a honeycomb under in-plane loading. Increasing the relative density or *h/l* changes its shape (reproduced with kind permission of Cambridge University Press).

dictates the global stress–strain curve of a honeycomb block. In the first, linear elastic stage, the cell walls simply bend elastically with small deflections.

The second stage may be governed by one of three different possible failure mechanisms of cell walls: elastic buckling, plastic collapse or brittle fracture. The first two mechanisms are analogous to those of a column under compressive loads; depending on the slenderness of the column (h/l here), a column could fail by Euler buckling or plastic yielding. Hence, cell walls with small values of h/l buckle elastically, while those with large values of h/l yield (or collapse plastically). Honeycombs, having brittle base materials with small critical strains, fail by brittle fracture as a result of excessive strains induced in the cell walls. This third mechanism is often accompanied by considerable fluctuations in the plateau stress.

Details of typical post-collapse behaviour of honeycomb cells are shown in Fig. 10.3 (Papka and Kyniakides, 1994 and 1998), for an aluminium honeycomb of 15 rows by 10 columns of cells. Successive crushing events as photographed in Fig. 10.3(b) are marked in the experimental load–displacement curve plotted in Fig. 10.3(a). Initially, the honeycomb deforms elastically and uniformly. The cell walls bend symmetrically about vertical axes through their centres. When the stress is about 110 kPa, the load–displacement curve becomes softer until an initial peak stress is reached at 121.9 kPa. Further crushing leads to deformation localised within one row of cells (photographs 2 and 3). The force reduces from the initial peak value and fluctuates as a result of plastic collapse of cell walls and geometrical constraint of neighbouring cells. Once a layer of cells is fully crushed with cell walls touching each other, this localised deformation propagates to an adjacent layer of cells (photographs 4–7).

These crushing events can be simulated using a finite element (FE) analysis package. Figure 10.4(a) depicts the FE and experimental load-displacement curves, showing good agreement. Figure 10.4(b) shows the successive crush zones derived from FE. In the FE model, beam elements were used for the cell walls and the solid wall material was idealised as bi-linear (i.e. elastic, linear strain-hardening) with a post-yield modulus of $E/100$ (E is the modulus of elasticity). A distinction may be necessary between the initial peak stress and the propagation stress, the latter being slightly lower. However here we do not distinguish the two.

The most relevant properties to energy absorption are ***plateau stress*** and ***densification strain*** (also known as ***locking strain***), ε_D. Theoretically, the densification strain should be equal to the porosity, p^*, or from Eq. [10.2]

$$\varepsilon_D = p^* = 1 - \frac{(2 + c/l)h/l}{2\cos\theta(c/l + \sin\theta)} \qquad [10.5]$$

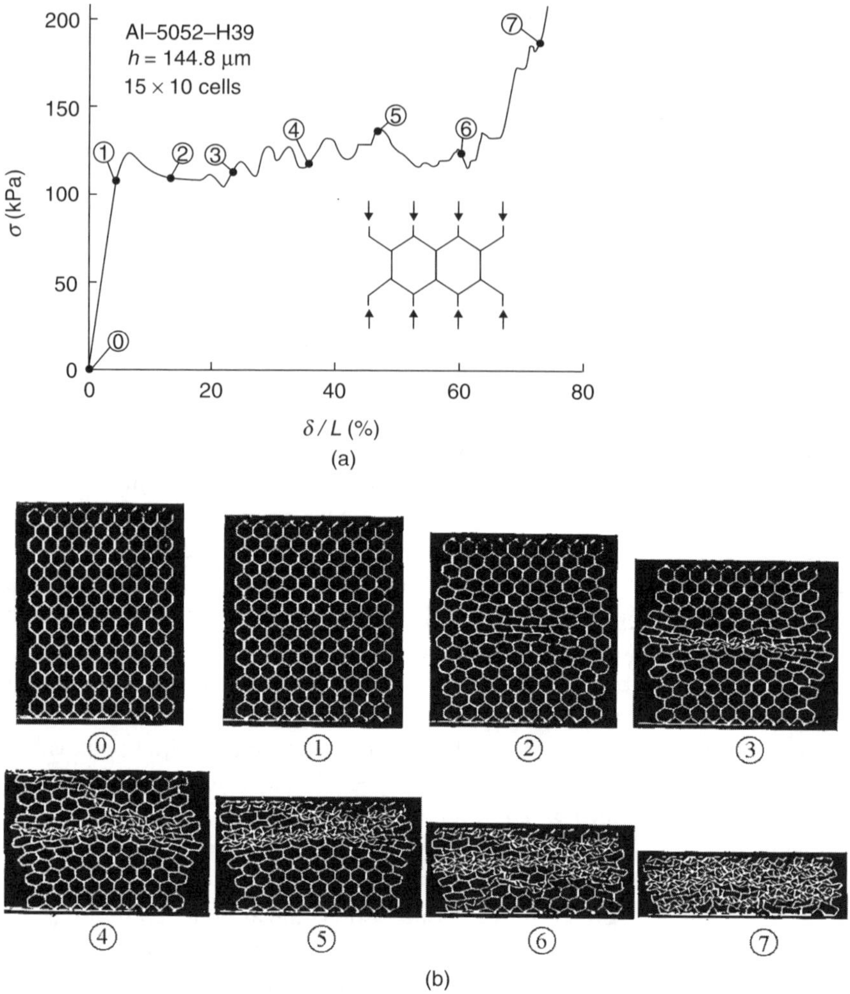

10.3 (a) Experimental load–displacement curve of an aluminium honeycomb crushed in the X_2 direction; (b) successive events of honeycomb under compression. (Papka and Kyriakides, 1998) (reproduced with kind permission of Elsevier).

But in practice, it was found that ε_D is less than that given by Eq. [10.5]. Taking the same empirical factor as that for foams, which will be mentioned later, the densification strain for honeycombs may be given as

$$\varepsilon_D = 1 - 1.4\frac{(2 + c/l)h/l}{2\cos\theta(c/l + \sin\theta)} \qquad [10.6]$$

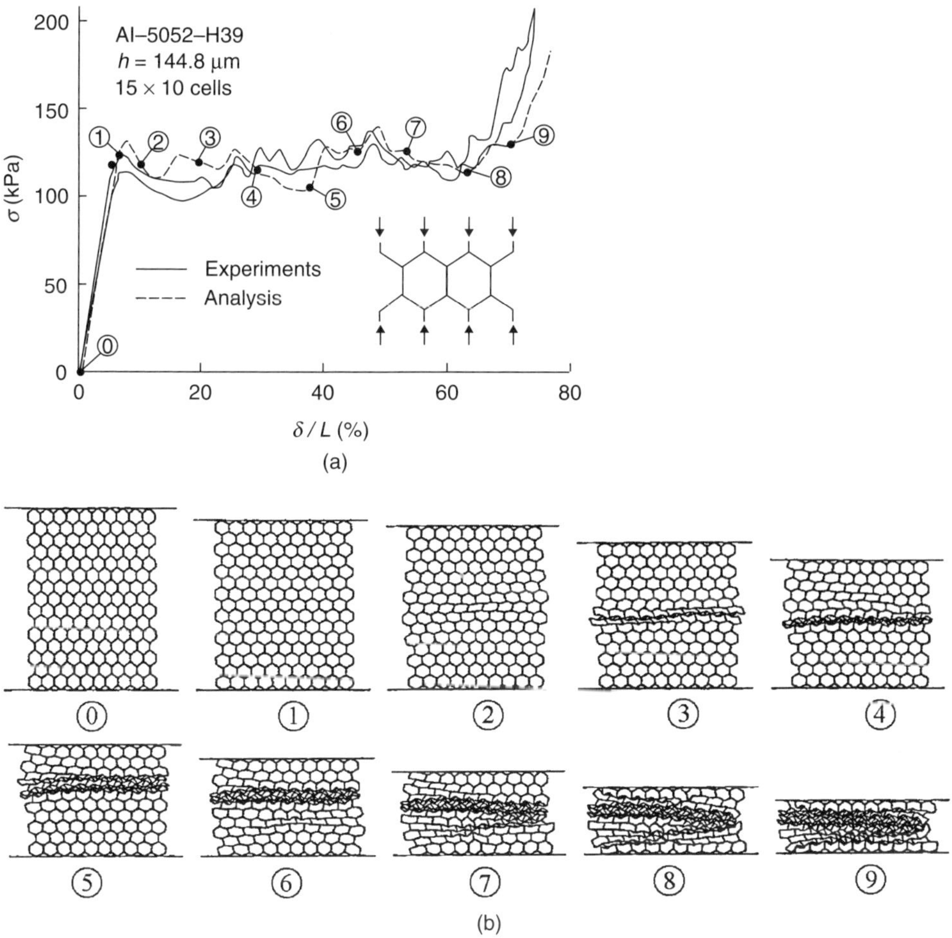

10.4 (a) Load–displacement curves from two experiments and a finite element analysis; (b) successive crushing events of the honeycomb as obtained from the finite element analysis (Papka and Kyriakides, 1998) (reproduced with kind permission of Elsevier).

10.1.2 Plateau stress under in-plane loading

As mentioned before, the plateau stress is governed by the cell failure mechanism. For small values of h/l, elastic buckling of cell walls occurs, as sketched in Fig. 10.5. In this case the vertical walls behave very much like columns under compressive loads. For external stress σ_2, the corresponding column force P is simply, from vertical force equilibrium

$$P = 2\sigma_2 bl\cos\theta \qquad [10.7]$$

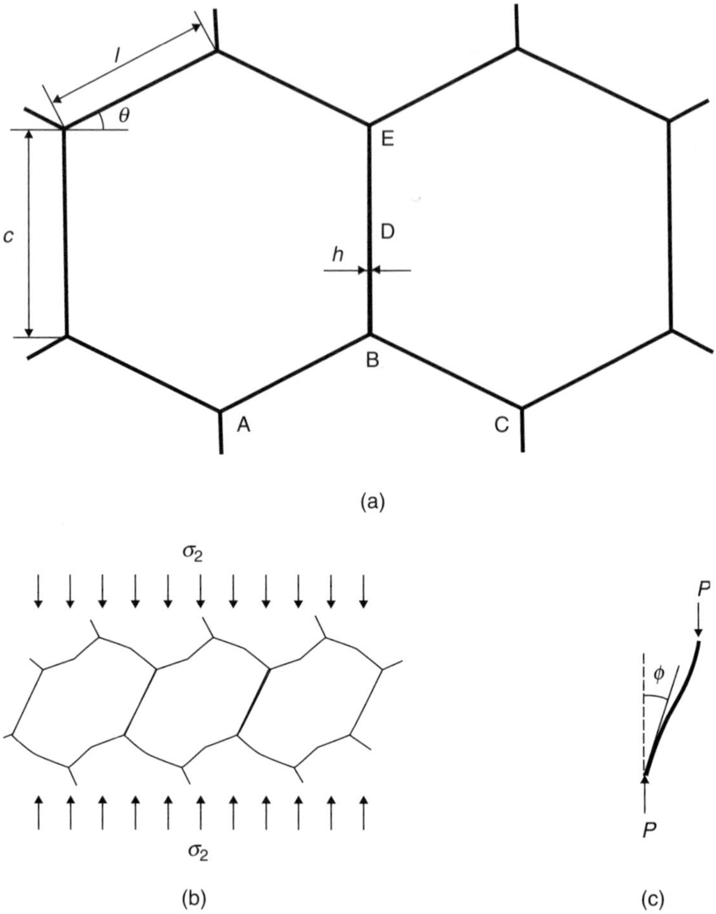

10.5 (a) Cell geometry; (b) and (c) failure of honeycomb cells by elastic buckling of cell walls such as BE, proposed by Gibson and Ashby, 1997.

where b is the breadth of a cell. The Euler buckling load of a column is (Timoshenko and Gere, 1961)

$$P_{cr} = \frac{n^2 \pi^2 E_s I}{c^2} \tag{10.8}$$

where I is the second moment of area, E_s is the elastic modulus of the cell wall solid and $I = bh^3/12$ for the vertical cell walls. The factor n describes the end constraint of the column. When $P = P_{cr}$, elastic buckling occurs. Hence, from Eqs [10.7] and [10.8], we have the critical stress

$$\frac{\sigma_{e2}}{E_s} = \frac{n^2\pi^2}{24}\frac{h^3}{lc^2}\frac{1}{\cos\theta} \quad [10.9]$$

Here subscript $e2$ denotes elastic buckling stress in the X_2 direction. Note that for an end free to rotate, $n = 0.5$. For an end constrained to rotate, $n = 2$. Its theoretical value can be derived for honeycombs (Gibson and Ashby, 1997). It is found that for regular hexagons ($l = c$, $\theta = 30°$), $n = 0.69$. Hence

$$\frac{\sigma_{e2}}{E_s} = 0.22\left(\frac{h}{l}\right)^3 \quad [10.10]$$

This indicates that the non-dimensionalised critical stress is proportional to h/l to the power of 3, when elastic buckling is the failure mechanism of the cell walls. Equation [10.10] agrees well with experimental data for elastomeric honeycombs. Note that elastic buckling in the X_1 direction does not occur as there are no cell walls lying in this direction or under pure compression.

Now, for honeycombs with relatively thick cell walls, plastic collapse of these walls will be the mechanism governing the plateau stress. Hence for each hexagonal cell, six plastic hinges are needed for it to become a collapse mechanism (Fig. 10.6). When beam AB rotates by a small angular increment ϕ under stress σ_1, point B moves inward with respect to point A by $l\sin\theta\phi$. The external work done by σ_1 should be equal to the plastic energy dissipated by hinges A, B, C and D, i.e.

$$2\sigma_{p1}(c + l\sin\theta)bl\sin\theta\phi = 4M_p\phi \quad [10.11]$$

where $\sigma_{p1}(c + l\sin\theta)b$ is the force at B due to σ_{p1}, in the direction of σ_1 and $M_p = (1/4)Y_s h^2 b$ with Y_s being the yield stress of the solid. Therefore

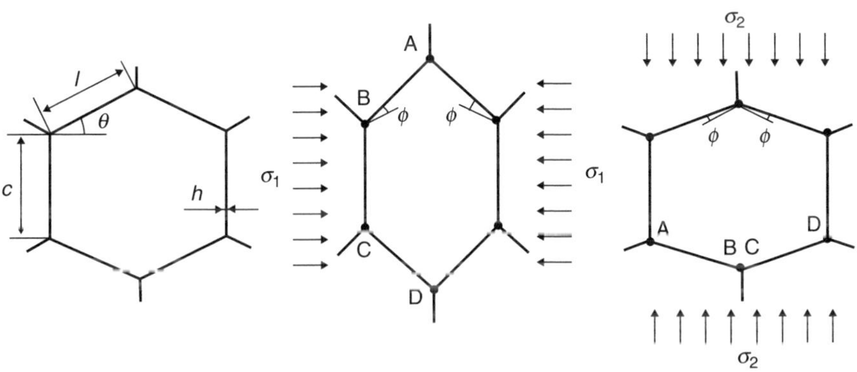

10.6 Failure of honeycomb cells by collapse of cell walls with localised plastic hinges, proposed by Gibson and Ashby, 1997.

$$\frac{\sigma_{p1}}{Y_s} = \left(\frac{h}{l}\right)^2 \frac{1}{2(c/l + \sin\theta)\sin\theta} \tag{10.12}$$

When $l = c$ and $\theta = 30°$

$$\frac{\sigma_{p1}}{Y_s} = \frac{2}{3}\left(\frac{h}{l}\right)^2 \tag{10.13}$$

Similar analysis gives σ_{p2} as

$$\frac{\sigma_{p2}}{Y_s} = \left(\frac{h}{l}\right)^2 \frac{1}{2\cos^2\theta} \tag{10.14}$$

Comparing Eq. [10.14] with Eq. [10.9] for elastic buckling, we can determine when elastic buckling should occur, i.e. $\sigma_{e2} \le \sigma_{p2}$. Hence, the critical thickness is given by

$$\left(\frac{h}{l}\right)_{cr} = \frac{12}{n^2\pi^2\cos\theta}\left(\frac{h}{l}\right)^2 \frac{Y_s}{E_s}$$

or

$$\left(\frac{h}{l}\right)_{cr} = 3\frac{Y_s}{E_s} \tag{10.15}$$

when $l = c$, $\theta = 30°$.

The theoretical analysis given above (Gibson and Ashby, 1997), using a plastic collapse argument, over-predicts the plateau stress when compared with experiment (see Fig. 10.7). A different theoretical expression is obtained when the external work done is expressed directly as a function of σ_1, without involving a force calculation. Thus, Eq. [10.11] becomes

$$\sigma_{p1}(c + 2l\sin\theta)bl\sin\theta\phi = 4M_p\phi \tag{10.16}$$

which leads to

$$\frac{\sigma_{p1}}{Y_s} = \left(\frac{h}{l}\right)^2 \frac{1}{2(c/l + 2\sin\theta)b\sin\theta} \tag{10.17}$$

For the case when $l = c$ and $\theta = 30°$

$$\frac{\sigma_{p1}}{Y_s} = \frac{1}{2}\left(\frac{h}{l}\right)^2 \tag{10.18}$$

Equation [10.18] gives values about 75 % of those from Eq. [10.13], which means that results from Eq. [10.18] would be much closer to the experimental data plotted in Fig. 10.7.

The above method of analysis can be used to study the crushing behaviour of honeycombs under bi-axial stress. Interested readers should refer to

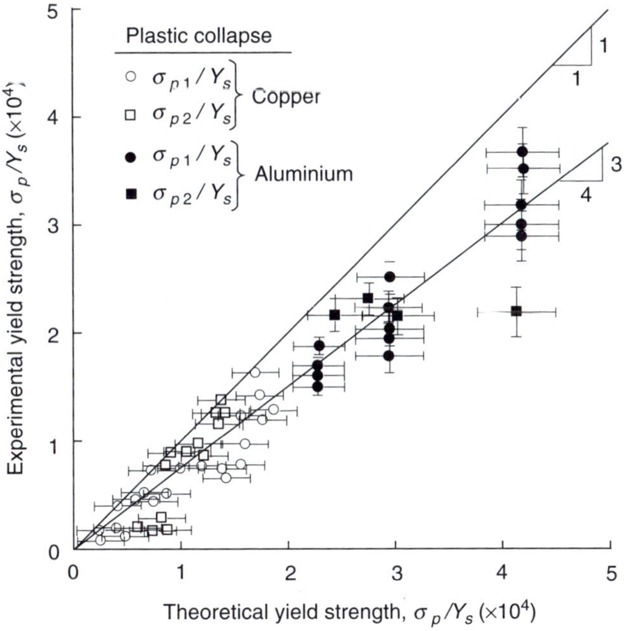

10.7 Comparison between theory Eq. [10.12] and experiment in terms of non-dimensionalised yield strength. When the theoretical results are modified by 75%, they agree with the experimental data much better.

Gibson and Ashby (1997) and Klintworth and Stronge (1988, 1989), for example.

10.1.3 Out-of-plane loading

When crushed in the out-of-plane (X_3) direction, the plateau stress is governed by either elastic buckling or plastic collapse, as for the in-plane loading case. Simple results can be obtained by studying the cell wall behaviour.

For elastic buckling, the cell walls can be treated as flat plates with suitable rotational constraints. The buckling load of individual cell wall plates is summed up, which leads to the overall stress (Gibson and Ashby, 1997)

$$\sigma_{e3} \approx \frac{2}{1-v_s^2} \frac{l/c+2}{(c/l+\sin\theta)\cos\theta} \left(\frac{h}{l}\right)^3 \quad [10.19]$$

or, for regular hexagons with Poisson's ratio $\upsilon_s = 0.3$

$$\sigma_{e3} = 5.2\left(\frac{h}{l}\right)^3 \quad [10.20]$$

For plastic collapse, an analysis similar to that for a rectangular tube under axial loading (Section 6.2), but with $\psi = \pi/6$ was performed by Wierzbicki (1983). This takes into account both stretching and bending deformation. The mean crushing stress is, for $l = c$ and $\theta = 30°$

$$\frac{\sigma_{p3}}{Y_s} = 5.6\left(\frac{h}{l}\right)^{\frac{5}{3}} \tag{10.21}$$

The power is 5/3, instead of being 1 or 2, reflecting the combined effect of bending and stretching. Analysis of plastic bending alone of the cell walls gives (Gibson and Ashby, 1997)

$$\frac{\sigma_{p3}}{Y_s} \approx \frac{\pi}{4}\frac{c/l+2}{4(c/l+\sin\theta)\cos\theta}\left(\frac{h}{l}\right)^3 \tag{10.22}$$

or

$$\frac{\sigma_{p3}}{Y_s} \approx 2\left(\frac{h}{l}\right)^2 \tag{10.23}$$

when $l = c$, $\theta = 30°$.

10.2 Foams

10.2.1 Cell structure, relative density, stress–strain curves and densification strain

In honeycombs, the cells are two-dimensional. Cellular materials with three-dimensional cells are called foams. When cells are connected by beam-type edges only, i.e. fluids can flow among the cells, they are called ***open-cell***. On the other hand, when a cell is fully enclosed with cell walls, so that there is no passage for a fluid to flow among the cells, this is called ***closed-cell***. Two examples are shown in Fig. 10.8. It is possible for a foam to have both open and closed cells. Polyhedral cells which can be packed to fill a space include triangular, rhombic and hexagonal prisms, the rhombic dodecahedron and the tetrakaidecahedron (Gibson and Ashby, 1997). However, for the purpose of the present discussion, the parameter relative density seems sufficient to characterise foams. For open-cell foams

$$\frac{\rho^*}{\rho_s} = C_2\left(\frac{h}{l}\right)^2 \tag{10.24}$$

and for closed-cell foams

$$\frac{\rho^*}{\rho_s} = C_3\frac{h}{l} \tag{10.25}$$

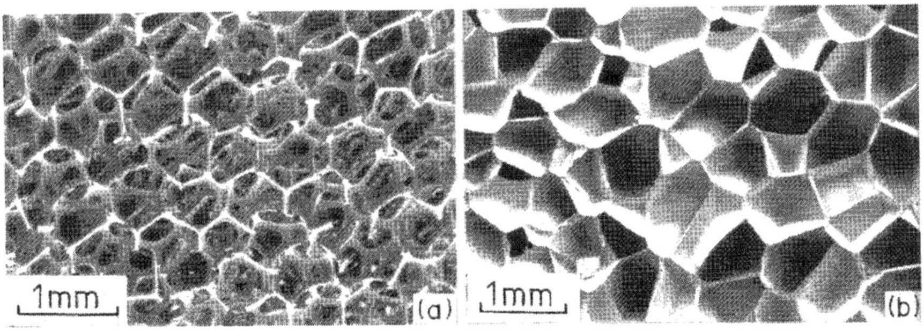

10.8 Example of open cell (left, polyurethane) and closed cell (right, polyethylene) foams (Gibson and Ashby, 1997) (reproduced with kind permission of Cambridge University Press).

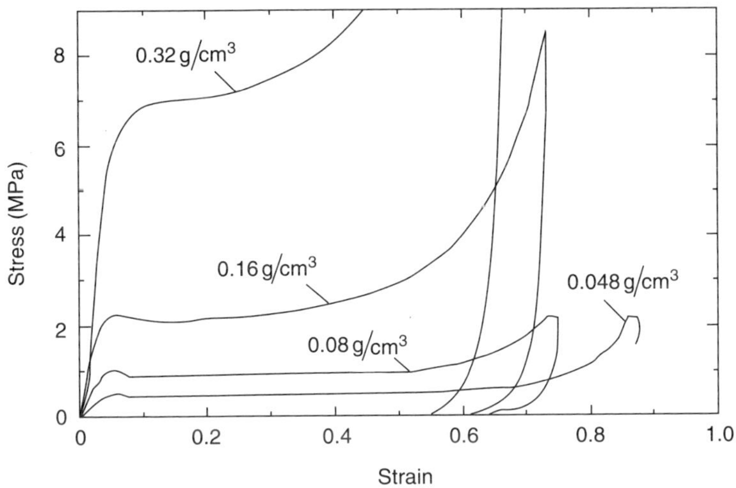

10.9 Stress–strain curves for closed cell rigid polyurethane foams of various densities (reproduced with kind permission of American Society of Civil Engineers).

Here, as for honeycombs in Eq. (10.1), C_2 and C_3 are numerical constants depending on the cell shape.

The response of foams and their theoretical treatment parallels what has already been stated for honeycombs. Typical compressive stress–strain curves are shown in Fig. 10.9 for closed-cell rigid polyurethane foams of various different densities (Maji *et al.*, 1995). Broadly, each curve has three stages: linear-elastic response, yielding with a plateau stress and densification when the stress increases rapidly with strain. As density increases, the initial elastic modulus and the plateau stress increase, but the densification strain reduces. As for honeycombs, the densification strain for both open- and closed-cell foams is

$$\varepsilon_D = 1 - 1.4\left(\frac{\rho^*}{\rho_s}\right) \qquad [10.26]$$

The coefficient 1.4 is obtained from a number of experiments.

10.2.2 Plateau stress

As for honeycombs, the plateau stress is governed by the failure mechanism of the foam cells: elastic buckling, plastic collapse or fracture. One additional contribution for closed-cell foams comes from compression of the air or fluid trapped within the cells, which enhances the average plateau stress.

The elastic buckling of open-celled foam may be studied by means of an idealised cell structure, Fig. 10.10(a) (Gibson and Ashby, 1997). The Euler buckling load of a strut is given by Eq. [10.8]. Hence, the corresponding nominal stress is

$$\sigma_e \propto \frac{P_{cr}}{l^2} \propto \frac{E_s I}{l^4} \propto E_s\left(\frac{h}{l}\right)^4 \qquad [10.27]$$

Note that $I \propto h^4$. Because $\rho^*/\rho_s \propto (h/l)^2$ for open-celled foams, we have

$$\frac{\sigma_e}{E_s} \propto \left(\frac{\rho^*}{\rho_s}\right)^2 \qquad [10.28]$$

Refinements to the above equation can be made when the corners occupy a significant portion of the volume. By fitting the analysis to experimental data, the equations are

$$\frac{\sigma_e}{E_s} = 0.05\left(\frac{\rho^*}{\rho_s}\right)^2 \qquad [10.29]$$

and more accurately

$$\frac{\sigma_e}{E_s} = 0.03\left(\frac{\rho^*}{\rho_s}\right)^2\left[1+\left(\frac{\rho^*}{\rho_s}\right)^{\frac{1}{2}}\right]^2 \qquad [10.30]$$

For closed-celled foams, the initial pressure within a cell, p_o, may cause a tension of $p_o - p_{atm}$ in the cell walls, p_{atm} being the atmospheric pressure. The external stress must overcome this tension first, before causing cell wall buckling. Hence, by modifying Eqs [10.29] and [10.30], respectively

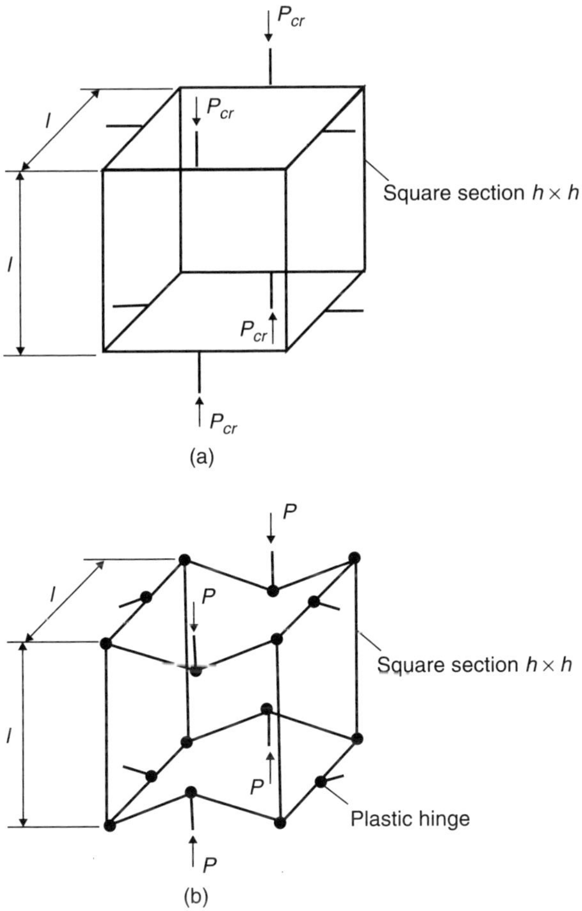

10.10 A highly idealised open cell under: (a) elastic buckling of the cell walls; (b) plastic collapse of cell walls.

$$\frac{\sigma_e}{E_s} = 0.05\left(\frac{\rho^*}{\rho_s}\right)^2 + \frac{p_o - p_{atm}}{E_s} \qquad [10.31]$$

and

$$\frac{\sigma_e}{E_s} = 0.03\left(\frac{\rho^*}{\rho_s}\right)^2\left[1+\left(\frac{\rho^*}{\rho_s}\right)^{\frac{1}{2}}\right]^2 + \frac{p_o - p_{atm}}{E_s} \qquad [10.32]$$

When the foam is further crushed, the fluid trapped within the cells exerts a larger pressure because of the decrease in cell volume, which can be evaluated from Boyle's law. Consequently, the stress is also related to strain ε

$$\frac{\sigma}{E_s} = 0.05\left(\frac{\rho^*}{\rho_s}\right)^2 + \frac{p_o - p_{atm}}{E_s(1-\varepsilon-\rho^*/\rho_s)} \qquad [10.33]$$

The idealised cell in Fig. 10.10(a) can also be used to analyse the collapse of the cell, Fig. 10.10(b). Because the fully plastic bending moment $M_p \propto Y_s h^3/4$, the force $P \propto M_p/l \propto Y_s h^3/l$. Consequently, the nominal stress σ_p is

$$\sigma_p \propto \frac{P}{l^2} \propto Y_s \frac{h^3}{l^3}$$

Because $\rho^*/\rho_s \propto (h/l)^2$ (Eq. [10.24]), we have

$$\frac{\sigma_p}{Y_s} \propto \left(\frac{\rho^*}{\rho_s}\right)^{\frac{3}{2}} \qquad [10.34]$$

Fitting this equation (and its refined form) obtained for open-cell foams to experimental results

$$\frac{\sigma_p}{Y_s} = 0.3\left(\frac{\rho^*}{\rho_s}\right)^{\frac{3}{2}} \qquad [10.35]$$

and

$$\frac{\sigma_p}{Y_s} = 0.23\left(\frac{\rho^*}{\rho_s}\right)^{\frac{3}{2}}\left[1+\left(\frac{\rho^*}{\rho_s}\right)^{\frac{1}{2}}\right] \qquad [10.36]$$

The plastic collapse of closed-cell foams involves not only bending of cell edges, but also stretching of cell walls; the latter contributes to the stress as $\propto \rho^*/\rho_s$. Let the volume fraction of the cell edges be ϕ; then the remaining fraction of the solid, $1 - \phi$, is due to the cell walls. The collapse strength of closed-cell plastic foams is

$$\frac{\sigma_p}{Y_s} = 0.3\left(\phi\frac{\rho^*}{\rho_s}\right)^{\frac{3}{2}} + (1-\phi)\frac{\rho^*}{\rho_s} + \frac{p_o - p_{atm}}{Y_s} \qquad [10.37]$$

Santosa and Wierzbicki (1998a) used a truncated cube as an idealised closed cell and performed analytical and finite element analyses. They obtained

$$\frac{\sigma_p}{Y_s} = 0.63\left(\frac{\rho^*}{\rho_s}\right)^{\frac{2}{3}} + 0.07\frac{\rho^*}{\rho_s} + 0.80\left(\frac{\rho^*}{\rho_s}\right)^2 \qquad [10.38]$$

They approximated the finite element results using

(a)

10.11 A typical CYMAT aluminium foam (a) with its stress–strain curve (b). Non-uniform deformation occurs (c) (Ruan *et al.*, 2002) (reproduced with kind permission of Elsevier).

$$\frac{\sigma_p}{Y_s} = 1.05\left(\frac{\rho^*}{\rho_s}\right)^{1.52} \qquad [10.39]$$

which has almost the same power as in Eq. [10.35], but the coefficient 1.05 is about three times the 0.3 in Eq. [10.35]. Other constitutive equations for a general stress–strain curve were proposed (e.g. Chang *et al.*, 1998).

10.2.3 Metal foams

Recently, metal foams have emerged as a new class of materials with great potential as energy-absorbing structures, among other applications. Metal foams based on aluminium or nickel are the most commonly used at present. Ashby *et al.* (2000) have summarised present knowledge on metal foams. Here we briefly present some results with respect to their energy-absorption performance.

A typical CYMAT aluminium foam and its stress–strain curve are shown in Fig. 10.11 (Ruan *et al.*, 2002). The characteristics discussed previously are present in this plot. Furthermore, because the cells are not uniformly distributed, localised deformation occurs, starting from the weakest location (Fig. 10.11(c)) (Ruan *et al.*, 2002). Note that in this figure, crushed zones A, B and C were of the same initial size.

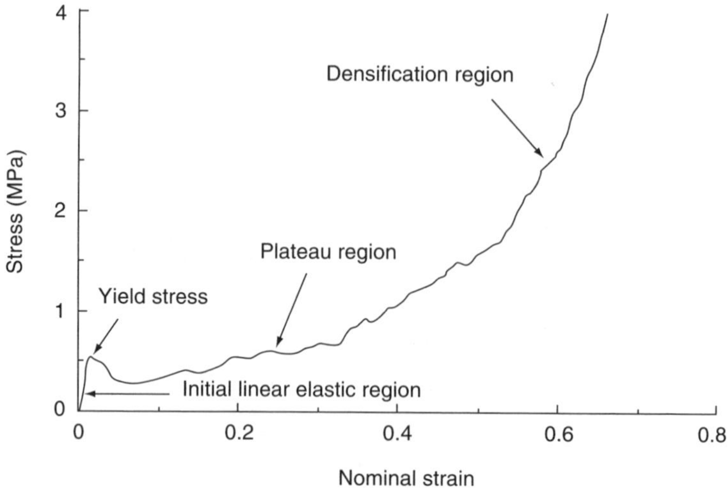

(b)

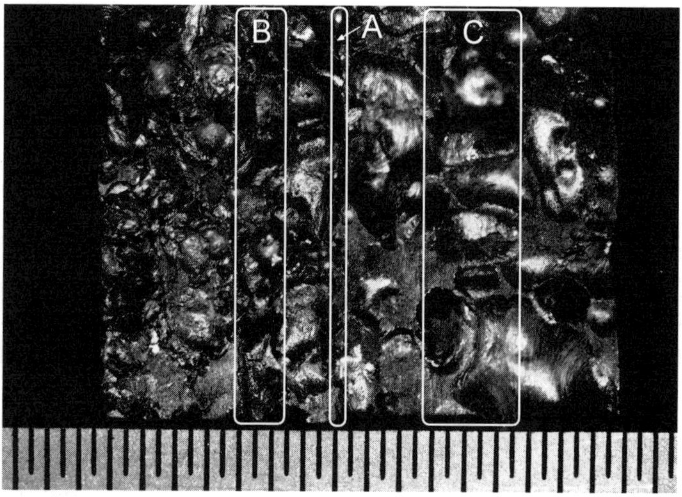

(c)

10.11 Continued.

In the design of structures for energy absorption, the peak load or stress is a major consideration. A peak or plateau stress which is too high will cause damage to goods in their packaging, or severe injury to passengers in vehicles, for example. Therefore, the energy-absorption performance of cellular material can be presented by plotting the specific energy against the plateau stress (Fig. 10.12, for metal foams) (Ashby *et al.*, 2000). Hence for a practical application, with maximum allowable stress specified, a metal

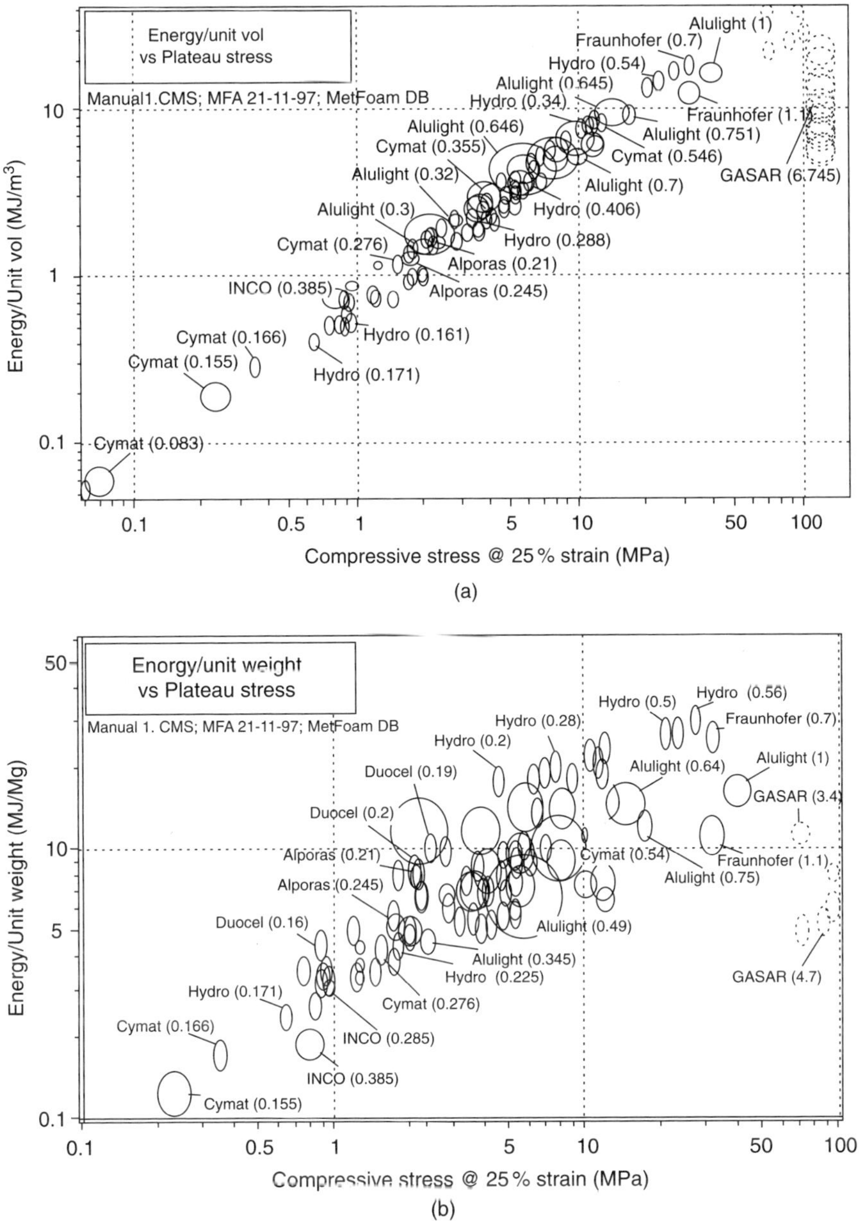

10.12 Energy absorption of metal foams plotted against the plateau stress (taken as the compressive strength at 25% strain): (a) energy per unit volume; (b) energy per unit weight. The value of density is given in Mg/m^3 (Ashby *et al.*, 2000) (reproduced with kind permission of Elsevier).

foam with the best specific energy absorption can be readily selected. More discussion on this type of plot will be given in Section 12.2.

10.3 Wood

Wood is a natural cellular material and can be used for energy absorption. It has three orthotropic planes (Fig. 10.13): radial, tangential and axial. Its microstructure along the axial direction is very different from that of the other two planes. Figure 10.14 shows those for cedar (Gibson and Ashby, 1997). It consists of highly elongated cells whose cross-section is often hexagonal. In the radial direction, arrays of cells which are smaller and more rectangular than others form radial rays (Fig. 10.13). This makes the compressive strength in the radial direction stronger than that in the tangential direction (by about 40 %).

Typical stress–strain curves are shown in Fig. 10.15(a) for axial stress and Fig. 10.15(b) for radial stress, for oak, redwood, pine and balsa (Reid and Peng, 1997). They exhibit general characteristics of the stress–strain curves of honeycombs and foams discussed earlier. One may note that an initial peak stress is present when compressed in the axial direction, as a result of buckling of the column-like fibres shown in Fig. 10.14. Again, the most relevant parameters in energy absorption are the plateau stress and the densification strain.

The axial collapse mechanism is very complicated. Nevertheless, most woods are of high density and plastic yielding occurs first. This leads to axial stress $\sigma_A \propto \rho^*/\rho_s$ (Gibson and Ashby, 1997) or

$$\frac{\sigma_A}{Y_s} = c\left(\frac{\rho^*}{\rho_s}\right) = 0.34\left(\frac{\rho^*}{\rho_s}\right) \qquad [10.40]$$

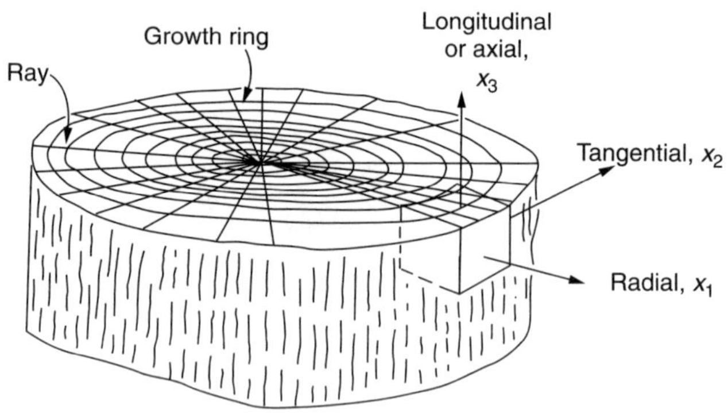

10.13 Definition of the axial, radial and tangential directions for a tree trunk (reproduced with kind permission of Cambridge University Press).

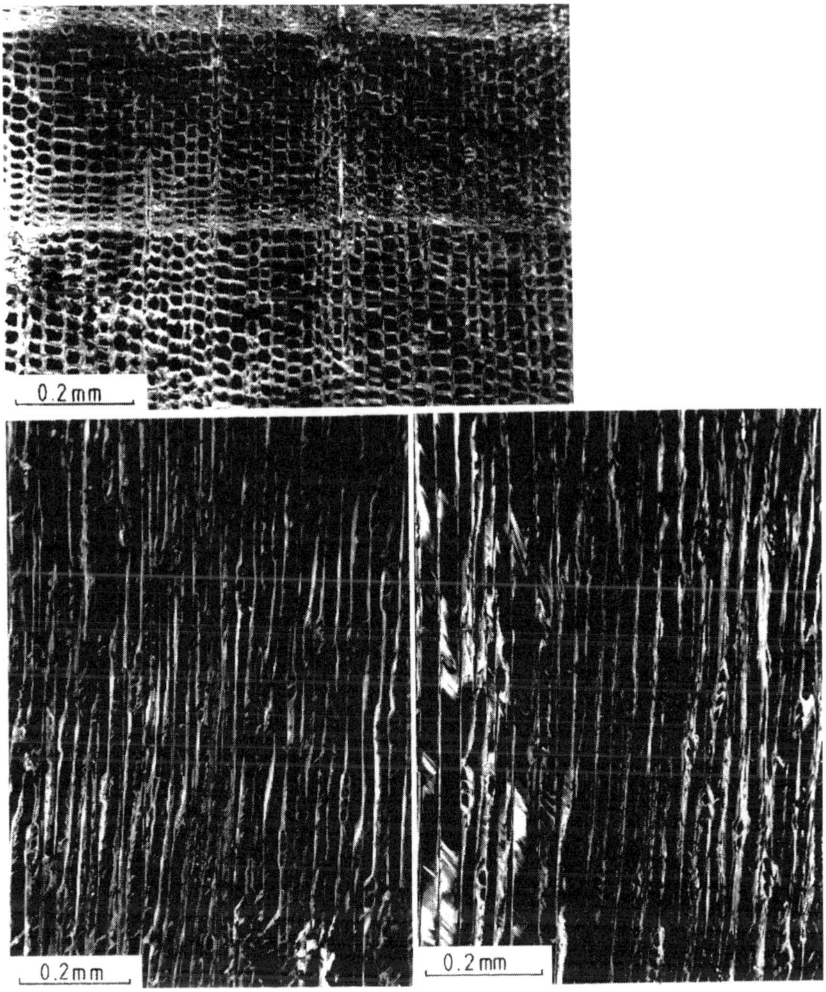

10.14 Photographs showing microstructures of cedar in the three orthogonal directions (Gibson and Ashby, 1997) (reproduced with kind permission of Cambridge University Press).

where Y_s is the yield strength of the cell wall solid. The coefficient 0.34 is obtained by fitting the above formula to experimental results. Alternatively, σ_A in MN/m^2 is

$$\sigma_A = 120\frac{\rho^*}{\rho_s} \qquad [10.41]$$

Reid *et al.* (1993) used a coefficient of 150 instead of 120 in the above equation to match their data.

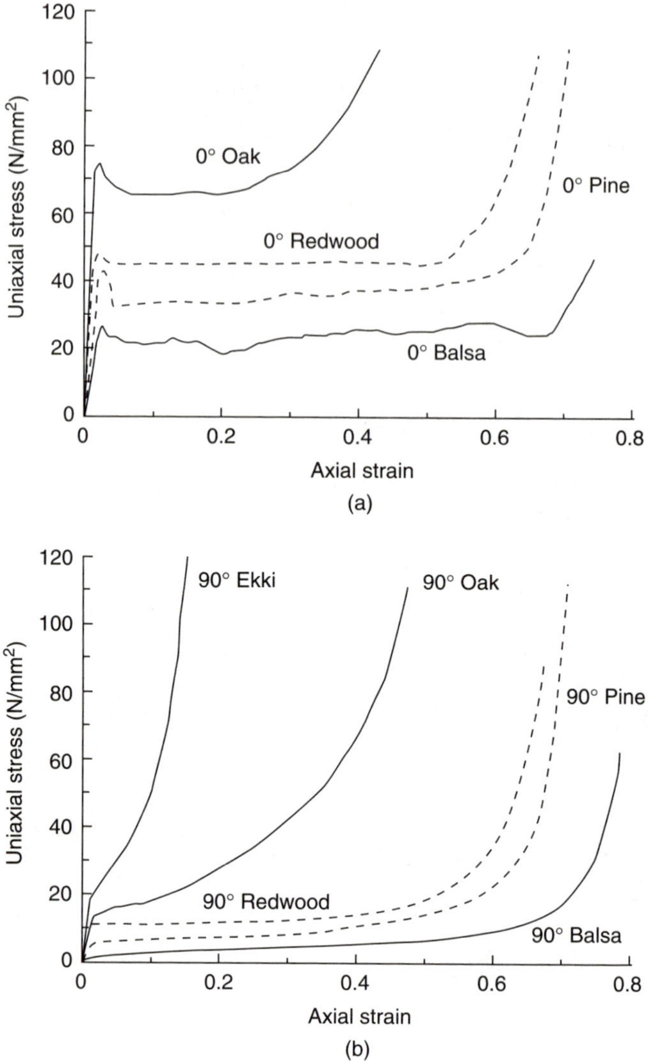

10.15 Uniaxial compressive stress–strain curves for several woods: (a) axial direction; (b) radial direction (reproduced with kind permission of Elsevier).

The collapse mechanism in the radial and tangential direction is largely dominated by plastic bending of cell walls. Hence as for honeycombs we have, for tangential strength

$$\frac{\sigma_t}{Y_s} \propto \left(\frac{h}{l}\right)^2 \qquad [10.42]$$

or

$$\frac{\sigma_t}{Y_s} \propto \left(\frac{\rho^*}{\rho_s}\right)^2 \qquad [10.43]$$

Again, empirically,

$$\frac{\sigma_t}{Y_s} = 0.14\left(\frac{\rho^*}{\rho_s}\right)^2 \qquad [10.44]$$

or

$$\sigma_t = 50\left(\frac{\rho^*}{\rho_s}\right)^2 \quad \text{MN/m}^2 \qquad [10.45]$$

and the radial stress is

$$\sigma_R = 1.4\sigma_t = 70\left(\frac{\rho^*}{\rho_s}\right)^2 \quad \text{MN/m}^2 \qquad [10.46]$$

The densification strain can be expressed in the same form as Eq. [10.26], but the coefficient (which was 1.4) is now 2 for low density woods, 1.3 for oak and 1.35 for pine and redwood (Reid *et al.*, 1993).

10.4 Impact response of cellular materials

10.4.1 Rigid, perfectly-plastic shock theory

As mentioned in Fig. 2.15 and discussed in Section 4.6, when the stress–strain curve is convex about the strain axis, the subsequent plastic stress wave travels at an increasing speed, leading to a shock wave front. This concept has been used in analysing the dynamic response of one-dimensional ring systems in Section 4.6. It is now clear that almost all of the cellular solids considered earlier have such a stress–strain curve and hence shock wave theory should apply here too (Reid and Peng, 1997; Ashby *et al.*, 2000).

Consider a mass G with velocity V_o impacting an initially stationary cylinder (Fig. 10.16(b)). The actual stress–strain curve is idealised as rigid, perfectly-plastic which locks at the densification strain, ε_D (Fig. 10.16(a)). A plastic wave front develops, travelling at velocity c_p. Upstream of this wave front, the material is stationary with stress σ_p. Downstream of the wave front, the material is compacted at the densification strain ε_D with a density $\rho_D = \rho/(1 - \varepsilon_D)$. The material also moves at the same velocity as the instantaneous velocity of the mass G, which decreases with time. The stress level for the dense material at the wave front jumps to σ_D, which varies with the instantaneous velocity v_D, as will be seen later. Consider the instant when the current length of the compacted cylinder is l. The corresponding initial

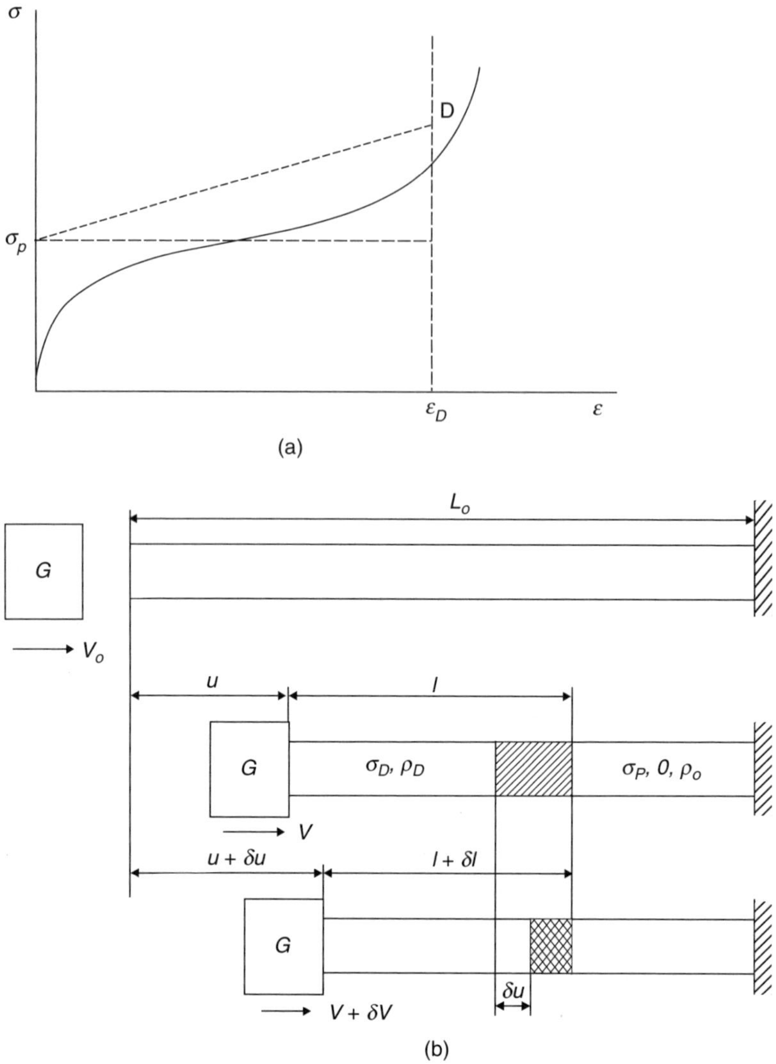

10.16 Rigid, perfectly plastic shock theory: (a) a typical stress–strain curve for cellular material, which is idealised into rigid, perfectly plastic followed by a complete locking; (b) sketch of a mass G impacting an initially stationary cylinder.

length is $l_o = l/(1 - \varepsilon_D)$. The plastic work done in compressing this length is $\sigma_p \varepsilon_D Al/(1 - \varepsilon_D)$. Here A is the cross-sectional area, which is assumed to be constant. Consideration of energy balance gives

$$\frac{1}{2}\left(G+\frac{\rho_o}{1-\varepsilon_D}Al\right)v_D^2+\sigma_p\varepsilon_D A\frac{1}{1-\varepsilon_D}=\frac{1}{2}GV_o^2 \qquad [10.47]$$

The plastic wave speed c_p is given by (Fig. 10.16(a))

$$c_p = \sqrt{\frac{(\sigma_D - \sigma_p)/\varepsilon_D}{\rho_o}} \qquad [10.48]$$

and the particle velocity is $v_D = c_p\varepsilon_D$ (see Eq. [2.54]). Conservation of momentum for an element at the wave front gives, for a time increment δt

$$(\sigma_D - \sigma_p)A\delta t = \frac{\rho_o A c_p \delta t v_D}{1-\varepsilon_D} \qquad [10.49]$$

Hence

$$\sigma_D = \sigma_p + \frac{\rho_o c_p v_D}{1-\varepsilon_D} \qquad [10.50]$$

Solving Eqs [10.47] and [10.50]

$$\sigma_D = \sigma_p + \frac{\rho_o}{\varepsilon_D}\frac{GV_o^2 - 2\sigma_p A l \varepsilon_D/(1-\varepsilon_D)}{G + \rho_o A l/(1-\varepsilon_D)} \qquad [10.51]$$

The above equation indicates that, as the length of the compacted cylinder l increases, the stress σ_D decreases. When $l = 0$, Eq. [10.51] gives the initial maximum stress:

$$\sigma_D = \sigma_p + \frac{\rho_o V_o^2}{\varepsilon_D} \qquad [10.52]$$

This suggests that the peak stress is enhanced under impact loading purely as a result of this inertia effect. The enhancement $(\sigma_D - \sigma_p)$ is proportional to the initial velocity squared.

Reid and Peng (1997) considered the case when both the mass G and the cylinder have the same initial velocity and then they impact onto a rigid flat anvil. The result for peak stress is identical to that given by Eq. [10.52].

Plastic shock theory is applicable to cellular materials such as honeycombs, foams and woods, provided that the impact velocity is sufficiently high and a plane plastic wave front develops. Theoretical prediction agrees well with the experimental results. Comparisons are given in Fig. 10.17 for two types of wood: balsa and pine (Reid and Peng, 1997), in terms of the ratio of σ_D/σ_p.

10.4.2 One-dimensional mass-spring model

The overall stress–strain curves of low density cellular materials may be reasonably idealised as rigid, perfectly plastic before densification, as treated in the previous section. Nevertheless, plastic crushing of these materials is progressive and the cells collapse layer by layer (Fig. 10.3). This

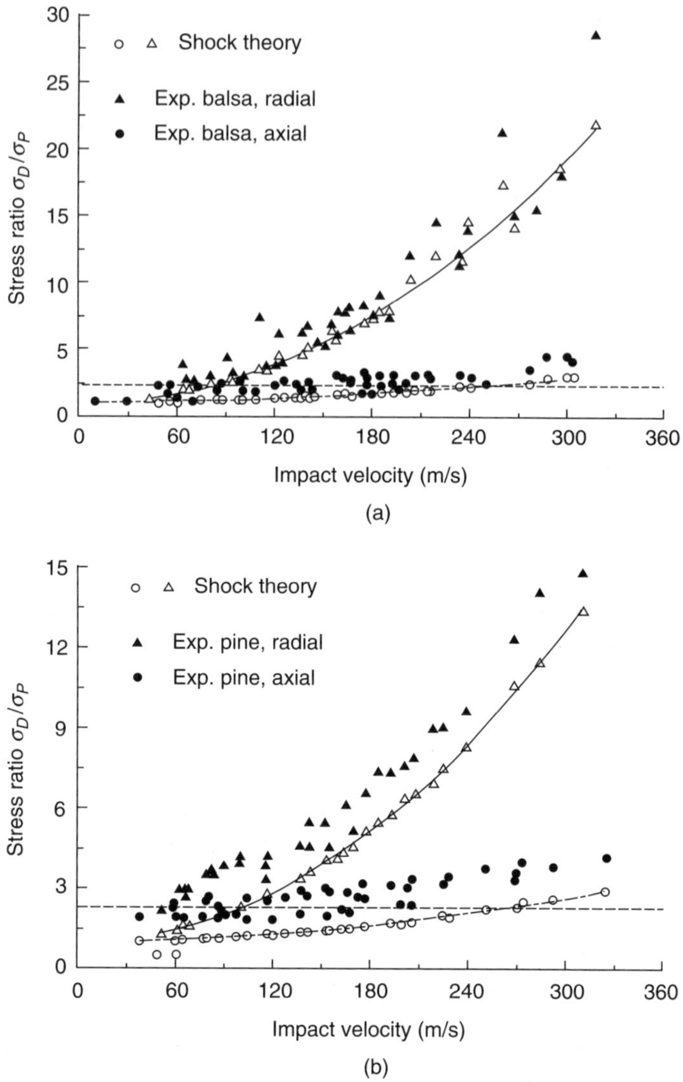

10.17 Comparison between the theory and experiment: (a) balsa; (b) pine (Reid and Peng, 1997) (reproduced with kind permission of Elsevier).

involves fluctuations in force, the degree of which depends on the cell geometry and post-collapse mode. In order to capture this localised deformation detail and to examine the dynamic effect on such a system, a one-dimensional mass-spring model was proposed (Shim *et al.*, 1990). This model may be regarded as a generalised version of the one for ring systems presented in Chapter 4.

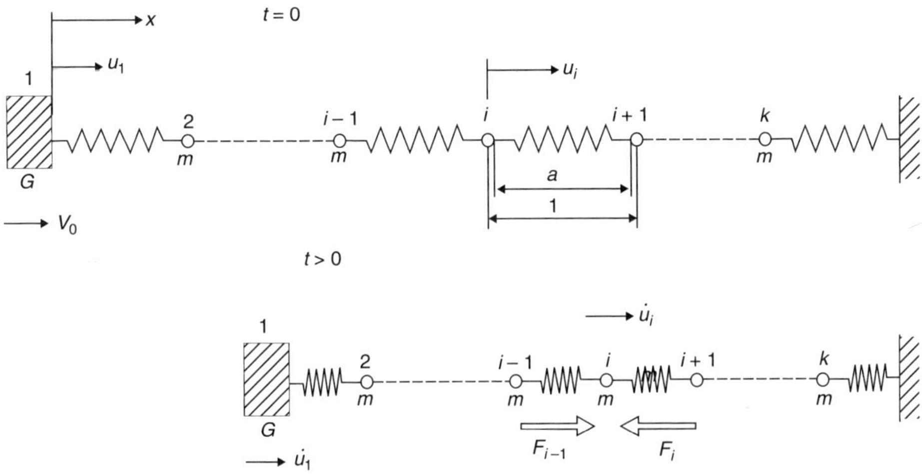

10.18 A one-dimensional mass-spring model (reproduced with kind permission of American Society of Mechanical Engineers).

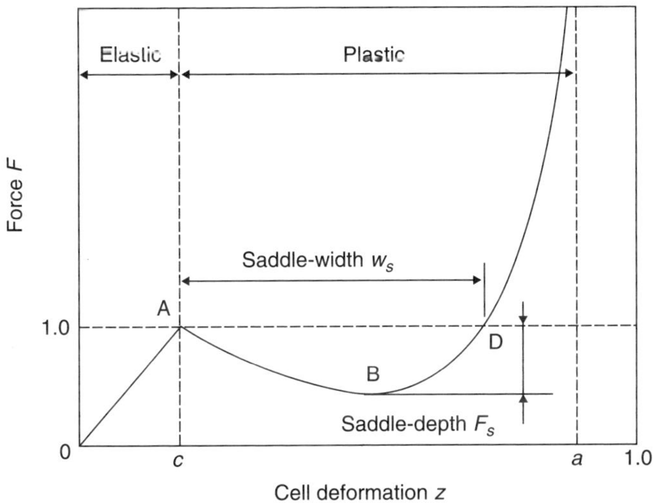

10.19 Force–displacement characteristic of the spring (reproduced with kind permission of American Society of Mechanical Engineers).

Consider the mass-spring system in Fig. 10.18, under impact of a mass G. The mass of each cell (such as a honeycomb cell) is lumped into a point mass m and a massless spring describing the crushing characteristic of each cell. For a general case, the spring characteristics are shown in Fig. 10.19; and may be described by the following:

$$\frac{F}{F_{01}} = \begin{cases} \dfrac{z}{c} & ;0 \le z < c \text{ (elastic)} \\ \exp\left[\dfrac{a(z-c)}{(a-z)^n} - b(z-c)\right] & ;c \le z \le a \text{ (post-yield)} \end{cases} \qquad [10.53]$$

where z is cell deformation. The shape of the non-dimensionalised force F/F_o and cell deformation is therefore governed by parameters a, b, c and n. Adjusting their values leads to different spring characteristics such as softening or hardening. The governing equations are

$$G\ddot{u}_1 + F(z_1) = 0 \qquad [10.54]$$

$$m\ddot{u}_1 + F(z_i) - F(z_{i-1}) = 0 \qquad (i = 1,2,3,\ldots,k) \qquad [10.55]$$

where the amount of cell crushing is $z_i = u_i - u_{i+1}$. The u_i terms are local coordinates for the springs. Global position coordinates x_i are given by $x_i = u_{(i-1)} + u_i$. Initial conditions are

$$\dot{u}_1(0) = V_o \qquad [10.56]$$

$$\dot{u}_i(0) = 0 \qquad i = 2,3,\ldots,k \qquad [10.57]$$

$$x_i(0) = i - 1 \qquad i = 1,2,3,\ldots,k \qquad [10.58]$$

After initial yielding, any unloading is assumed to be purely inelastic.

The above equations can be solved numerically to obtain the non-dimensional mass position, cell deformation and impact mass deceleration, see Figs 10.20 (a), (b) and (c), respectively. The number of cells k = 10, m/G = 0.1, V_o = 5, saddle-width w_s = 0.507 (see Fig. 10.19).

In this model, the mass m is assumed to be constant. Gao *et al.* (2003a and b) recognised that the effective mass of a cell of general shape and the collapse mechanism can vary, see e.g. Section 7.2. Hence they proposed a variable mass-spring model in order to capture the dynamic response of cellular materials. The inertia effect of the cells (micro-inertia) was discussed by Stronge and Shim (1988).

10.4.3 Dynamic crushing of honeycombs

Crushing modes

The above shock model applies to the case when a plane plastic wave front develops and propagates. This occurs when the impact velocity is sufficiently high. At low velocities, deformation may be localised too, but within a band which is inclined to the impact face. D. Ruan *et al.* (2003) studied, by means of FE analysis, the in-plane response of honeycombs subjected to impact by a massless rigid plate of constant velocity during crushing. The FE model

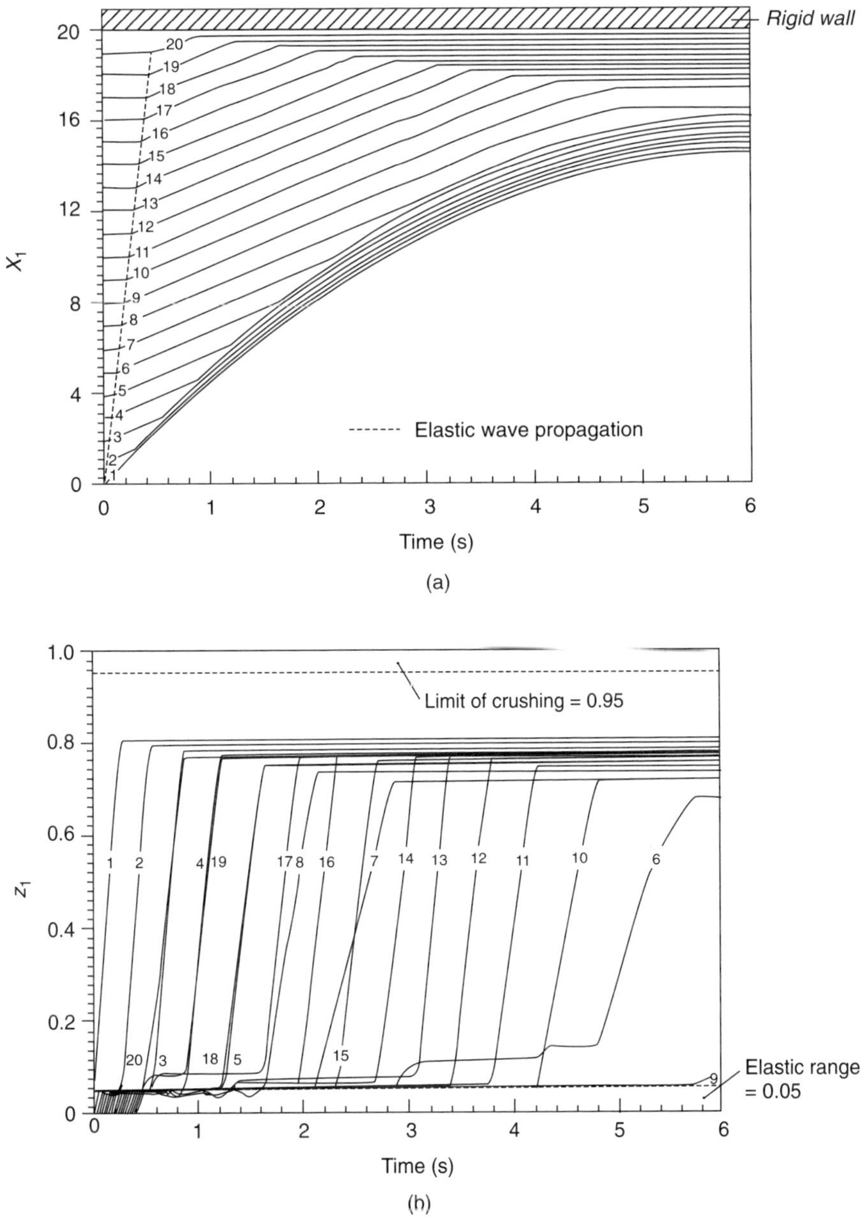

10.20 Results from the mass-spring model: (a) mass position; (b) cell deformation; (c) mass deceleration – the reference deceleration is $a_o = F_o/G$ (Shim *et al.*, 1990) (reproduced with kind permission of American Society of Mechanical Engineers).

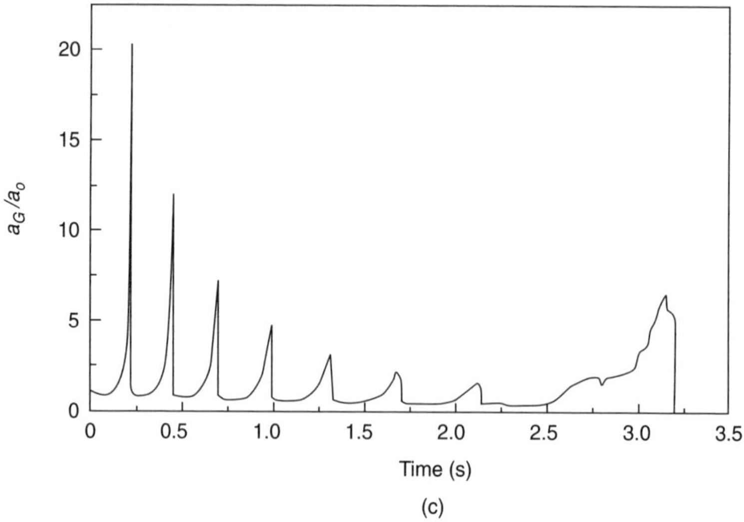

10.20 Continued.

used has 16 cells in the X_1 direction and 15 cells in the X_2 direction. The cell size is $s = 4.7$ mm and $l = 2.7$ mm (corresponding to $\theta = 30°$) for all cases; the cell wall thickness h varies from 0.08 mm to 0.5 mm.

ABAQUS/EXPLICIT was employed for the dynamic analysis. The cell wall material was assumed to be elastic, perfectly plastic with a Young's modulus of 69 GPa and a yield stress of 76 MPa. Each edge of the cell wall was modelled by three shell elements (type S4R). The model consisted of a total of 2280 shell elements. Each hexagonal cell was defined as a single self-contact surface. Self contact was also defined between the outside faces of the cell, which might contact other cells during crushing.

The impact velocity v of the rigid plate varied from 3.5 m/s to 280 m/s in order to study the effect of loading rate. When crushed in the X_1 direction, all degrees of freedom of the left edge of the specimen were fixed and the top and bottom edges were free. A horizontal constant velocity (along the X_1 direction) was applied to the right face of the striking plate. Similarly, for impact in the vertical (X_2) direction, all degrees of freedom of the bottom edge were fixed and the left and right edges were free. A vertical constant velocity was applied to the top face of the rigid striking plate.

Figures 10.21 and 10.22 show deformation modes in the X_1 direction with h = 0.2 mm, for the two cases where impact velocity v = 3.5 m/s and v = 70 m/s, respectively. For v = 3.5 m/s, initial localisation occurs when the displacement of the right edge is small (Fig. 10.21(a)) and this produces an 'X' shaped band starting from the struck end. With an increase of displacement, a second localised 'X' band is developed from the fixed edge, and this intersects with the first localisation band to form a rhombus at the

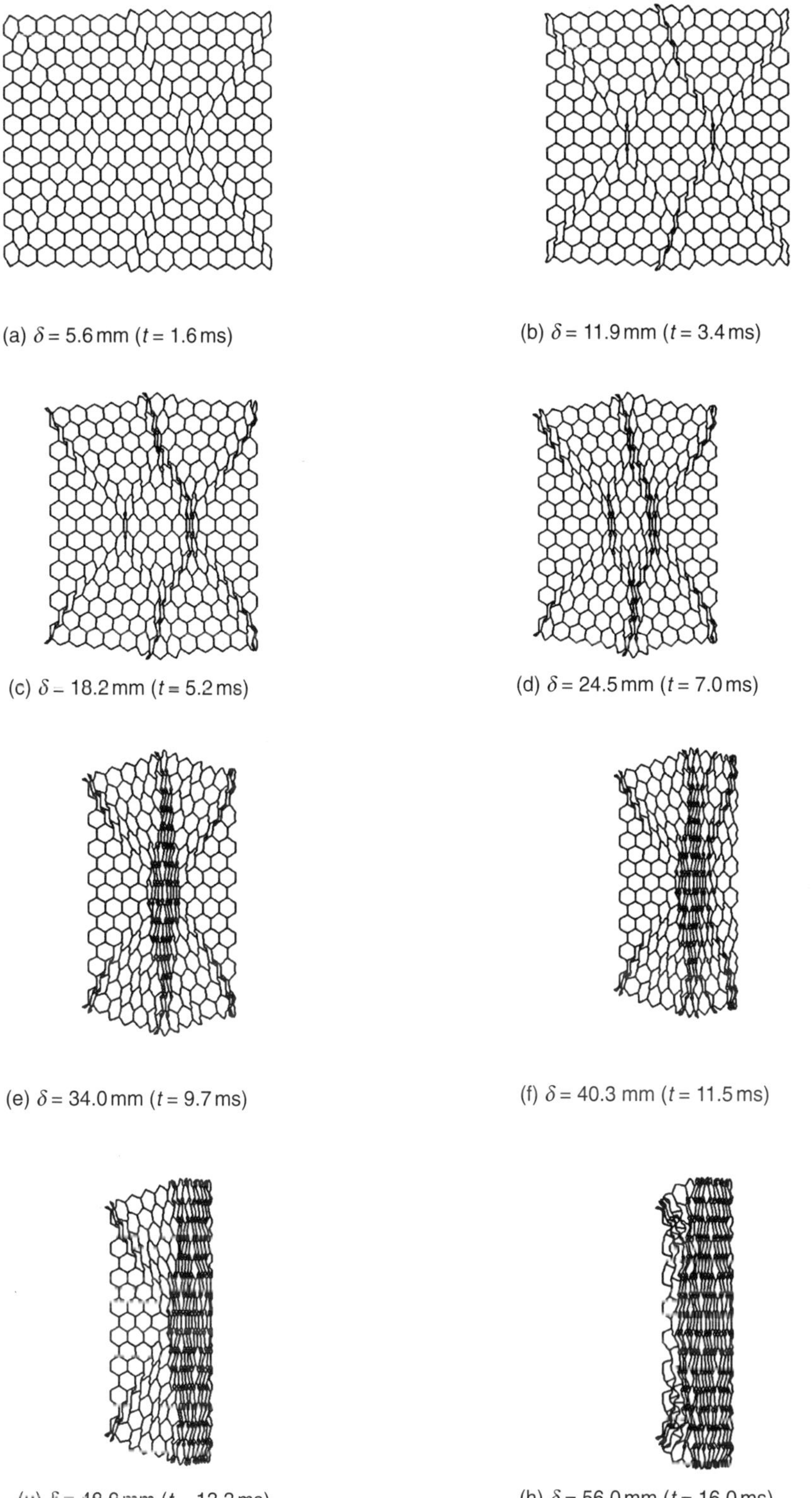

10.21 Crushing of a honeycomb in the X_1 direction, $h = 0.2$ mm, $v = 3.5$ m/s. Deformations are localised initially within an 'X' shaped band (D. Ruan *et al.*, 2003) (reproduced with kind permission of Elsevier).

centre of the specimen (Figs 10.21(b) and (c)). As the crushing proceeds, more localised bands occur with one more layer of cells crushed along the 'X' bands (Fig. 10.21(d)). After that, localisation takes place within the central rhombus (Figs 10.21(e) and (f)). Finally, when deformation within the rhombus is exhausted, more localised bands occur near the loading edge (Fig. 10.21(g)), until the honeycomb is completely crushed (Fig. 10.21(h)). The 'X' shaped localisation band was also observed in our quasi-static experiments.

When the impact velocity is even higher, for example at $v = 70\,\text{m/s}$, no obvious localised bands have been found within the block through the whole crushing process (Fig. 10.22). Only a localised transverse band perpendicular to the impact is observed at the loading edge and it continues to propagate, layer by layer, to the fixed edge. This deformation is in the manner of plane wave propagation.

At intermediate impact velocities (not shown here), a number of cells near the right edge of the model are slightly crushed within a 'V' shaped block, and no obvious localised deformation band is observed at the beginning of deformation. Afterwards, a localised deformation band occurs near the loading edge. The band is in the shape of an 'X' but is slightly slimmer than that for a lower speed impact. With further deformation, more localised bands develop progressively. When the displacement increases to about 40 mm, a new oblique localised band occurs, but from the fixed edge, and this band propagates towards the existing ones, which seem to have stopped developing. Then the bands interact with each other and the sample is totally crushed.

Mode classification map

All the three observed deformation modes in the X_1 direction may be classified into three types. Type one is the 'X' shaped deformation mode shown in Fig. 10.21(a). The feature of this mode is that the 'X' shaped localised bands can be observed clearly when the honeycomb is crushed by a deformation as small as 5.6 mm. Type three is 'I' mode, as shown in Fig. 10.22. In this mode, there is no obvious oblique localised band through the whole crushing process and only vertical bands normal to the loading direction are found. Type two is a transitional mode between the 'X' and 'I' modes, and is called 'V' mode. The localised bands are oblique, but they do not form a complete 'X' shape when the honeycomb is crushed by a displacement of 5.6 mm. Figure 10.23(a) is a sketch of these three modes. The 'I' and 'X' modes are distinctive and the 'V' mode is less so.

Based on this convention, we plot logarithmically, the deformation modes in the X_1 direction of honeycombs with different cell wall thickness in Fig. 10.23(b). Deformation modes are all of 'X' mode at low velocities and 'I'

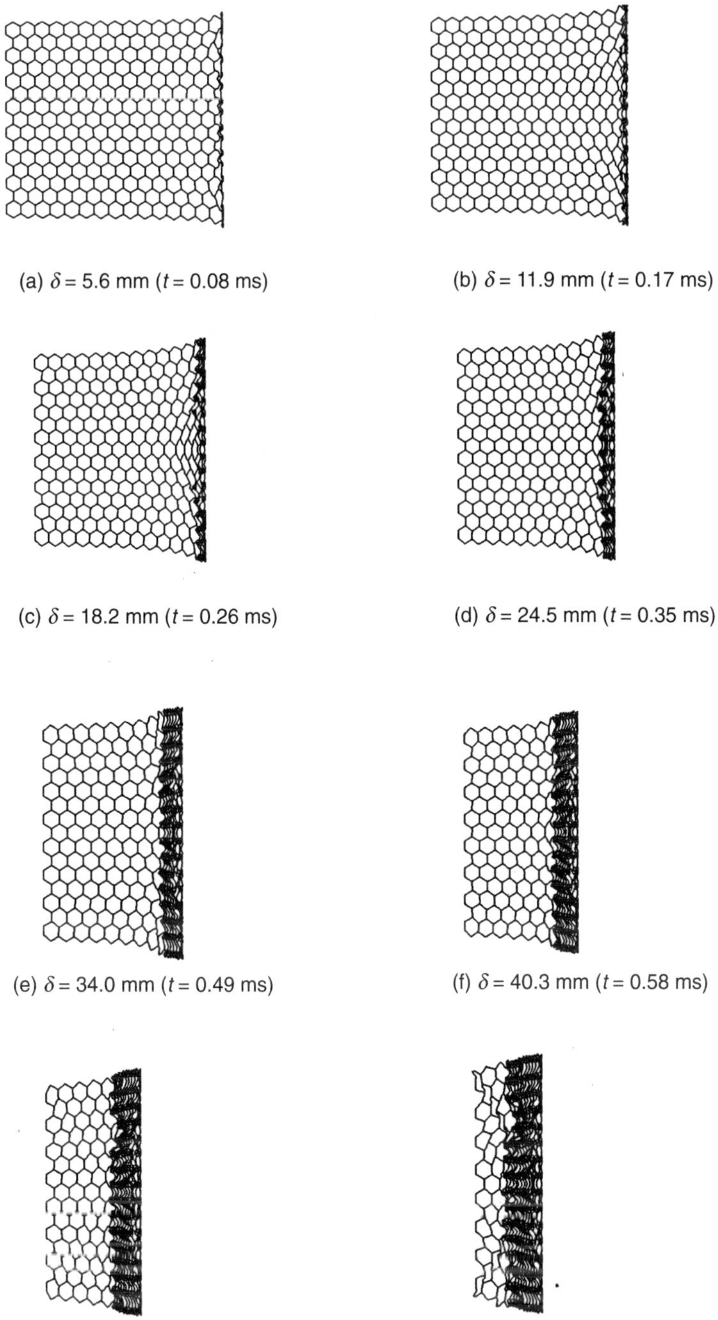

(a) $\delta = 5.6$ mm ($t = 0.08$ ms)
(b) $\delta = 11.9$ mm ($t = 0.17$ ms)
(c) $\delta = 18.2$ mm ($t = 0.26$ ms)
(d) $\delta = 24.5$ mm ($t = 0.35$ ms)
(e) $\delta = 34.0$ mm ($t = 0.49$ ms)
(f) $\delta = 40.3$ mm ($t = 0.58$ ms)
(g) $\delta = 46.6$ mm ($t = 0.67$ ms)
(h) $\delta = 56.0$ mm ($t = 0.80$ ms)

10.22 Crushing of a honeycomb in the X_1 direction, $h = 0.2$ mm, $v = 70$ m/ε (D. Ruan *et al.*, 2003) (reproduced with kind permission of Elsevier).

mode at high velocities. The critical velocity at which the deformation mode switches from one type to another depends on the cell wall thickness.

From a dimensional analysis, a non-dimensional critical velocity $v/(\sigma/\rho)^{1/2}$ is often adopted for the dynamic response of solids. We follow a similar argument for the whole honeycomb specimen; here the dimensionless group is only dependent upon another group, h/l, assuming that the size of the honeycomb specimen is immaterial. But since $\sigma \propto (h/l)^2$ (Eq. [10.13]) and $\rho \propto h/l$ (Eq. [10.3]), we would expect $v \propto \sqrt{h/l}$, which corresponds to a straight line in Fig. 10.23(b). From this figure, v_{c1} is almost independent of h/l, but $v_{c2} \propto \sqrt{h/l}$. The empirical equations for the two critical velocities are, respectively

$$v_{c1} = 14 \qquad (\text{m/s}) \qquad [10.59]$$

and

$$v_{c2} = 277\sqrt{h/l} \qquad (\text{m/s}) \qquad [10.60]$$

Plateau stress

Our main interest is in the plateau stress, which is important for energy absorption. The values of dynamic plateau stress are calculated as the ratio of plateau force to loading cross-sectional area. Theoretical static plateau stresses are calculated following Gibson and Ashby (1997) (Eq. [10.12]), whilst we multiply their formula for plateau stress by a factor of 1.15 to account for the plane strain condition of the cell walls. For the same cell wall thickness, dynamic plateau stresses are higher than the theoretical static values. The shock theory equation [10.52] is applied here.

Figures 10.24(a)–(c) show the change of plateau stress ($\sigma_D - \sigma_p$) with impact velocity for various wall thicknesses, where calculated dynamic plateau stresses from FE are plotted as diamond symbols. For all values of cell wall thickness, the plateau stress increases with impact velocity. When velocities are higher than a certain value, plateau stress ($\sigma_D - \sigma_p$) shows a good correlation to velocity by a square law, corresponding to a slope of 2 on the double logarithmic plot. The slope is lower than 2 when h is relatively small, say 0.08 mm and larger than 2 when h is relatively large, 0.3–0.5 mm in this example calculation.

A final empirical equation is obtained as follows

$$\frac{\sigma_D}{Y_s} = 0.8\left(\frac{h}{l}\right)^2 + \left[62\left(\frac{h}{l}\right)^2 + 41\left(\frac{h}{l}\right) + 0.01\right] \times 10^{-6} v^2 \qquad [10.61]$$

This is the equation for dynamic plateau stress when the velocity is high, i.e. for 'I' mode. Similar observations may be made for the plateau stress in the X_2 direction.

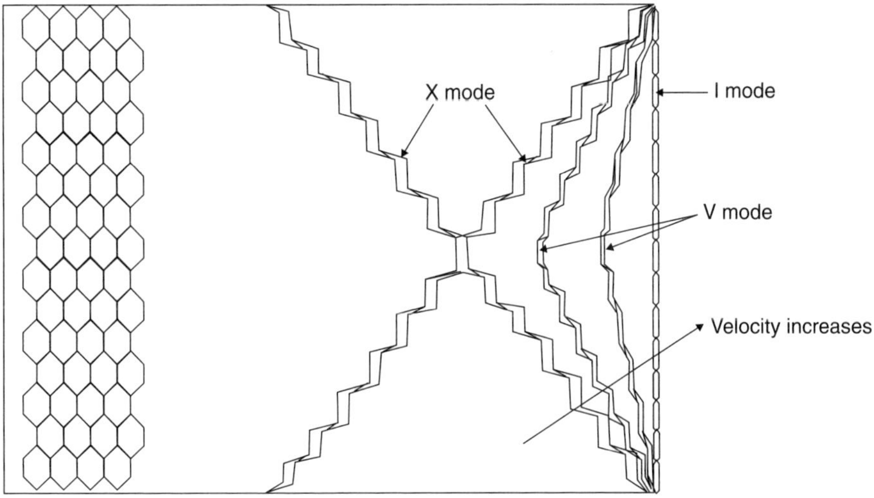

(a)

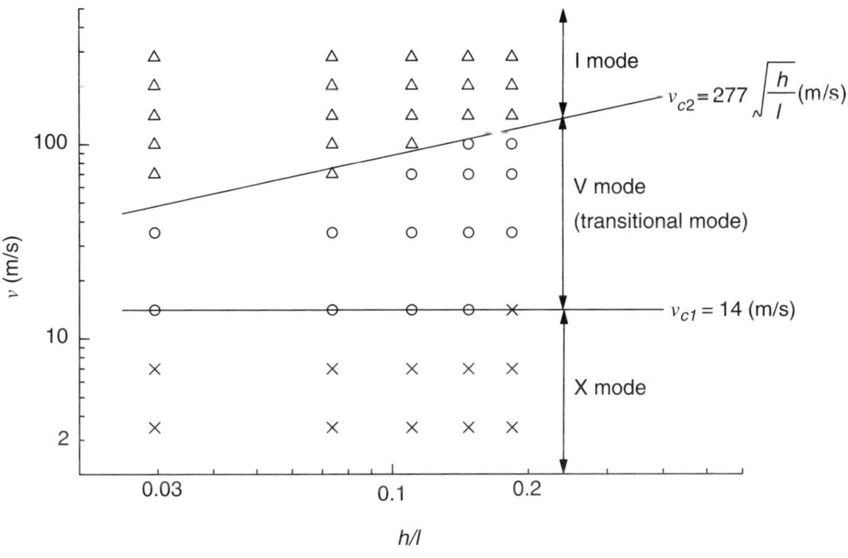

(b)

10.23 (a) Sketch of the three types of deformation modes (reproduced with kind permission of Elsevier); (b) deformation mode map in terms of *h/l* and *v*. *Δ* = 'I' mode, O = 'V' mode, X = 'X' mode (D. Ruan *et al.*, 2003) (reproduced with kind permission of Elsevier).

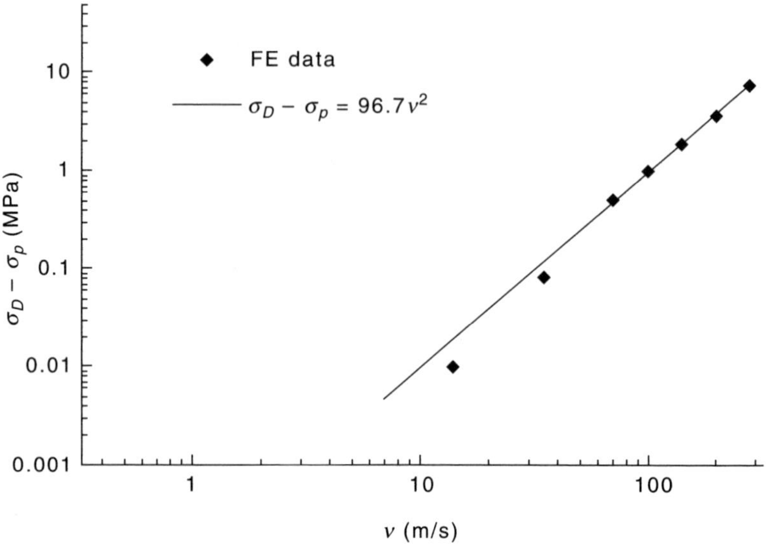

(a)

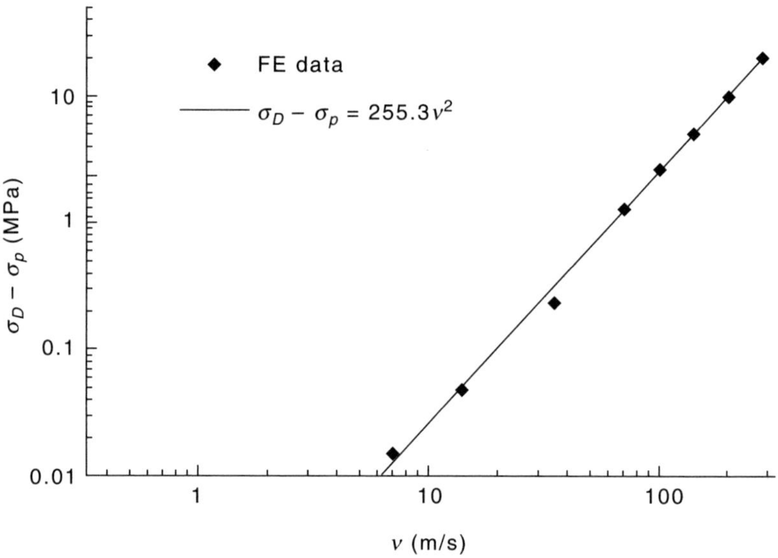

(b)

10.24 Change of plateau stresses in X_1 direction with the impact velocity: (a) $h = 0.08$ mm; (b) $h = 0.2$ mm; (c) $h = 0.3$ mm (D. Ruan *et al.*, 2003).

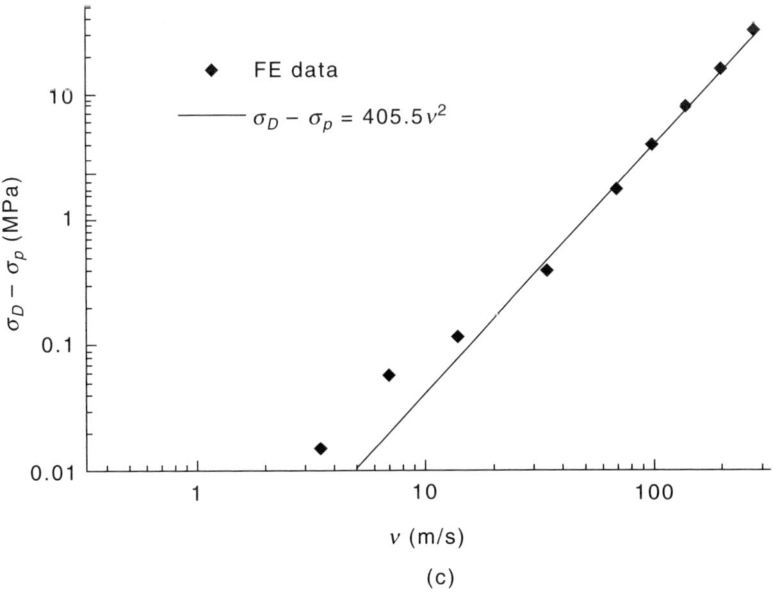

10.24 Continued.

10.4.4 Strain-rate effect

Strain-rate has a primary effect on the solid material of the cell walls, just in the same way as it has on conventional solids, as discussed in Chapter 2. This generally enhances the yield stress of the cell walls and hence increases the overall collapse stress.

For open cell foams, strain-rate plays another important role. When the foam is crushed, fluid initially trapped in the cells escapes. This requires additional external work to overcome the friction between the fluid and the cell edges. At higher strain-rates, this work is larger, leading to an increase in the plateau stress. For a block of linear dimension L, Gibson and Ashby (1997) obtained the resulting contribution to stress as

$$\sigma_g = \frac{C\mu\dot{\varepsilon}}{1-\varepsilon}\left(\frac{L}{l}\right)^2 \qquad [10.62]$$

Hence, the contribution from the fluid is proportional, through a constant $C \sim 1$, to the viscosity of the fluid μ, the strain-rate $\dot{\varepsilon}$ and to the reciprocal of cell size l.

A number of investigations have been made concerning the strain-rate effect of cellular materials. For example, Zhao (1997) and Gilchrist and Mills (2001) studied polymeric foams, while dynamic crushing of honeycombs has been reported by Wu and Jiang (1997), Zhao and Gary (1998),

Baker *et al.* (1998) and Ruan *et al.* (2001). Dynamic crushing of metal foams has been conducted by Mukai *et al.* (1999), Deshpande and Fleck (2000) and Reid *et al.* (2001). Dynamic response of porous materials has been investigated by Wang *et al.* (2001 and 2003) and Lu *et al.* (2001).

10.5 Cellular textile composites

10.5.1 Background

As shown in Section 10.1, cellular structures or cellular materials usually contain a large amount of space and display high specific-energy-absorption capacity, so that they are usually very good for energy absorption. On the other hand, textile structures consist of assemblies of fibres with given orientations and low packing density. They can be deformed easily into required configurations. Therefore, by combining the excellent characteristics of textile structures and cellular materials, cellular textile composites can provide ideal candidates for meeting various energy-absorption requirements as stated in Section 1.3.2 (Xue *et al.*, 2000a).

Yu *et al.* (2000) first conducted a systematic investigation into the energy-absorption capacity of cellular textile composites. A new class of cellular textile composite with high specific energy-absorption capacity has been developed and studied thoroughly. A brief introduction to this class of composite material will be given below.

10.5.2 Two configurations studied for grid-domed textile composites

To fabricate cellular textile composites, typically, the 1×1 interlock double jersey knitted fabrics made of multi-filament textured nylon yarn (DJ-N) were first produced. The originally flat fabrics were then formed into three-dimensional cellular textile composites by a two-step method. The fabric was first shaped into a grid-domed sheet with cells comprising specially designed domes. These textile preforms were then coated with polyester resin, followed by a heat setting treatment at 200 °C for five minutes and finally cured at room ambient conditions.

The mechanical properties of the grid-domed textile composite depend on the material system selected (i.e. type and size of fibres, fabric architecture, resin material and add-on ratio, as well as the fabrication process) and the cell configuration. In particular, two configurations have been examined in detail. In Configuration 1, each cell consists of a truncated conical shell and a hemispherical dome (Fig. 10.25(a)); in Configuration 2 each cell consists of a flat-topped truncated conical shell only (Fig. 10.25(b)).

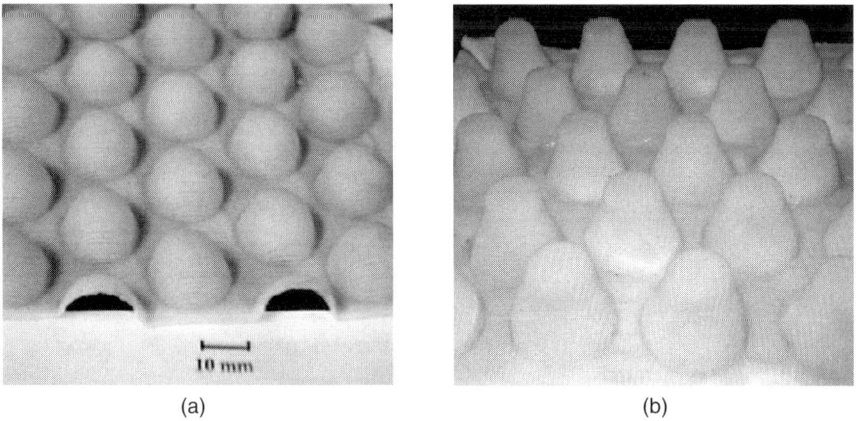

10.25 Grid-domed textile composites: (a) Configuration 1; (b) Configuration 2 (Xue *et al.*, 2000b).

Both experiment and analytical analysis revealed (Yu *et al.*, 2000) that the large deformation process of the grid-domed composite of Configuration 1 consisted of three stages, i.e. local inversion of the hemispherical dome, global plastic collapse of the truncated spherical cap and large plastic deformation of the truncated conical shell. Most of the energy-absorption was contributed by the membrane-force dominated large plastic deformation stage. This implies, therefore, that the flat-topped truncated conical shell composites of Configuration 2 should be more efficient in energy absorption.

Both samples with Configurations 1 and 2 were tested under identical quasi-static and impact conditions. The sample size and detailed loading conditions can be found in Xue *et al.* (2000b). Figures 10.26(a) and (b) depict the load–displacement curves under quasi-static compression and under impact loading, respectively, for samples with the two configurations. It is seen that the elastic deformation stage for samples of Configuration 2 was much shorter than that for samples of Configuration 1. For samples of Configuration 2, as deformation progressed, the rate of load increase diminished and plastic deformation became dominant. The force required for deformation remained almost constant and the deformation stroke was long and in a stable mode.

It is obvious from these results that, compared with Configuration 1, grid-domed textile composites of Configuration 2 demonstrate higher energy absorption capacity, more stable deformation mode, lower peak force and almost constant force magnitude during the large deformation process. Hence grid-domed composites of Configuration 2 should be ideal candidates for energy absorbers.

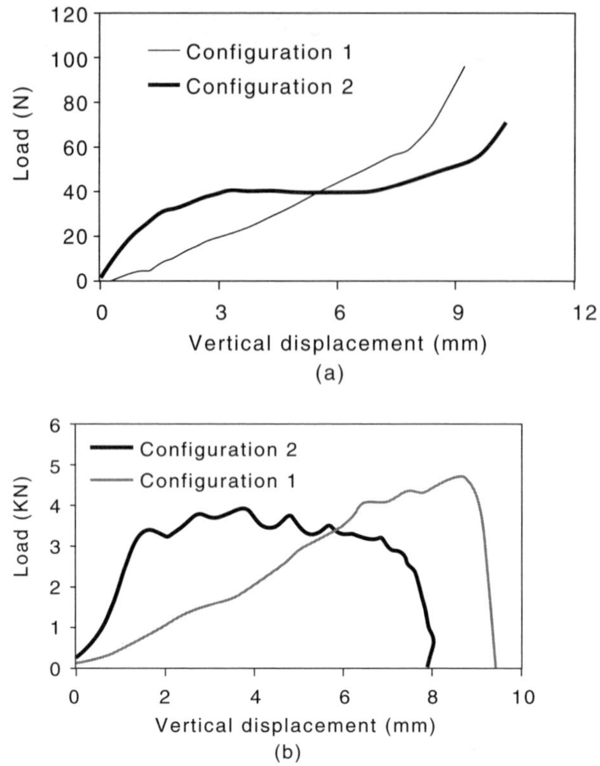

10.26 Load–displacement curves for samples of the two configurations: (a) under quasi-static compression; (b) under impact loading (Xue *et al.*, 2000b).

10.5.3 Effects of cell geometry and cell distribution upon energy-absorption capacity

Cell geometry

It was found from an experimental study (Xue *et al.*, 2000b) that for grid-domed textile composites of Configuration 2, an effective parameter to control the energy-absorption performance was the diameter ratio of the cell's top to the cell's bottom, or the semi-apical angle of the truncated conical cell. If the top diameter is too small compared with the bottom diameter, or the semi-apical angle of the cell is too large, this would be detrimental to the energy-absorption capacity of the grid-domed composite. For instance, it was found that for samples with the same percentage of resin add-on (170 %), the same bottom diameter, 18 mm, and the same cell height, 13.8 mm, the optimal top diameter within the examined parameter range was 10 mm and the corresponding semi-apical angle about 16°.

Cell distribution

As observed in the tests, the cell top and flat base of the grid-domed composites did not deform noticeably when the conical wall underwent a large deformation. Therefore, the area of conical wall can be regarded as the effective area in absorbing energy under axial quasi-static compression and impact loading. To verify this observation, two sets of samples were produced with the same level of resin add-on. Both sets of samples possessed the same total area of cell walls, but different cell numbers and cell size.

The test results indicated that changing the size of cells did not affect the magnitude of the load and the energy-absorption capacity of the samples provided that the total effective area of the samples remained unchanged. It is thus concluded that the effective cell wall area of cells dominates the energy-absorption capacity of grid-domed textile composites when other conditions remain the same. Therefore, one should aim to increase the effective area within the specified total area so as to enhance overall energy-absorption capacity. Cell density (i.e. the number of cells per unit area) is one of the parameters related to the effective area. Increasing cell density will result in an increase in the effective area for energy absorption within the same specified projected area, thus raising the energy-absorption capacity of the material. However, further increasing cell density is restricted practically by formability of the composite fabrics.

10.5.4 Theoretical models of a flat-topped conical shell under axial compression

Macroscopic deformation modes observed in experiments

The macroscopic deformation modes of thin-walled structural components subjected to axial loading greatly affect their energy-absorption capacity. In order to observe the deformation mode of a flat-topped conical shell under axial compression, a specifically designed compression test was carried out with a series of photographs taken from various angles. The front view and side view of a typical deformed cell are shown in Fig. 10.27. It is evident that a two-lobe diamond-pattern mechanism was formed. During the deformation process, the following deformation characteristics were observed.

(1) The circumferences of the top and bottom circles remained almost unchanged, whilst the cell top simply descended vertically during the shell large deformation process.
(2) The flat-topped conical shell collapsed plastically into a two lobe diamond-pattern.

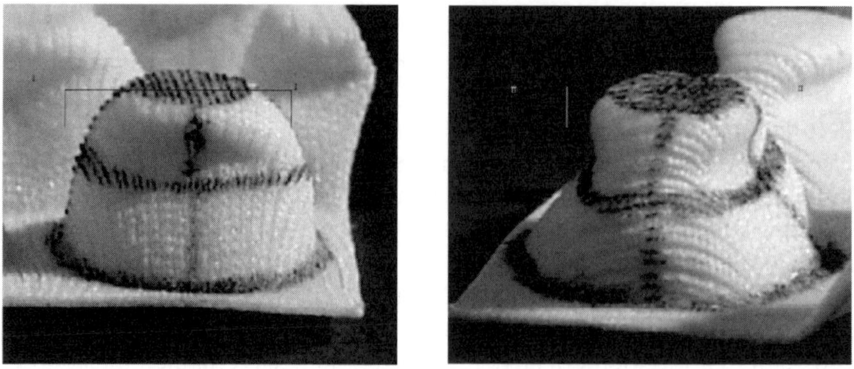

(a) (b)

10.27 A deformed cell: (a) from front view; (b) from side view (Xue *et al.*, 2001b) (reproduced with kind permission of Elsevier).

(3) Plastic hinge lines formed along the top and bottom circles of a cell, as well as along a horizontal circumferential loop at about two thirds of the height, resulting in two rhombuses on the front and back surfaces of the cell.

Basic assumptions of theoretical models

Because of the complexity of the elastic-plastic transition coupled with changes in cell geometry, it is difficult to formulate a single analytical model to simulate the entire elastic-plastic large deformation process of a flat-topped conical shell under axial compression. However, it is possible to construct two separate models suitable for different stages of the process and capable of predicting the major characteristics of the shell under axial compression. That is, an elastic model can be applied to the early stage of the shell's deformation and a rigid-plastic model can be employed when the shell experiences large deformation with negligible elastic deformation. Since the energy absorption capacity is primarily associated with large deformation, the following discussion will mainly focus on the rigid-plastic model.

Based on the experimental observations described above, assumptions are made as follows:

(1) In the elastic model, the material is assumed to be linearly elastic; in the large deformation model it is assumed to be rigid-plastic.
(2) The top circle of the cell remains unchanged and simply descends vertically during the deformation process while the bottom circle of the cell is unchanged and fixed on the base.

(3) When a horizontal plastic hinge line forms, the circle deforms in plane into an elongated circle, whilst the perimeter of the horizontal plastic hinge line along the circumferential direction remains constant (so as to reduce membrane deformation).
(4) At least one generator of the conical shell remains unchanged in length, whilst all other generators are under compression.
(5) The energy dissipated by shear deformation is neglected.

Formulation of theoretical models

Initially the shell behaves elastically. According to an equilibrium analysis of the thin-walled conical shell (Gould, 1999) and comprehensive geometric considerations (Xue *et al*., 2001b), the load–displacement relationship can be expressed as

$$P = 2\pi h\left(a + \frac{1}{2}H\tan\alpha\right)E\bar{\varepsilon}\cos\alpha$$
$$= \frac{\pi h\left(a + \frac{1}{2}H\tan\alpha\right)E\Delta\cos^3\alpha}{H}\left(\frac{2}{3}\sin\alpha\sqrt{\frac{6\Delta}{H}} + 1\right) \quad [10.63]$$

where E is Young's modulus of the material, h the wall thickness of the conical shell, H the height, α the semi-apical angle, Δ the vertical displacement at the top and P the vertical load applied.

With an increase in vertical displacement, the flat-topped conical shell is deformed plastically. The deformed configuration of the flat-topped conical shell is sketched in Fig. 10.28(a). Plastic hinge lines are formed along the top and bottom circles, along a horizontal circumferential loop at height $(1 - \lambda)H$ with λ being a parameter to be determined later, as well as along the sides of rhombus ADBF. The geometry of the horizontal circumferential hinge line is sketched in Fig. 10.28(b), where the length of straight line AB is supposed to be equal to that of arc A′B′ based on assumption 3 above, so the magnitude of b must satisfy the following equation

$$b = b_o(1 + \phi - \sin\phi) \quad [10.64]$$

where angle ϕ serves as a process parameter. Because of the orthotropic properties of the material, the outward portions of the horizontal hinge line, i.e. arcs AA_1 and BC, take place where the wale direction of the fabric is perpendicular or nearly perpendicular to the generator of the conical shell. The inward portions of the horizontal hinge line, i.e. lines AB and A_1C, take place where the wale direction of the fabric is parallel or almost parallel to the generator of the conical shell. Obviously, the shape of the horizontal hinge line AA_1CB, as an elongated circle, evolves with increasing ϕ.

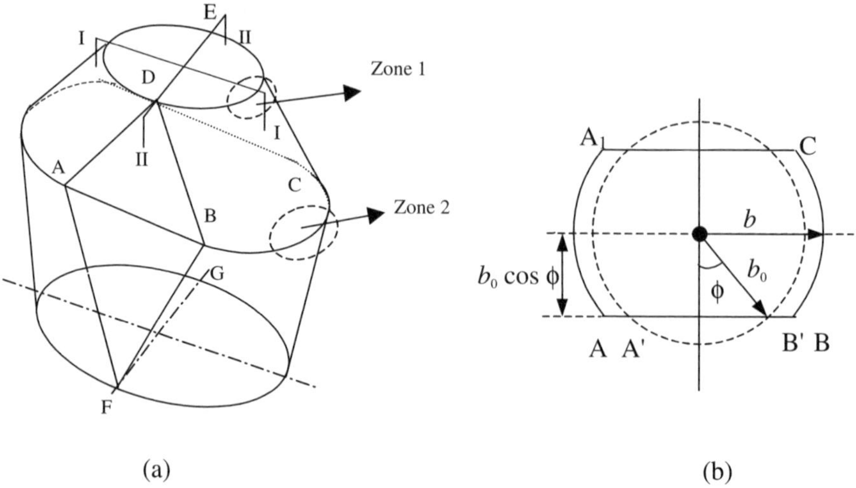

10.28 Rigid-plastic deformation model of a flat-topped conical shell: (a) deformed shape with plastic hinge lines; (b) elongated circle as the horizontal circumferential hinge line (Xue *et al.*, 2001b).

By considering the geometry of the deformed shapes of cross-sections I–I and II–II during large plastic deformation, all geometric quantities, such as the current length of various generators, can be expressed in terms of ϕ and λ (refer to Xue *et al.*, 2001b, for details). When the generator length on cross-section II–II is assumed to remain constant during large deformation of the shell, calculations confirm that the generator on cross-section I–I is under compression.

As the material is assumed to be rigid-plastic in this large deformation model, the work done by the compressive force is dissipated in two parts: by bending along the plastic hinge lines, and stretching of wall segments between hinge lines. The first part of energy dissipation can be written as

$$W_b = (W_b)_{top} + (W_b)_{bottom} + 2(W_b)_{AB} + 2(W_b)_{BC} + 4(W_b)_{DA} + 4(W_b)_{AF} \qquad [10.65]$$

which is proportional to the plastic bending moment per unit width of the shell wall, M_o.

The second part of energy dissipation is due to stretching of the shell segments between the plastic hinge lines, while the change in the wall thickness is assumed to be negligible during stretching. The initial side surface area of the flat-topped conical shell before compression is

$$S_o = \frac{\pi H}{\cos \alpha}(2a + H \tan \alpha) \qquad [10.66]$$

During large deformation, the side surface area changes. The areas of segments ABD, AFB, BCDE and BCFG can be approximately expressed in terms of their side lengths and angles; hence they depend on ϕ and λ (refer to Xue *et al.*, 2001b). By considering all of these areas, we find

$$W_m = \sigma_y h \sum |S_i - S_{io}| \tag{10.67}$$

where S_i is the area of the segments after deformation and S_{io} is the initial area corresponding to S_i. σ_y denotes the flow stress of the material, which may vary with equivalent strain due to strain-hardening of the material.

Thus, the total energy dissipation during large plastic deformation of the flat-topped conical shell is

$$W = W_b + W_m \tag{10.68}$$

which should be equal to the external work done by the axial compressive force P during the deformation process. That is

$$W(\phi,\lambda) = W_b + W_m = \int_0^{\Delta} P \mathrm{d}\Delta \tag{10.69}$$

which leads to

$$\begin{aligned} P &= \frac{\partial W(\phi,\lambda)}{\partial \Delta} = \frac{\partial W(\phi,\lambda)}{\partial \phi} \cdot \frac{\partial \phi}{\partial \Delta} \\ &= \left(\frac{\partial W_b(\phi,\lambda)}{\partial \phi} + \frac{\partial W_m(\phi,\lambda)}{\partial \phi} \right) \cdot \frac{\partial \phi}{\partial \Delta} \end{aligned} \tag{10.70}$$

where the relationship between ϕ and Δ can be obtained as

$$\frac{\partial \Delta}{\partial \phi} = b_o \sin\phi \left[\frac{(a - b_o \cos\phi)}{h_1} + \frac{(c - b_o \cos\phi)}{h_2} \right] \tag{10.71}$$

where h_1 and h_2 are two geometric quantities depending on a, b_o and c; whilst $\partial W_b(\phi, \lambda)/\partial \phi$ and $\partial W_m(\phi, \lambda)/\partial \phi$ can be determined from Eqs [10.65] and [10.67]. After P is minimised in terms of λ the load–displacement curve of a flat-topped conical shell under axial compression as well as its energy-displacement relationship can be obtained.

Numerical examples and comparison with experiment

In numerical examples, the cell parameters chosen were the same as those of the tested samples. That is, the radius of the top circle is $a = 5$ mm, the radius of the bottom circle is $c = 9$ mm, the shell height is $H = 13.8$ mm, the shell thickness is $h = 1$ mm and the semi-apical angle of the shell is $\alpha = 16.16°$. The material properties are described in Xue *et al.* (2001b). The load–displacement curves predicted by the two theoretical models and that obtained from experiment are compared in Fig. 10.29(a). Here 'linear

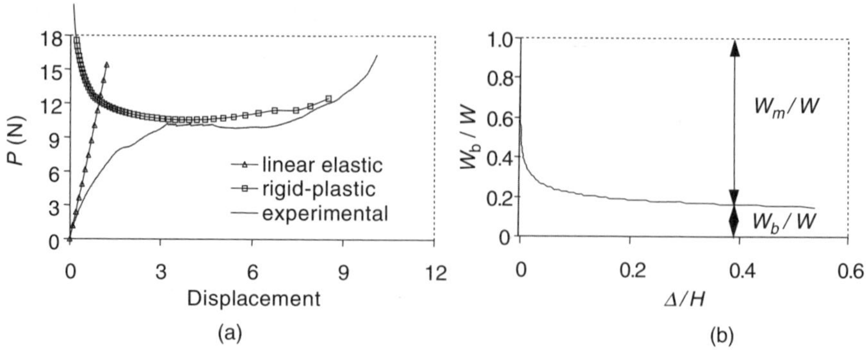

10.29 Predictions of the theoretical models: (a) the predicted load–displacement curves, compared with experiment; (b) the partitioning of the energy dissipation, predicted by the rigid-plastic model (Xue *et al.*, 2001b).

elastic' pertains to numerical predictions from the elastic model, whilst 'rigid-plastic' pertains to predictions from the rigid-plastic model.

It is evident that the behaviour of the real flat-topped conical shell textile composite under compression can be represented by the elastic model in the early stage and by the rigid-plastic model for the later stage, respectively. During the transition from elastic deformation to large plastic deformation, the load–displacement curve should undergo a smooth path between the two theoretical curves depicted in Fig. 10.29(a). This implies that the discrepancy between the theoretical predictions and the experimental result in the transition stage should be much smaller than that exhibited in Fig. 10.29(a).

Because the deformation modes proposed are merely kinematically admissible for a flat-topped conical cell, they do not necessarily satisfy the equilibrium equation. Therefore, these theoretical models only provide upper bounds for the load–displacement characteristics, as evidently exhibited in Fig. 10.29(a).

Figure 10.29(b) shows the proportion of energy dissipated by bending deformation, W_b, within the total energy dissipation, W. Clearly, W_b/W gradually reduces with increasing vertical displacement, implying that energy dissipation due to cell membrane deformation will be dominant when deflection of the cell is large. For the cell geometry given here, the membrane deformation dissipates about 85 % of the total energy when the vertical displacement reaches half of the cell height.

Strain distributions

The distribution of strain in the shell wall is depicted in Fig. 10.30(a), where ε_m denotes average strain in a wall segment resulting from membrane

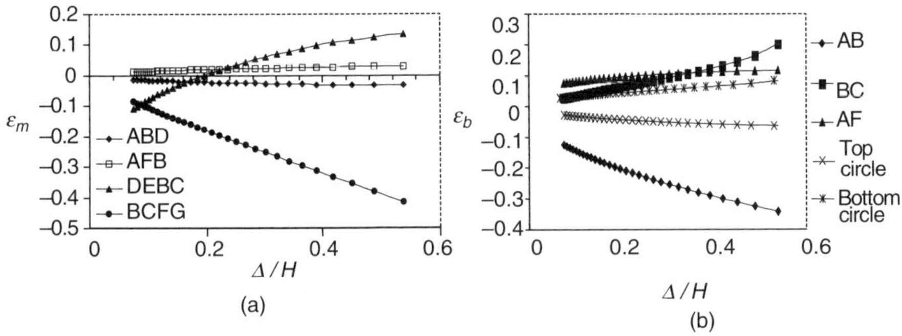

10.30 Distributions of strains, predicted by the rigid-plastic model: (a) the membrane strains in the shell segments; (b) the bending strains along the plastic hinge lines.

deformation. ABD marked in Fig. 10.30(a) pertains to segment ABD shown in Fig. 10.28(a), and so on. From Fig. 10.30(a), it is found that the strains in segments DEBC and BCFG are much larger than those in segments ABD and AFB. The magnitude of the compressive strain in segment BCFG increases with increasing vertical displacement. In segment DEBC, the material is compressed when vertical displacement Δ is less than $0.2H$, then extended as the cell experiences even larger deformation.

The distribution of strain resulting from bending deformation along the plastic hinge lines is depicted in Fig. 10.30(b), where ε_b is the maximum bending strain along the subject hinge line. The effective length of the plastic hinge is taken as being three times the wall thickness. AB marked in Fig. 10.30(b) pertains to plastic hinge line AB shown in Fig. 10.28(a), and so on. Strain ε_b is defined as positive when the outer surface of the shell is extended. It is seen that the largest (absolute) maximum bending strain is along plastic hinge line AB; this strain also increases with increasing vertical displacement.

10.5.5 Mesoscopic failure mechanisms observed by SEM

For composite materials, internal material failure generally is initiated before any change in macroscopic appearance or behaviour is observed. For flat-topped conical shells made of a nylon/polyester material system subjected to axial compression, mesoscopic failure mechanisms were *in situ* examined by SEM (refer to Xue *et al.*, 2001a). The sample coated with gold was put between two parallel steel plates and compressed at a crosshead speed of 0.3 mm/min. Images were taken at every 1 mm increase in crosshead displacement and hence the microscopic failure process in the observed zones was recorded.

Two distinct regions, with different microscopic characteristics, were focused on: first the region near the top of the shell marked as Zone 1 in Fig. 10.28(a), where the material was mainly subjected to compression; secondly the region near the plastic hinge line BC marked as Zone 2 in Fig. 10.28(a), where the material mainly underwent bending deformation. The mesoscopic deformation mechanisms governing these two zones are quite different.

Consider Zone 1, where the material was mainly compressed; relevant SEM images are shown in Fig. 10.31. When the local strain approached 12 %, which is the breakage strain of the matrix material (polyester), some cracks were initiated. As deformation proceeded, these cracks propagated. Cracks propagated preferentially through the weakest regions of the composite structure, i.e. through resin-rich regions or interfaces between fibres and matrix. In this process, the interface between fibre and matrix material failed, resulting in debonding. When a crack propagated in a direction normal to a fibre, breakage of the fibre eventually occurred. Thus, the main microscopic features of energy dissipation in Zone 1 during the compression process, can be characterised as (a) development and propagation of cracks in matrix; (b) fibre/matrix interface debonding; and (c) fibre fracture.

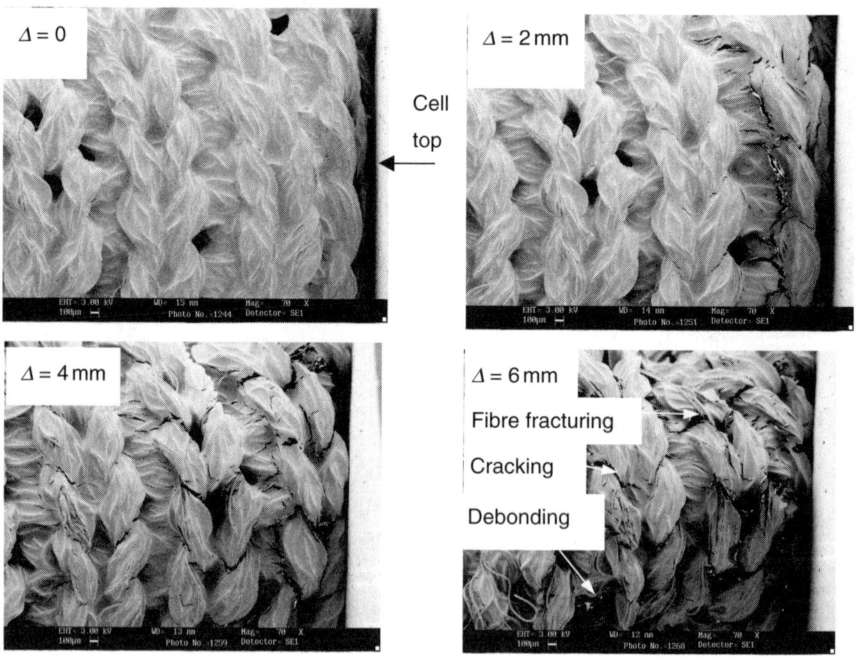

10.31 SEM images showing the evolution of microscopic failure mechanisms at Zone 1 (70×) (Xue *et al.*, 2001a) (reproduced with kind permission of Elsevier).

Local out-of-plane buckling and a change of fabric architecture were observed in Zone 2, i.e. the middle part of the shell wall, where there is significant bending strain. The cracks initiated in the near-top region, Zone 1, propagated towards the middle part of the conical shell without causing shell separation. By comparing the images of Zones 1 and 2, it is apparent that the strain in Zone 1 increased more quickly than that in Zone 2; consequently, cracking in Zone 1 occurred earlier than in Zone 2.

As seen in Section 10.5.4, for the flat-topped conical shell three-dimensional textile composite examined, the cells collapsed plastically into a diamond pattern as shown in Fig. 10.28(a). This plastic collapse dominated the total energy absorption. The energy-absorption capacity of the material can be obtained by integrating the area under the load vs displacement curves shown in Fig. 10.26. It is observed that the load–displacement curve does not fluctuate greatly or drop sharply. This implies that the contribution to total energy-absorption capacity from mesoscopic damage, i.e. matrix cracking, fibre fracturing and the fibre/matrix debonding, is not very significant. The effect of mesoscopic damage on macroscopic material response can be observed only when it has accumulated to a certain level.

10.5.6 Concluding remarks

When designing cellular textile composites for energy-absorption purposes, a very large plastic (i.e. unrecoverable) deformation with a constant reactive force is the targeted performance. Accordingly, the material system and fabric architecture are specially selected to achieve this purpose, so that the textile composite has better mechanical properties and higher energy-

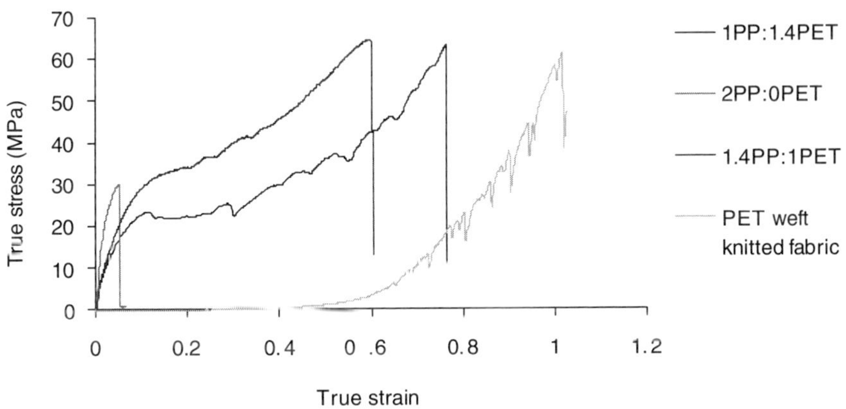

10.32 Typical tensile stress–strain curves of the thermoplastic (PET/PP) textile composites with different fibre contents (along the wale direction).

absorption capacity than that of any of its components, fibre or matrix. This is true for both thermoset and thermoplastic textile composites. Figure 10.32 shows typical tensile properties of the fibre (PET), the matrix (PP) and the PET/PP thermoplastic textile composite (refer to Yu *et al.*, 2001). It is evident from this figure that, although the matrix is very brittle, and while the fibre can only sustain stresses at a low level under large elongation, the textile composite exhibits an excellent combination of appropriate stress level and great deformability. This provides a clear example of the outstanding energy-absorption capacity of the textile composite.

The design of the cell configuration and cell pattern is another issue critical to the achievement of higher energy-absorption capacity in grid-domed textile composites. After a careful experimental study of the effects of geometric parameters, an optimal design of the cell geometry and cell distribution may be achieved. Another powerful tool for optimising the cell design is the theoretical model developed, which analytically describes the large deformation process of a flat-topped conical cell and is capable of predicting the energy dissipation characteristics.

It is demonstrated that, in comparison with conventional energy-absorbing materials (e.g. polyester and polyethylene foams), the new class of grid-domed cellular textile composites possess several advantages. They have higher specific energy-absorption capacity, a more stable deformation mode and almost constant force magnitude during the large deformation process under quasi-static and impact loading. This class of composites has found applications as energy-absorbing liners of various safety helmets protecting cyclists, motorbike riders and players of extreme sports (Tao *et al.*, 2003).

11
Composite materials and structures

Composite materials and structures have gained much attention over the last three decades. They are now widely used in the aerospace industry, for example. In addition to their excellent performance with high specific strength and specific stiffness, they possess good energy-absorption behaviour. This chapter discusses the energy-absorption characteristics of thin-walled tubes made of fibre-reinforced composites, metal tubes externally reinforced with fibres and sandwich panels.

11.1 Factors that influence the energy-absorption characteristics

As one may anticipate, the energy-absorption behaviour of composite materials and structural components is affected by a number of factors. These factors may be broadly classified into composite materials and properties, fabrication conditions, geometry and dimensions of the structural components, and test conditions. Fibre-reinforced composite materials are governed by the fibre material, matrix material, fibre/matrix interface and fibre content. Along with the fibre stacking sequence, fibre orientation and fibre form (whether unidirectional, woven fabric or braided fabric) are all important factors. Geometry includes both the cross-sectional shape of a tube (square, rectangular or circular) and lengthwise shape (tapered or constant). Geometry may also involve a trigging system such as chamfering of a tube end to initiate collapse. Testing conditions specify the loading direction with respect to the components (axial or transverse) and loading rate (static or dynamic). For composites, most properties are highly temperature-dependent and thus temperature is also an important factor.

Most studies on the energy absorption of composites are done with circular tubes; very few can be performed with composites alone. We therefore discuss the axial crushing behaviour of circular tubes first.

11.2 Axial crushing of circular tubes

11.2.1 Energy dissipating mechanisms and characteristics

Most composite tubes are made from high strength, high stiffness fibres (glass, carbon and Kevlar®), embedded in rigid cross-link thermosetting resins such as polyester and epoxy. Unlike ductile metals and themoplastics, the fibres and resins are brittle and they fail by fracture after an initial elastic deformation.

Fracture strain for typical glass fibres is about 1.5–2.0 % and for polyester resins it is between 1.5 and 3.5 %. Superficially, it may appear that they would thus absorb less energy than conventional metals. However, they actually perform much better when comparison is made in terms of the specific energy absorbed (energy per unit mass). Figure 11.1 illustrates typical values of the specific energy for some metals and polymer composite materials, namely, carbon fibres in a thermoplastic polyetheretherketone (PEEK) matrix (carbon/PEEK), carbon fibres in epoxy matrix, glass fibres in epoxy matrix, and chopped strand glass fibre mat-reinforced polyester composites (SMC) (Ramakrishna and Hamada, 1998). The high value of energy absorbed for carbon fibres in a PEEK matrix (almost 200 kJ/kg) is about twice that typically obtained for carbon-epoxy composites. This was attributed to the high fracture toughness of the PEEK matrix inhibiting crack growth, the fracture of a large number of fibres within the crush zone, and the large number of splits in the fronts generated by the resulting splaying mode of failure exhibited (this mode will be described in detail later).

Extensive studies have been conducted into the energy-absorption behaviour of composite materials and structures (e.g. Thornton, 1979; Gupta, *et al.* 1997); these have been reviewed by Mamalis *et al.* (1997) and

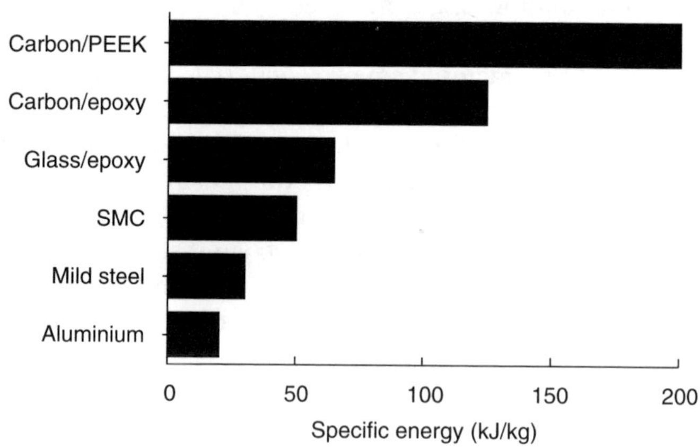

11.1 Typical values of specific energy absorption for some materials (Ramakrishna and Hamada, 1998).

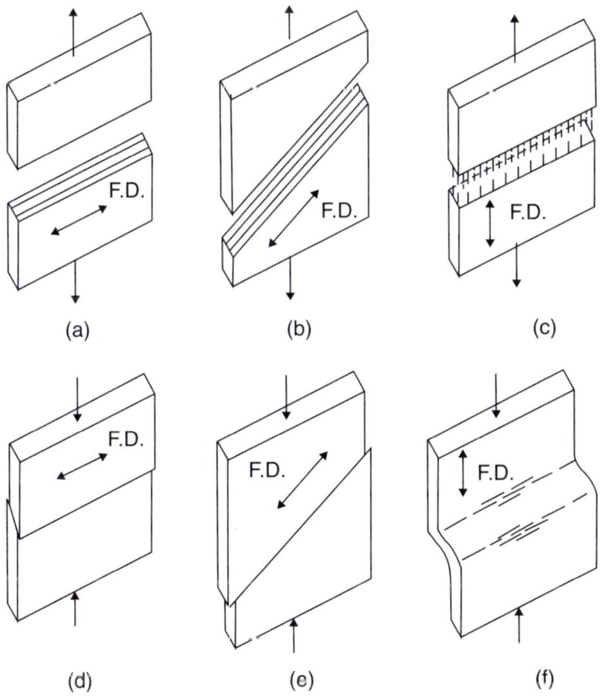

11.2 Failure modes of laminas under uniaxial tension or compression (Hull and Clyne, 1996).

Carruthers *et al.* (1998). The book by Mamalis *et al.* (1998) describes in detail the energy-absorbing behaviour of thin-walled structural components under different loading conditions, including axial crushing. The following discussion largely follows the work by Hull (1983, 1991) and Farley and Jones (1992) into the crushing mechanisms of the overall composite tubes and the detailed micro-cracking of their walls.

Figure 11.2 illustrates some of the failure modes observed in unidirectional laminas tested in uniaxial tension and compression (Hull and Clyne, 1996). They are self explanatory. The key to a particular mode lies in the direction of the principal stresses relative to the fibre orientation and the properties of the fibres, resins and fibre matrix interface. In most cases, cracks occur parallel to the fibre direction. Failure of the fibres is either tensile fracture (Fig. 11.2(c)) or micro-buckling under compression (Fig. 11.2(f)). These basic modes are helpful in describing the failure modes of composite tubes.

Depending on tube geometry, composite material and loading conditions, there are two possible modes by which a tube under axial compression may fail globally, in addition to the overall Euler-type buckling exhibited by

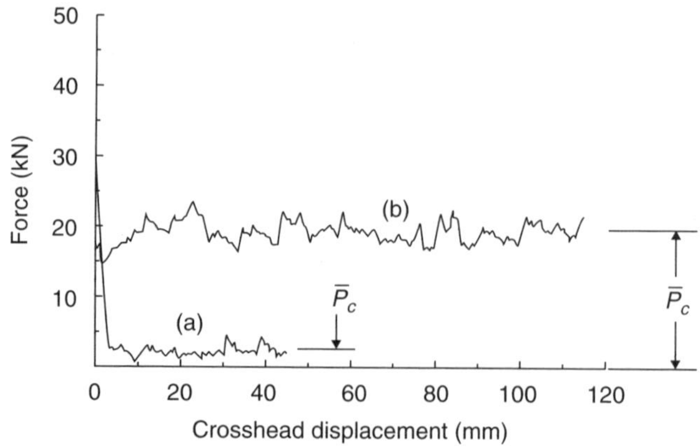

11.3 Load–displacement curves for a composite tube by central failure (a); and progressive crushing (b) (Hull, 1983).

metal tubes (Fig. 6.2). It can fail by sudden fracture around the centre of a tube or by a progressive type of crushing. Figure 11.3 shows typical force–displacement curves for these two types of failure.

Figure 11.4 shows four examples of a composite material tube failure by centre fracture (Hull, 1983). The four tubes are all made of polyester-glass fibres, but with different arrangements and volume fraction of the glass fibres. Briefly, Fig. 11.4(a) is for a sheet moulding compound (SMC) tube made by a hot-press moulding process. The short fibres lie randomly within the plane of the tube wall, leading to an isotropic property within this plane. Fracture was nucleated by shear in the fibre bundles and propagated rapidly through the resin-rich regions. In Fig. 11.4(b), the woven cloth tube's warp and weft fibres provide the stiffness and strength in the axial and hoop directions. Failure was initiated by local compressive buckling of the axial fibres (see Fig. 11.2(f)), which resulted in a through-wall shear fracture. The shear failure in the resin parallel to the fibres (Figs 11.2(a)–(e)) accounts for the mode of the filament-wound tube in Fig. 11.4(c). The pultruded tube in Fig. 11.4(d) mainly consists of multiple layers of woven cloth and random fibres. Column-type buckling of the outer fibres was the failure mode, accompanied by shear of the inner layers.

Central failure of a composite material tube leads to a sudden drop in the axial force after an initial peak (Fig. 11.3(a)). This is catastrophic and the energy absorbed is very low. However, by designing a suitable triggering mechanism, such as chamfering one end of the tube, the same tube may instead fail by progressive crushing at a much higher and almost constant load. This is a most desirable feature of an energy absorber, as discussed in Chapter 1.

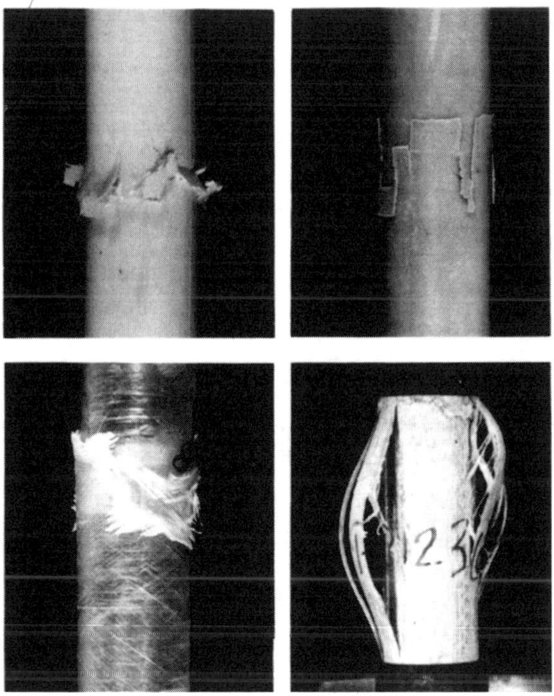

11.4 Centre failure of tubes made of four composites: (a) sheet moulding compound; (b) woven cloth; (c) helically wound; (d) pultruded (Hull, 1983).

A typical load–displacement curve is shown in Fig. 11.5 for a brittle fibre-reinforced plastic (FRP) tube under progressive crushing, as sketched in Fig. 11.6. Stage I of the load–displacement curve corresponds to crushing of the chamfered end and the subsequent formation of a crush zone, which progresses in stage II. Once the effective stroke is exhausted, the debris within the tube starts to compact, leading to a rapid increase in load again in stage III.

Micro-cracking and fracture of tube walls have been studied by several researchers, notably Hull (1991) and Farley and Jones (1992). Two extremes exist for the crushing mechanism at a micro-level for FRP tubes under progressive crushing and most specimens exhibit a combination of these two modes. One mode is called splaying (or lamina bending) and the other is fragmentation (or transverse shearing).

Splaying progressive crushing is illustrated in Fig. 11.7, with details shown in Fig. 11.8. Very long inter-laminar and parallel-to-axis cracks are the main feature, with few or no fibres cracking. The development of such a mecha-

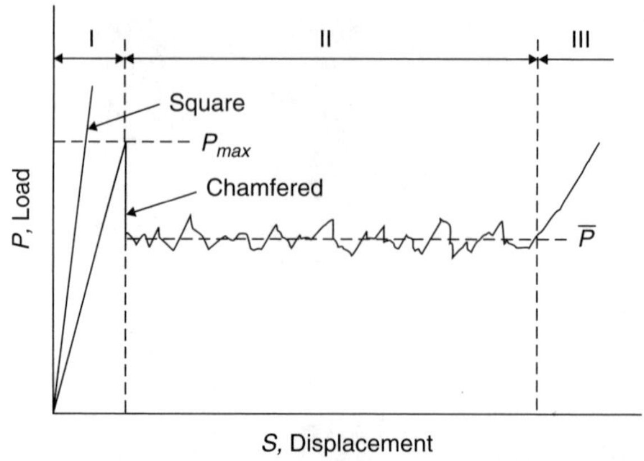

11.5 Load–displacement curve for tubes with ends chamfered (Hull, 1991) (reproduced with kind permission of Elsevier).

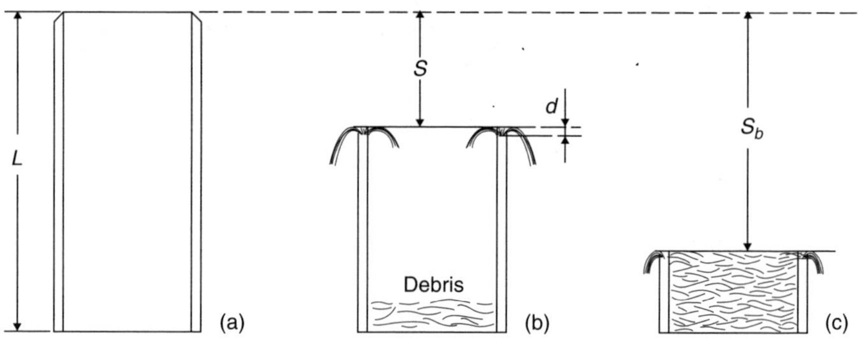

11.6 Sketch of a chamfered tube under axial crushing: (a) before loading; (b) during progressive crushing; (c) fully crushed with debris contained inside (Hull, 1991) (reproduced with kind permission of Elsevier).

nism is sketched in Fig. 11.9. The inner hoop-wound layer is crushed first with shear cracks (Fig. 11.9(a)), followed by the separation of the cracked hoop layers from the axial layers (Fig. 11.9(b)). Kinking and buckling of the fibres then takes place and a well-defined zone occurs with the formation of cracks around the middle of the axial layers (Fig. 11.9(c)). Afterwards, the crushed materials act like a wedge forcing the axial material to the inside and outside of the tube. Another example involving axial splitting and splaying is shown in Fig. 11.10, for a woven glass cloth-epoxy resin tube.

Figure 11.11 shows an example of the progressive crushing by ***fragmentation mode***, for a tube made from woven E-glass cloth and a general-purpose epoxy resin. This mode is characterised by formation of

11.7 Photograph of a splaying mode (Keal, 1983).

11.8 Details of a splaying mode showing debris wedge and wall crack (Keal, 1983).

fragments in the crush zone (Fig. 11.12), as a result of short inter-laminar and longitudinal cracks.

Details of the crush zone are shown in Fig. 11.13 and a sketch of the fragmentation sequence is given in Fig. 11.14. Briefly, for the (0/90/90/0) tube with a chamfer (Fig. 11.14(a)), crushing of the chamfer occurs first with splitting and comprehensive buckling in both hoop and axial directions (Fig. 11.14(b)), leading to the initial fragmentation. In the hoop direction,

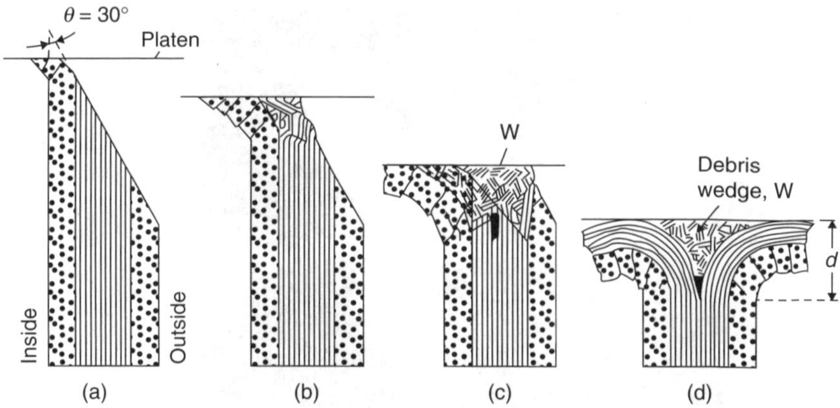

11.9 Sketch of progressive formation of a splaying mode crush zone (Hull, 1991) (reproduced with kind permission of Elsevier).

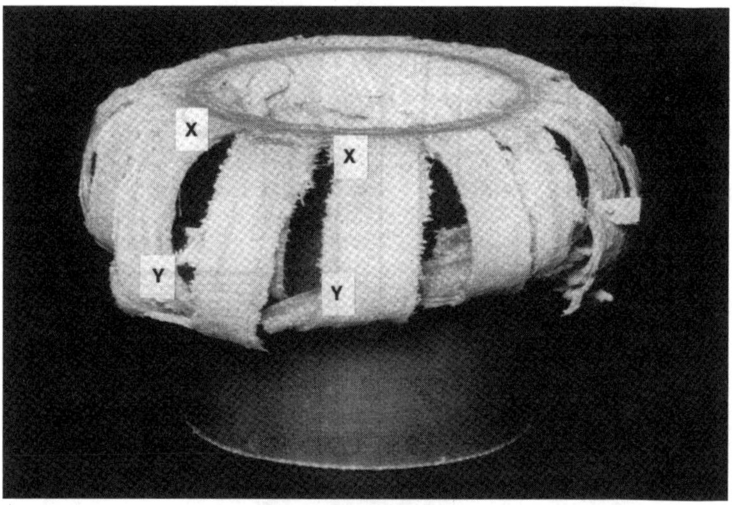

11.10 Another example of splaying mode involving a woven glass cloth-epoxy resin tube (Hull, 1991) (reproduced with kind permission of Elsevier).

compressive buckling on the inside of the tube and tensile fracture on the outside of the tube occur. This process repeats itself for the full wall section (Figs 11.14(c) and (d)).

Whether a splaying or fragmentation mode occurs depends on factors such as the laminate configuration and the failure strengths of the individual laminar failure modes. A particular mode is the result of competition

11.11 Fragmentation mode by a woven glass cloth tube (Hull, 1991) (reproduced with kind permission of Elsevier).

11.12 Details of internal fragments from the crush zone (Berry, 1984).

between parameters which favour either of these two modes. One key factor is the relative strength in the axial and hoop directions.

For tubes made from a series of glass cloth-polyester prepreg materials, when the number of the hoop fibres (H) is large compared with the axial

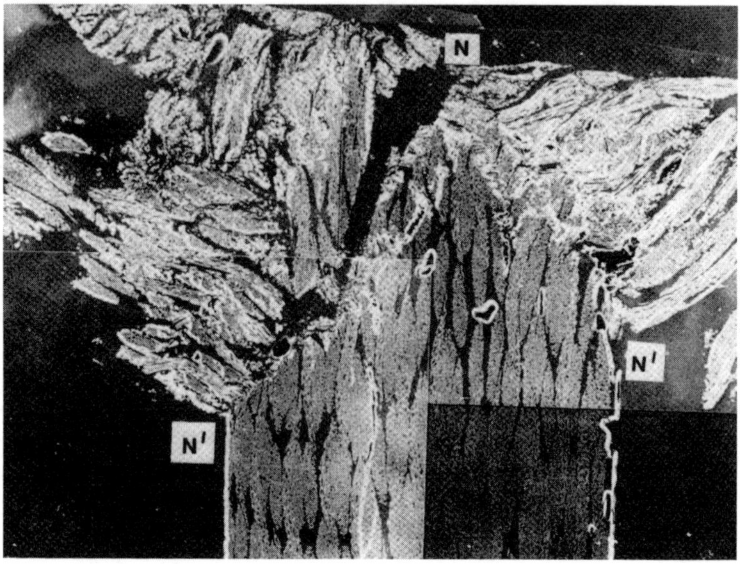

11.13 Photograph of a detailed crush zone (Berry, 1984).

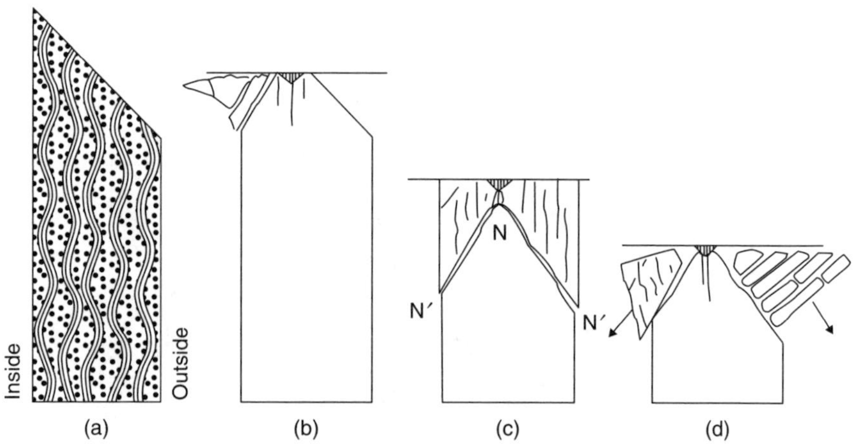

11.14 Sketch of the fragmentation sequence (Berry, 1984).

ones (A) (say H:A between 4:1 and 8.5:1), micro-fragmentation occurs. On the other hand, when the hoop constraint is weak (H:A between 1:7 and 1:8.5) no axial fibre fracture and splaying occurs. Gradually increasing the hoop resistance leads to a sharper micro-bending (with a small radius) and eventually fracture of the axial fibres. The effect of the hoop-to-axial ratio on the specific energy absorbed is shown in Fig. 11.15 for two different

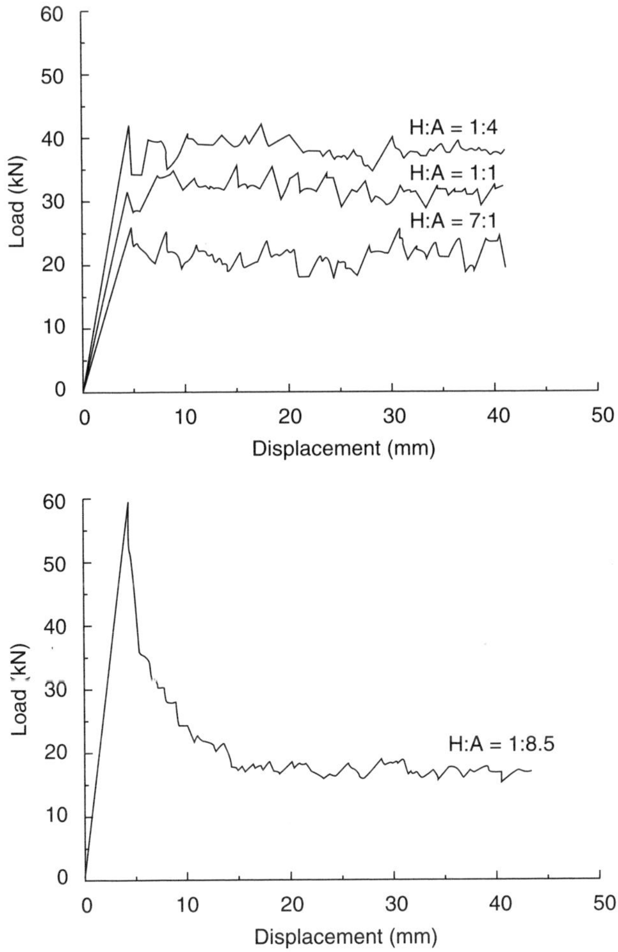

(a)

11.15 (a) Load–displacement curves for glass cloth-polyester resin tubes of different values of H:A; (b) effect of H:A ratio on specific energy absorption (•, 4 mm/s; o, 4 m/s) (Berry, 1984).

loading rates (4 mm/s and 4 m/s). Note that all the tubes have approximately 10 layers of cloth in a wall thickness of 2 mm.

Table 11.1 presents the results for a range of carbon fibre-reinforced epoxy tubes (Farley, 1986). Higher matrix failure strain leads to a fragmentation mode, with a superior specific energy performance to that of a splaying mode.

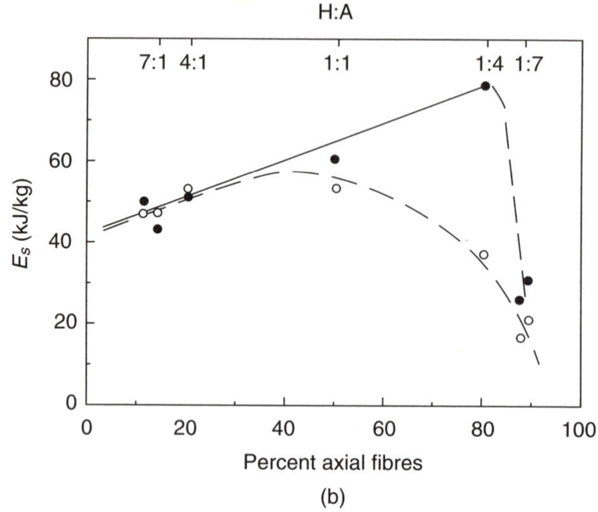

11.15 Continued

Table 11.1 Values of specific energy absorption and corresponding mode for carbon-reinforced epoxy tubes (Farley, 1986)

Matrix failure strain	Lay-up	Specific energy absorption (kJ/kg)	Failure mode
0.020	$[0/\pm 15]^4$	125	Fragmentation
0.010	$[0/\pm 15]^4$	94	Splaying (lamina bending)
0.020	$[0/\pm 45]^4$	85	Fragmentation
0.010	$[0/\pm 45]^4$	69	Splaying (lamina bending)
0.020	$[0/\pm 75]^4$	74	Fragmentation
0.010	$[0/\pm 75]^4$	54	Splaying (lamina bending)

11.2.2 Effect of ply orientation

The effect of winding angle for filament-wound glass fibre-polyester resin tubes is shown in Fig. 11.16, where ϕ is the angle between the fibre direction and the longitudinal axis of the tube. The tubes have four layers of fibres at $\pm\phi$ and the volume fraction of fibres is about 0.45. The maximum value of specific energy occurs at $\phi = \pm 65°$. For $\phi = 90°$, through-wall shear takes place in a fragmentation mode at the crush front and the hoop layers detach from the wall. It must be stressed that the trend in Fig. 11.16 is only true for this particular material. For graphite/epoxy tubes and Kevlar®/epoxy tubes, the effect of fibre orientation is different (Fig. 11.17). Other studies (Kindervater, 1990) broadly show a similar trend.

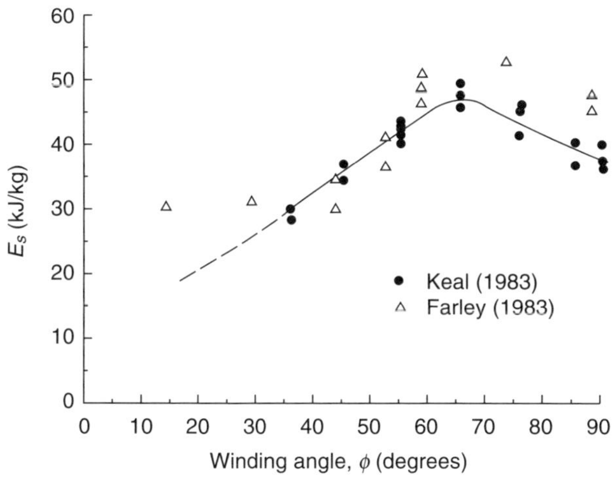

11.16 Effect of fibre orientation on specific energy absorption for glass fibre-polyester resin tubes (D = 50 mm, h = 3–4 mm).

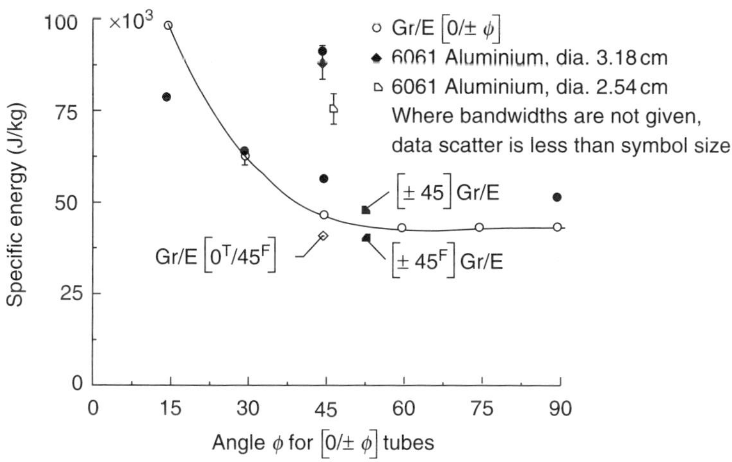

(a)

11.17 Effect of fibre orientation on specific energy absorption for FRP tubes: (a) graphite/epoxy tubes; (b) Kevlar/epoxy tubes; (c) glass/epoxy tubes (Farley, 1983).

11.2.3 Effect of diameter to thickness ratio (D/h)

Figure 11.18 illustrates the effect of the inner diameter to thickness ratio, D/h, on the specific energy for $[\pm 45]_N$ carbon-epoxy tubes (Farley, 1986). Clearly, increasing D/h leads to a large fall in the specific energy, as a result

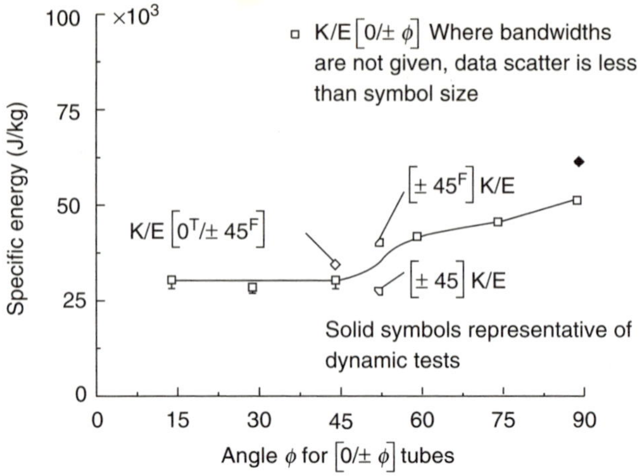

(b)

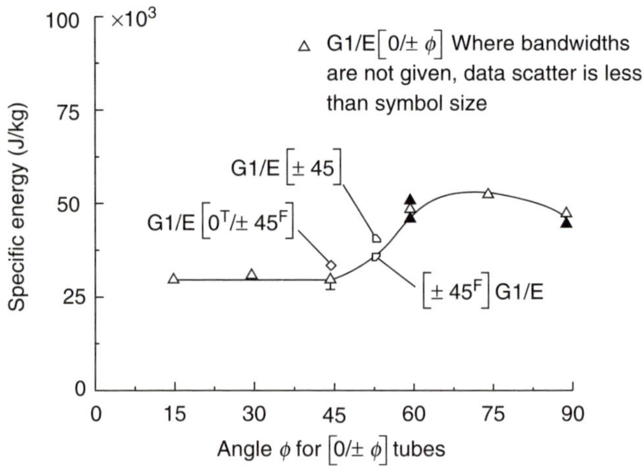

(c)

11.17 Continued

of an increase in the length and number of inter-laminar cracks with lower failure strength. A similar trend exists for glass-polyester conical frusta (Mamalis *et al.*, 1991c). However, carbon fibre/PEEK tubes behave differently. Their specific energy seems to be affected mainly by the absolute value of h rather than D/h; in the study by Ramakrishna and Hamada (1998), the specific energy was a maximum when h = 2~3 mm for D = 35.5–96.0 mm. Thornton and Edwards (1982) characterise the tubes by considering the relative density, which is defined as the ratio of the volume of the tube to that of a solid of the same external dimensions. Stable collapse

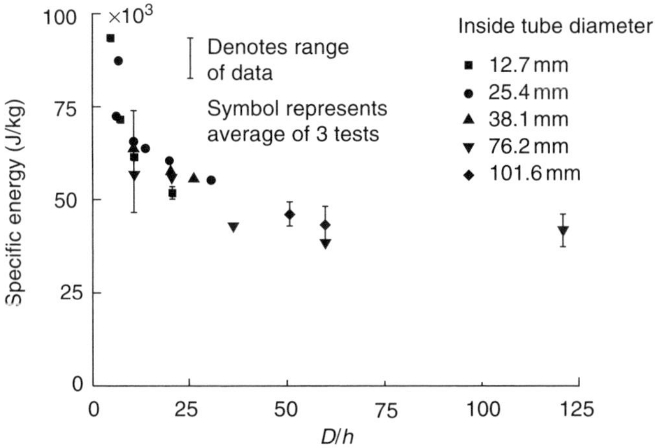

11.18 Effect of inside diameter to thickness ratio on specific energy for carbon epoxy tubes (Farley, 1986).

occurred when the relative density was above 0.025 and 0.06 for carbon and glass FRP tubes, respectively, and below 0.15 for aramid circular tubes. Fairfull and Hull (1987) studied the effect of specimen dimensions on the energy-absorption capacity for glass cloth/epoxy tubes.

11.3 Axial crushing of tubes with other geometries

Square and rectangular tubes absorb less energy than comparable circular ones, when subjected to axial crushing (Thornton and Edwards, 1982; Mamalis *et al.*, 1992 and 1996a). In particular, it has been reported that square and rectangular tubes have respectively 0.8 and 0.5 times the specific energy absorption of similar circular sections (Kindervater, 1990). This observation is similar to the trend for metal tubes, although the exact mechanism is different.

Two other sections studied are a 'near-elliptical' shell (Farley and Jones, 1992) and a rail beam (Mamalis *et al.*, 1996a), Fig. 11.19. For carbon-epoxy and aramid-epoxy tubes of a 'near-elliptical' section, the crushing mode near the ends of the major axis was predominantly high energy brittle fracture while it was mainly lower energy lamina bending for those away from these regions. The energy-absorption capability increased as the included angle decreased; namely, as the tube became more elliptical, its energy-absorption capability increased. In fact, an increase of between 10 % and 30 % in the specific energy absorption was obtained by reducing the included angle from 180° (circular) to 90°. This could be because a larger

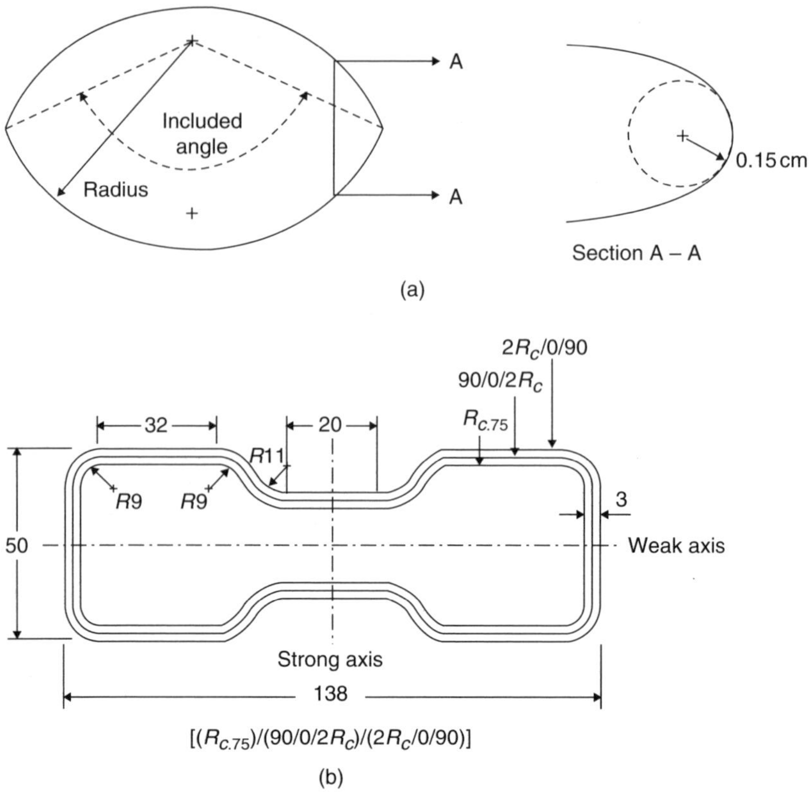

11.19 Tubes of other cross-section: (a) a 'near-elliptical' shell (Farley and Jones, 1992); (b) a rail beam (Mamalis *et al.*, 1996a).

portion of material fails by progressive crushing when the included angle is smaller, leading to better energy-absorption performance.

The hour-glass cross-section automotive frame rail considered (Fig. 11.19(b)) is made of a glass fibre/vinylester composite material. The specific energy of the progressively collapsed specimens is almost constant as the ratio of thickness/axial length increases. Also, this section has higher values of specific energy than comparable square tubes.

Energy absorption of conical shells made of composite materials has been studied (as were metallic cones). Some collapse modes observed are shown in Fig. 11.20 (Mamalis *et al.*, 1991a and 1996b). Broadly speaking, two modes exist for conical frusta: progressive crushing and central failure, similar to the case of circular tubes (Fig. 11.4). The specific energy absorption decreases as the semi-angle of the composite frusta increases. Further, when the semi-angle is larger than a critical value, lying between 15° and

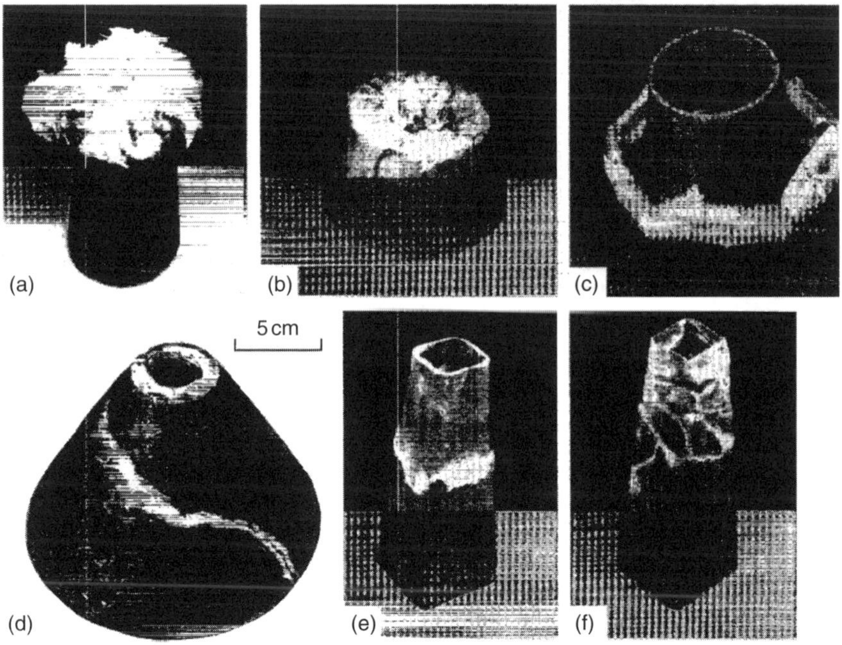

11.20 Various collapse modes of axially crushed conical shells (Mamalis *et al.*, 1991c, 1996b).

20°, unstable collapse occurs. Besides, conical shells do not need a collapse trigger mechanism, unlike circular tubes. Finally, axially loaded square frusta absorb less specific energy than comparable circular ones. The effect of foam-filling was studied for composite frusta (Gupta and Velmurugan, 1999) and for circular tubes with braided composite walls (Harte *et al.*, 2000).

11.4 Tubes under bending

Energy absorption for circular and square/rectangular tubes made of fibre-reinforced glass vinylester and polyester composites subjected to bending has been studied. Typical moment–rotation curves and the deformed specimens are shown in Fig. 11.21 for circular tubes and in Fig. 11.22 for square ones. The fracture characteristics on the tension side are very different from those on the compression side, which mainly exhibit local buckling of the fibres. Furthermore, the end clamping conditions greatly affect the moment–rotation curve (Fig. 11.21(b)). The insertion of a plug in the clamped end was found to reduce energy dissipation owing to shortening of the post-crushing regime. Clamping devices with round edges tend to

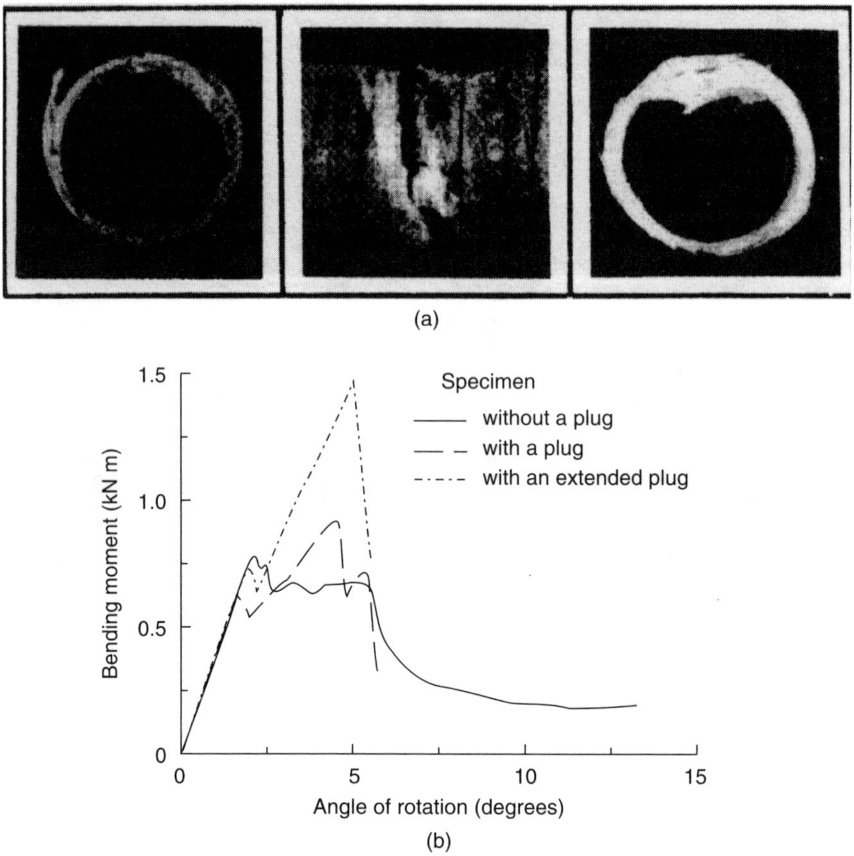

11.21 Bending collapse of a circular tube: (a) left, front and right views of the crush region; (b) moment–rotation curves of a tube with the same dimension but different clamping conditions (D = 56 mm, h = 2.3 mm, L = 256 mm) (Mamalis *et al.*, 1998) (reproduced with kind permission of CRC Press).

reduce stress concentrations and delay crack initiation and propagation, leading to a higher bending moment and energy absorption.

Both tubes for the curves in Fig. 11.22(b) have approximately the same dimensions ($40 \times 51 \times 2.7$ mm) with two layers, but are bent about a different axis. When bent about its strong axis, the energy absorption is better. Also, the energy absorption and peak moment increase with the tube thickness (not shown in this figure). It is interesting to note that rectangular tubes were found to have better crashworthy characteristics for large deformation under bending than circular tubes of similar dimensions. Bending tubes of hourglass shape (Fig. 11.19) was reported by Mamalis *et al.*).

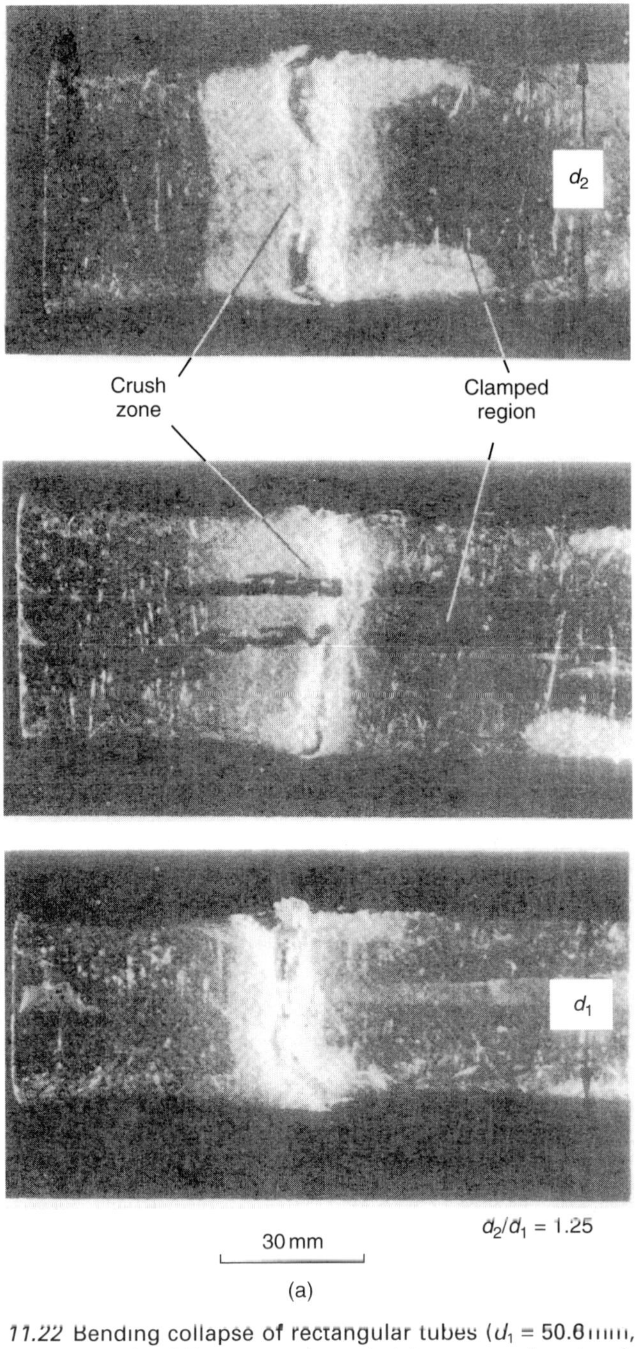

(a)

11.22 Bending collapse of rectangular tubes (d_1 = 50.6 mm, d_2 = 41.0 mm, h = 2.7 mm, two layers): (a) compression, tension and side views; (b) moment–rotation curves for bending about two different axes (Mamalis *et al.*, 1998) (reproduced with kind permission of CRC Press).

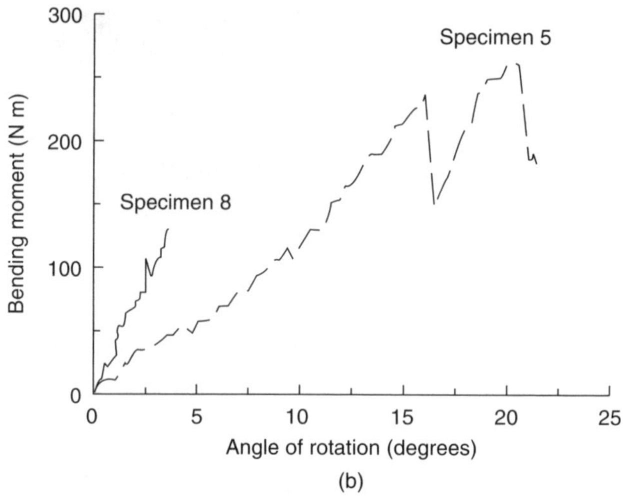

11.22 *Continued*

11.5 Comments on the crushing of composite tubes

Most of the presentation above relates to experimental observations of the global crushing behaviour (load-displacement curves and macro-mechanism of deformation/fractures), as well as detailed micro-level discussion of the energy-absorption mechanisms. Quantitative assessment or prediction of various energy components is very difficult for these complex processes. Nevertheless, interested readers should refer to the book by Mamalis *et al.* (1998) for some analytical treatment of tubular components under axial loading or bending.

Two other important factors have not yet been mentioned: loading rate and temperature. Present findings are inconclusive: some authors have reported an increase in specific energy absorption with an increased loading rate, while others have noted that this decreases or is insensitive to loading rate. The key to understanding this is to assess the relative sensitivity of each mechanism to the loading rate and then estimate the overall response. For example, friction can be a major mechanism and this generates a considerable amount of heat and, therefore, the magnitude of energy absorbed by this mechanism will be likely to vary with loading rate.

Temperature affects fracture toughness, the compressive strength of fibres and resins as well as friction. Woven glass fibre fabric/epoxy tubes tested at lower temperatures absorb a larger amount of energy than at room temperature (Ramakrishna and Hamada, 1998). A similar trend was found for glass cloth/epoxy tubes and glass fibre/polyester tubes, but not for carbon fibre/epoxy tubes, which show constant specific energy absorption for temperatures up to 150°, followed by a rapid decrease (Thornton *et al.*, 1985). Carbon/PEEK thermoplastic composite tubes when crushed progressively fail by splaying mode at temperatures between –60°C and 150°C. Whereas at lower temperatures (–100°C and –80°C) they failed catastrophically by axial splitting of tube walls leading to a lower value of specific energy (a result of the fact that the fracture toughness at –60°C is half of that at 100°C). At higher temperatures (>20°C), the specific energy decreases, too, owing to the decrease of compressive strength of composite materials, the reduced friction forces and smaller amount of fibre fracture with increasing temperature.

11.6 Axial crushing of composite wrapped metal tubes

Fibre/epoxy composites may be combined with metal tubes to absorb a greater amount of energy. As discussed in Chapter 6, when axially crushed, circular metal tubes fold both outwards and inwards. By partially constraining the tube wall's outward movement, the tube may be more efficient in energy absorption. Some examples of collapse modes are shown in Fig. 11.23 for metal tubes externally wrapped in glass/epoxy composites with a different number of layers and wrapping directions (Song *et al.*, 2000). When the inner metal tube of the compound tube is ductile, it deforms in asymmetric mode (diamond mode in Fig. 6.2), regardless of metal thickness and ply orientation (Fig. 11.23(a)). Brittle inner metal tubes undergo multiple fractures (Fig. 11.23(b)), accompanied by fibre fragmentation. When the ply angle is small, almost parallel to the tube axis, axial delamination occurs for the composites and they separate from the inner metal tubes (Fig. 11.23(c)). Finally, a brittle metal inner tube wrapped with thick composites at a large angle (almost circumferential only) may fail by unstable fracture collapse (Fig. 11.23(d)).

Theoretical analysis for such tubes failing in asymmetric mode is complex. However, note that for metal tubes the energy absorption is independent of the mode (Chapter 6). Hence by analogy, analysis for axisymmetric collapse of a compound tube should provide some guidance, following exactly the procedure for metal tubes only (Section 6.3). Consider the folding mechanism shown in Fig. 11.24 for an axisymmetric mode. Assume that the composite does not contribute any additional energy when the tube wall is bent with the composite in tension and eventually fracturing. The only additional considerations will then be bending energy for the composite under compression and the stretching energy of the composite.

The fully plastic bending moment of the combined metal and composite may be estimated, following Mamalis *et al.* (1991b), assuming that the

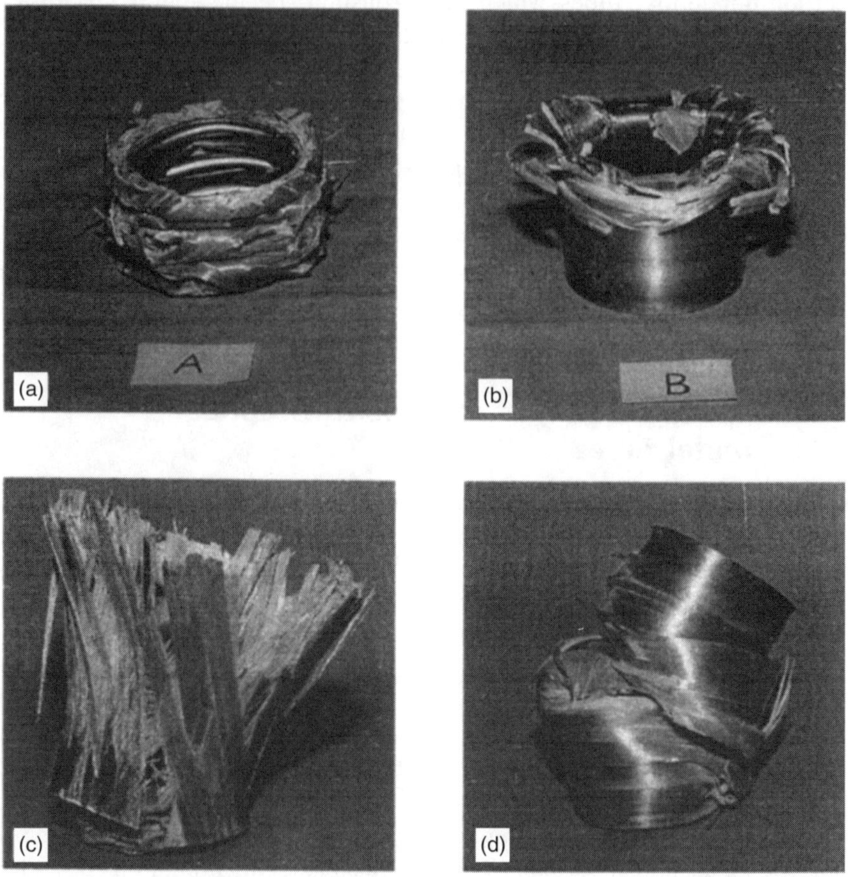

11.23 Failure modes of metal tubes externally reinforced with glass/epoxy composites (Song *et al.*, 2000) (reproduced with kind permission of Elsevier).

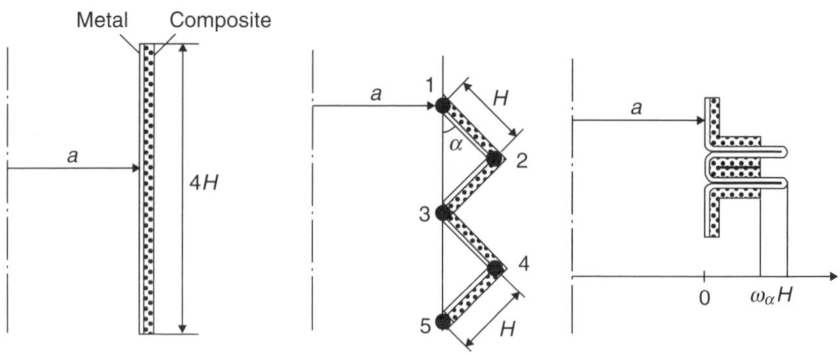

11.24 Collapse mode of a composite/metal wall assumed by Hanefi and Wierzbicki (1996).

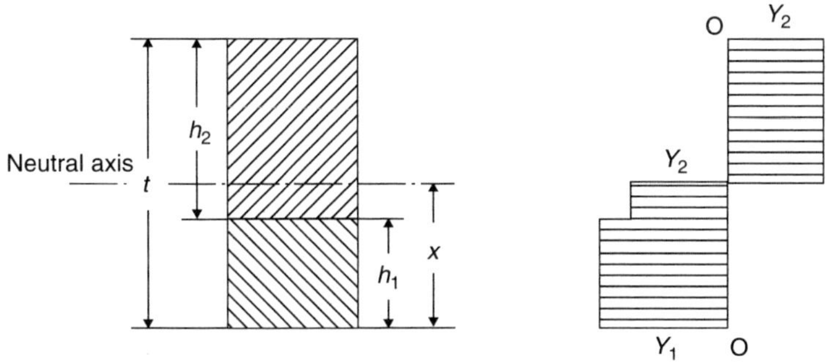

11.25 Stress state of a fully yielding section consisting of two materials (Mamalis *et al.*, 1991b).

behaviour of the composite is perfectly plastic under compression. Refer to the stress state shown in Fig. 11.25 for two materials of thickness h_1 and h_2, and yield stress Y_1 and Y_2, respectively. From the condition that the resultant force over the whole cross-section is zero, the neutral axis is determined by

$$x = \frac{Y_2(h+h_1) - Y_1 h_1}{2Y_2} \tag{11.1}$$

Taking moments about this neutral axis, the fully plastic bending moment is:

$$\begin{aligned} M_o &= Y_2 \frac{(h-x)^2}{2} + Y_2 \frac{(x-h_1)^2}{2} + Y_1 h_1 \left(x - \frac{h_1}{2} \right) \\ &= \frac{Y_2 h_2^2}{4} \left[1 + 2\frac{Y_1 h_1}{Y_2 h_2} + 2\frac{Y_1 h_1^2}{Y_2 h_2^2} - \left(\frac{Y_1 h_1}{Y_2 h_2} \right)^2 \right] \end{aligned} \tag{11.2}$$

Assume that the real value of the fully plastic moment for the compound tube is an average of Eq. [11.2] and that for bare metal

$$M_o = C \frac{Y_m h_m^2}{4} \tag{11.3}$$

where

$$C = \frac{1}{2} \left[2 + 2\frac{Y_c h_c}{Y_m h_m} + 2\frac{Y_c h_c^2}{Y_m h_m^2} - \left(\frac{Y_c h_c}{Y_m h_m} \right)^2 \right]$$

Subscripts c and m denote composite and metal, respectively.

Thus the total bending energy for hinges 1 to 5 in Fig. 11.24 is

$$W_b = \pi^2 D\, C Y_m h_m^2 \tag{11.4}$$

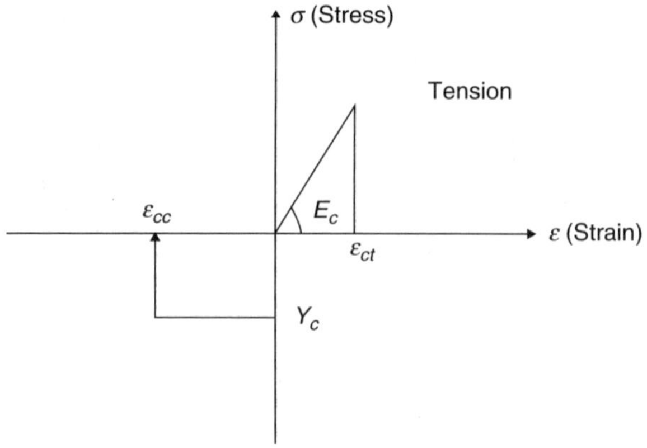

11.26 Idealised stress–strain relationship for composites.

The stretching energy for metal tube is, as before

$$W_m = \frac{\pi D}{2} Y_m h_m H^2 \tag{11.5}$$

Stretching energy for the composite, ignoring the fractured portion, is

$$W_c' = \pi D h_c E_c \varepsilon_{ct}^2 H \tag{11.6}$$

for the assumed stress–strain relationship for composites shown in Fig. 11.26. If the composite is compressed circumferentially when folding inwards, the energy is then

$$W_c'' = 2\pi h_c Y_c H^2 \tag{11.7}$$

The total membrane energy for the composite is assumed to be

$$W_c = W_c' + W_c'' = 2\pi h_c H\left(HY_c + \frac{D}{2} E_c \varepsilon_{ct}^2 \right) \tag{11.8}$$

From work balance, the mean force is

$$P_m = \frac{1}{4H}(\pi^2 D Y_m C h_m^2 + 4\pi h Y_{eq} H^2 + \pi D h_c E_c \varepsilon_{ct}^2 H) \tag{11.9}$$

where the equivalent stress $Y_{eq} = 1/h[Y_m h_m + (Y_c h_c)/2]$. The value of H is found by letting $dP_m/dH = 0$; then substituting H back into Eq. [11.9] leads to the average crushing force. This analysis is due to Hanefi and Wierzbicki (1996) and agrees fairly well with experiment (Wang *et al.*, 1992). Further modification was subsequently made which incorporates the effect of wrapping angle (X. Wang and Lu, 2002).

11.7 Composite sandwich panels

Laminates and composite sandwich panels are widely used, for example in aircraft. Their resistance to impact by a projectile is important in cases such as bird strike. The impact velocity (and hence loading rate) can be much higher than those cases previously discussed. Their energy-absorbing behaviour is dependent upon the material and dimensions of the skins as well as the core materials. Here we present studies of laminates first, followed by those of sandwich panels.

11.7.1 Penetration energy of composite skins

We discuss two types of composite laminates which can be used as skins of sandwich panels: E-glass woven laminates and Kevlar® laminates. Figure 11.27 shows specimens after static test of laminates which were made by hand lay-up using two weights of E-glass woven roving: 800 g/m^2 and 1500 g/m^2 (Roach *et al.*, 1998). The matrix was a polyester resin. Square specimens (200 × 200 mm^2) were fully clamped and indented with a flat-faced 20 mm diameter cylinder. Static tests were stopped at the maximum load of penetration. The corresponding static energy is shown directly related to the delamination area (Fig. 11.28), for different thicknesses. Similar trends exist when specimens are impacted with the indenter (Fig. 11.29 for a velocity of 60 m/s). The delamination zone is circular for static loading, but less so for impact loading. The effect of impact velocity seems negligible in the range studied.

Figure 11.30 shows load–indenter displacement curves for Kevlar®/polyester laminates, with diameter 114 mm and fully clamped at the edge (Zhu

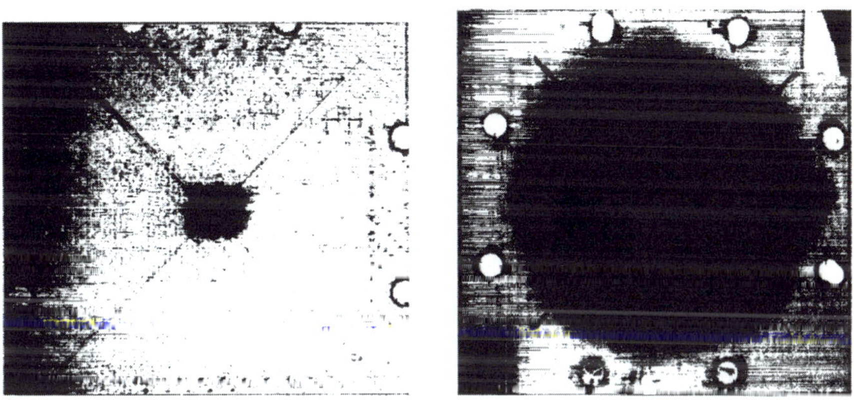

11.27 Photograph of E-glass woven laminates after static indentation (Roach *et al.*, 1998) (reproduced with kind permission of Elsevier).

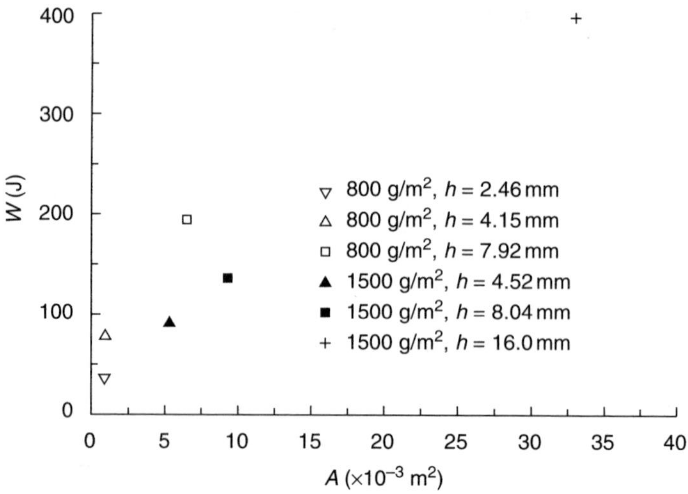

11.28 Plot of static energy versus delamination area for E-glass woven laminates of different weights (Roach *et al.*, 1998) (reproduced with kind permission of Elsevier).

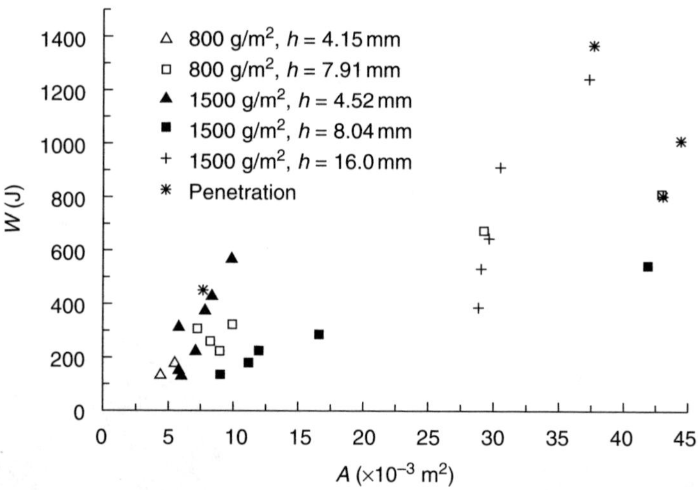

11.29 Same as Fig 11.28, but for a velocity of 60 m/s (Roach *et al.*, 1998) (reproduced with kind permission of Elsevier).

et al., 1992a). The 12.7 mm diameter indenter has a conical tip of 60°. Laminated plates are arranged in 0/90 and 0/45 lay-ups, with 5–24 layers (3.1–15 mm thick). Photographs of the damage for a 6.35 mm thick Kevlar/polyester laminate (Fig. 11.31) show initial global plate deflection, Fig. 11.31 (a), successive fibre failure and bulging at the distal side, Figs 11.31 (b)–(d), with

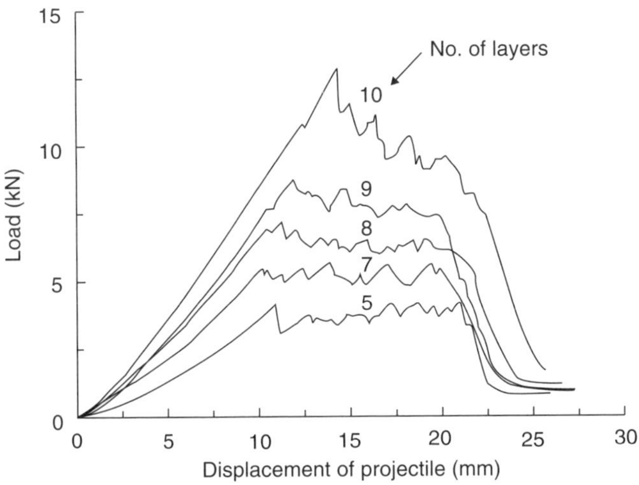

11.30 Quasi-static load–displacement curves of Kevlar/polyester laminates with different thicknesses by a 12.7 mm diameter steel cylinder with a 60° tip angle (Zhu *et al.*, 1992a)

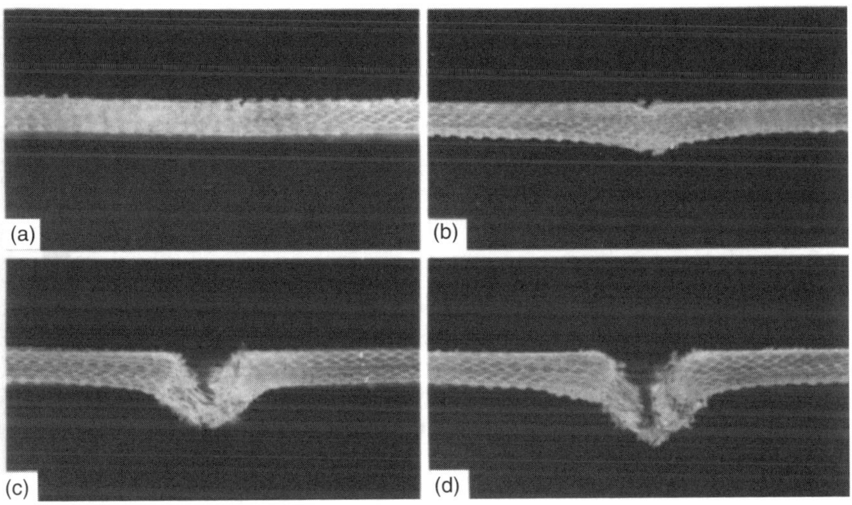

11.31 Photographs of the damage in four 6.35 mm thick laminates at four different indenter positions: (a) 6.35 mm; (b) 12.7 mm; (c) 19.05 mm and (d) 25.4 mm. The indenter and specimen are the same as in Fig. 11.30 (Zhu *et al.*, 1992a) (reproduced with kind permission of Elsevier).

delamination. The first peak in load (Fig. 11.30) corresponds to initiation of the fibre failure. Increasing the indenter diameter does not change the initial slope of the load–displacement curve, but the peak force reached is roughly proportional to the diameter of the projectile. The cone angle has

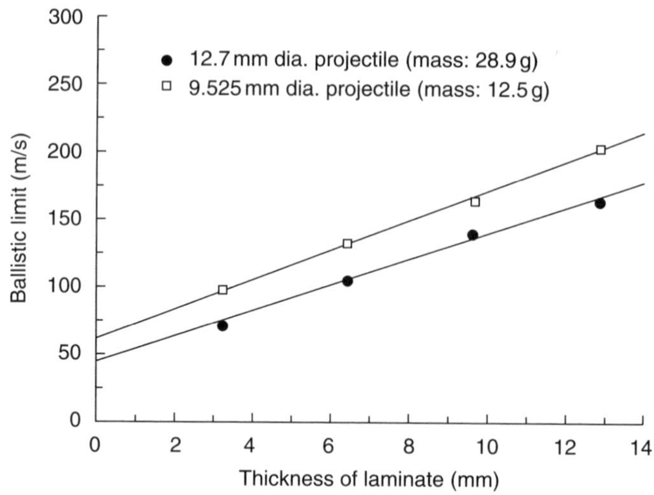

11.32 Ballistic limits of Kevlar/polyester composites with different thicknesses (Zhu *et al.*, 1992a).

a similar effect, with an increase in peak force (about 40 %) and energy absorption when the cone angle varies from 60° to 120°.

Impact tests produce more localised damage. The ballistic limit, which is defined as the velocity when the projectile is either stuck in the target or else exits with negligible velocity, increases with the total plate thickness (Fig. 11.32). An analytical model was proposed by Zhu *et al.* (1992b), which agrees with test results.

11.7.2 Composite sandwich panels

Good energy absorption can be achieved by employment of sandwich components: energy is dissipated by bending, stretching and fracture of the skins and by localised crushing of the core. Honeycombs and foams have excellent energy-absorption characteristics (Chapter 10) and they are commonly used as a sandwich core. These sandwich panels can be supported at the back fully, or (simply or fully) supported at the edges, depending on the application. When the impact energy is low, plastic deformation of both the skins and core occurs without any skin perforation or tearing. Delamination between the skin and core may take place. For high velocity ballistic impact, deformation is more localised and penetration takes place.

Bonding between the skin and core plays an important role, as evidenced by the static load–displacement curves (Fig. 11.33) for sandwich panels fully supported at the back face and indented with a 74 mm diameter cylindrical

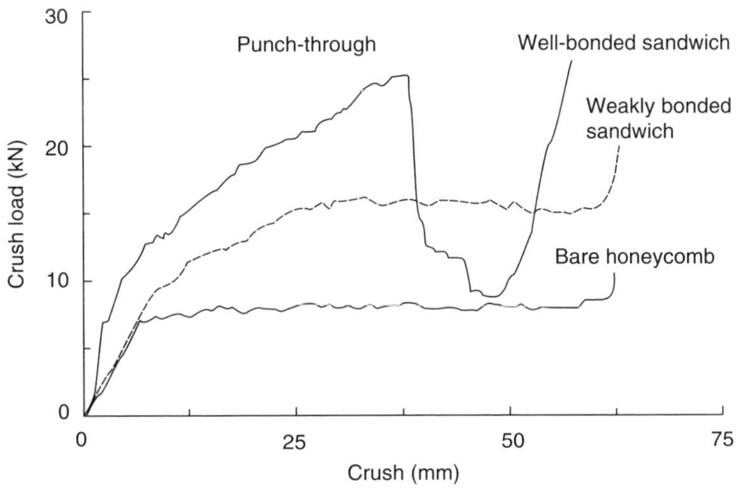

11.33 Quasi-static load–displacement curves for a sandwich panel fully supported at the back and indented with a radius cylindrical shell (Goldsmith and Sackman, 1992).

shell with a convex face (radius 458 mm) (Goldsmith and Sackman, 1992). The skins were 0.81 mm thick aluminium plates and the core was aluminium honeycomb (cell diameter 3.18 mm, thickness 0.0254 mm and density 72 kg/m^3). The crushing load for weakly bonded sandwich was almost twice that of the bare honeycomb. This increase is due to bending of the top skin and crushing of additional core material immediately around the indenter. A properly bonded sandwich has a much higher load, but the punch-through failure of the top skin occurs earlier than the weakly bonded sandwich. The crushing load and energy absorption increase with core density. For skins of the same thickness, aluminium faces produce much higher load and energy absorption than plastics such as fibreglass and Lexan®, which have similar performance.

For static load and low velocity impact, buckling of the core occurs uniformly over the sandwich thickness. High velocity impact may initiate buckling at both the front and distal faces as a result of stress wave propagation and reflection.

Similar load–displacement curves are produced by a sandwich panel which is simply supported and loaded at the centre compared with a comparable panel fully supported at the back face before punch-through is achieved (Fig. 11.34). Further, the dynamic curve for a velocity of 25 m/s is almost identical to its static counterpart, indicating a negligible dynamic effect for this velocity range. The deformed specimen (Fig. 11.35), which is similar to the static one, confirms this observation.

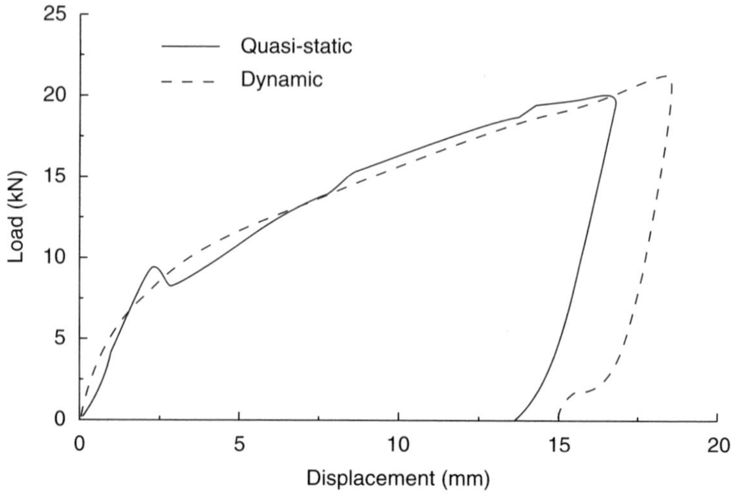

11.34 Same as in Fig. 11.33, but the panel is simply supported at the edges.

11.35 Photograph of a deformed simply supported sandwich panel struck with a projectile at 1219 m/s. (Goldsmith and Sackman, 1992) (reproduced with kind permission of Elsevier).

Perforation of sandwich panels occurs when the impact energy is sufficiently large. Examples of this are shown in Fig. 11.36 for Coremat® panels and Aerolam panels. Analysis of this complicated process is difficult. Nevertheless, some simplified mechanism may be used to estimate the

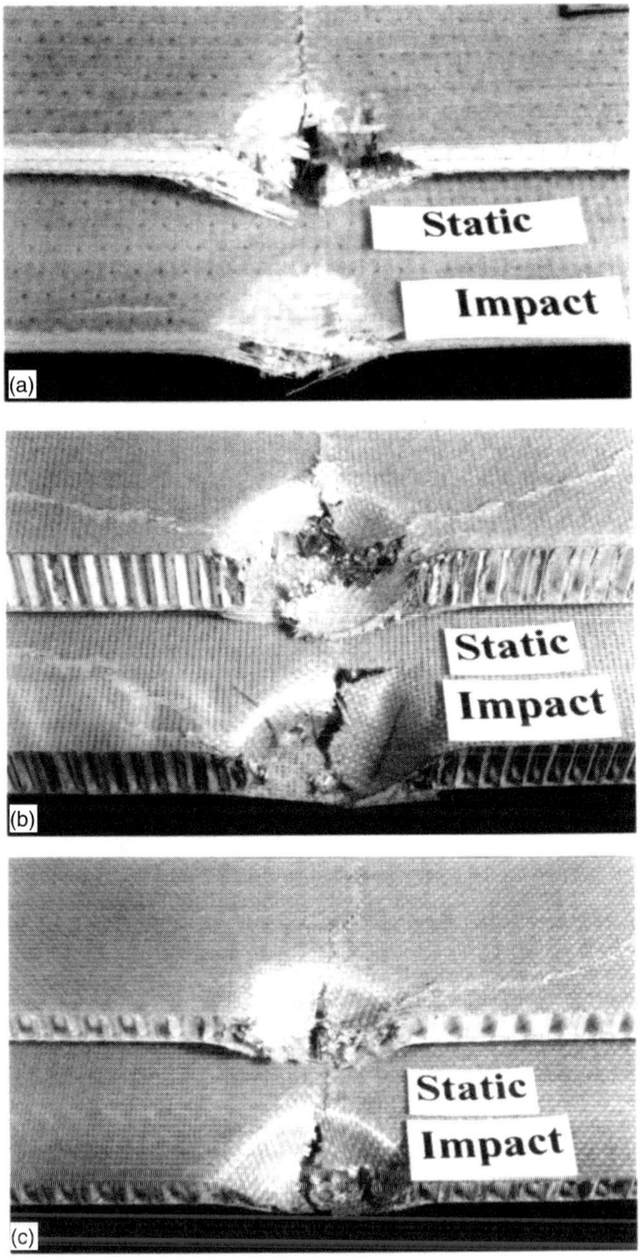

11.36 Photographs of sandwich panel cross-sections after static and impact perforation: (a) Coremat (impact mass/drop height: 20 kg/3 m); (b) 26 mm Aerolam (impact mass/drop height: 20 kg/3 m); (c) 13 mm Aerolam (impact mass/drop height: 10 kg/3 m) (see Mines *et al.* (1998) for details of fabrication) (reproduced with kind permission of Elsevier).

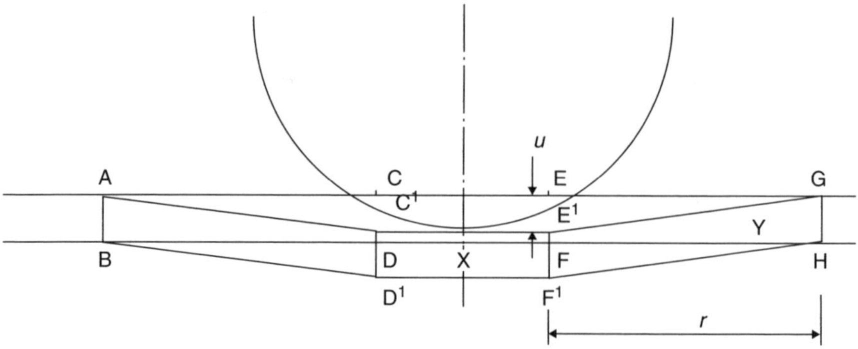

11.37 Idealised deformation mode by Mines *et al.* (1998).

energy absorption, e.g. Fig. 11.37 for Coremat® panels. The central region X is under compression while region Y undergoes shear. Thus, the energy in X can be assessed from the stress–strain curve of the core material and the volume in X. The energy associated with the shear dominated region Y can be estimated in a similar way. The shear stress–strain curve may be obtained from a uniaxial compression stress–strain curve, recognising that

$$\tau_{xy} = \frac{\sigma}{\sqrt{3}} \quad \text{and} \quad \gamma_{xy} = \frac{\varepsilon}{\sqrt{3}} \qquad [11.10]$$

The shear strain is related to the central displacement u as

$$\gamma_{xy} = \frac{u}{r} \qquad [11.11]$$

This analysis requires that the size of regions X and Y, and the deflection at perforation, are known, which can only be obtained from experiment.

Dimensional analysis by Reid and Wen (2000)

Owing to the complex nature of this problem, a dimensional analysis is used to obtain empirical equations of, for the static case, the energy absorbed to fracture, E_{frac}. This can be assumed to be the sum of local penetration energy E_l and energy corresponding to the global deformation E_g, i.e. (Reid and Wen, 2000)

$$E_{frac} = E_l + E_g \qquad [11.12]$$

Energy for global deformation is assumed to be of the form

$$\frac{E_g}{\sigma_u D^3} = A\left(\frac{L}{D}\right)^{\beta_1}\left(\frac{T}{D}\right)^{\beta_2} \qquad [11.13]$$

where σ_u is the failure stress of the laminates in tension, D is indenter diameter and T is total laminate thickness. A, β_1 and β_2 are constants to be determined experimentally.

For hemispherical-ended indenters, the local energy is

$$E_l = \frac{\pi D^2 T \sigma_u \varepsilon_{frac}}{8} \tag{11.14}$$

where ε_{frac} is fracture strain of the laminates.

For flat-faced punches

$$E_l = \frac{\sigma_u T D^2 \pi}{4}\left(\frac{K\sigma_c}{\sigma_u}\right)\left(\frac{C}{T}\right)\varepsilon_d + \frac{\sigma_u T D^2 \pi}{2}\left(\frac{\tau_{13}}{\sigma_u}\right)\left(\frac{T}{D}\right)\left(\frac{h}{T}\right)^2 + \frac{\sigma_u T D^2 \pi \varepsilon_{frac}}{8}\left(\frac{h}{T}\right) \tag{11.15}$$

where K is a factor usually taken to be 2. σ_c and ε_d are the compressive strength and densification strain of the core, respectively C and h are thickness of the core and skin, respectively. τ_{13} is the shear strength of the skin laminates. The three terms on the right hand side of Eq. [11.15] represent the energies absorbed by core crushing, shear plugging of the upper skin and fragmentation of the lower skin, respectively. Combining Eqs [11.12]–[11.15] gives the static energy to fracture of a sandwich panel.

A dynamic enhancement factor ϕ can be used again to characterise the dynamic effects. Empirically,

$$\phi = 1 + \beta\left(\frac{V_i}{V_o}\right) \qquad (V_i < V_o) \tag{11.16a}$$

$$\phi = 1 + \beta \qquad (V_i > V_o) \tag{11.16b}$$

where β is an empirical constant and velocity V_0 is given by

$$V_o = \sqrt{\frac{E_l}{\rho_l}}\varepsilon_{frac} \tag{11.17}$$

with E_l and ρ_l being the in-plane Young's modulus and density of the laminates, respectively. This velocity (≈80 m/s) indicates a transition of energy dissipation mechanisms between low and high velocity regions.

For high impact velocities, the failure of a sandwich panel may be regarded as being wave dominated. This is evident from experimental observation that for high velocities, debonding between the skins and core is less, but delamination within the skin laminates occurs within a larger area. In this case, the dynamic stress enhancement may be regarded as being the same as that in Eq. [11.16]. Namely, for the static linear elastic limit in through-thickness compression of the laminates, σ_e, the dynamic resistance is

$$\sigma = \left(1 + \Gamma\sqrt{\frac{\rho_l}{\sigma_e}}\right)\sigma_e \qquad [11.18]$$

where Γ is an empirical constant (2, 1.5 and $2\sin\theta/2$ for flat-faced, hemi-spherical-ended and conical-nosed projectiles, respectively) and θ is the cone angle. Therefore, the energy dissipated is

$$E = \frac{\pi D^2 T\sigma}{4} = \frac{\pi D^2 T}{4}\left(1 + \Gamma\sqrt{\frac{\rho_l}{\sigma_e}}V\right)\sigma_e \qquad [11.19]$$

The ballistic limit can be worked out by equating this energy to the kinetic energy of a projectile.

12

Case studies

Through four typical examples, this chapter demonstrates how to apply the theoretical models and fundamental studies of energy-absorbing elements illustrated in the previous chapters to real engineering problems, e.g. how to select materials and how to determine structural parameters so as to match the energy-absorption requirement to the structure/device concerned, together with other considerations in engineering design.

12.1 Rockfall protective net

12.1.1 Rockfall and its protection

When buildings, dams or roads are constructed in mountain or hill areas, an important safety concern is how to protect these structures from rockfall. Rockfall is the downslope movement of boulders (from natural slopes) or rock blocks (from cut faces) that, if not properly restrained, has the potential to destroy or damage structures along its path or create an obstacle to public transportation networks. The rock's motion down a slope may be triggered by disturbance caused by the construction work itself, or by rainfall, gale or earthquake after the construction.

Depending on the geological conditions, some sites may require permanent protective barriers/fences made of concrete or rock, and/or ditches, while others may only need a protective net, which is relatively light and moveable from one site to another after being used during the construction period. Figure 12.1(a) shows a typical rockfall protective net. In general, an ideal rockfall protective net should absorb the energy from a rockfall safely, no matter where the collision takes place on the net; it should be easy to install and maintain and the impact to posts should not result in a collapse of the structure.

12.1 (a) A rockfall protective net made of connected metal rings and (b) the protective net arrests the rolling rock and dissipates its kinetic energy by plastic deformation of the rings.

12.1.2 Energy-absorption capacity of a protective net

The protective net shown in Fig. 12.1(a) is made of metal rings and each ring is connected by four identical rings in its neighbourhood. When a rock rolls down from a slope, the protective net will arrest the rolling rock and dissipate its kinetic energy by plastic deformation of the rings, as shown in Fig. 12.1(b).

As sketched in Fig. 12.2, when the net is stretched, a ring of radius r in the stretched region will gradually change from a circular to a square shape, as a result of bending caused by the concentrated forces applied at the four corners along the diagonal directions from the neighbouring

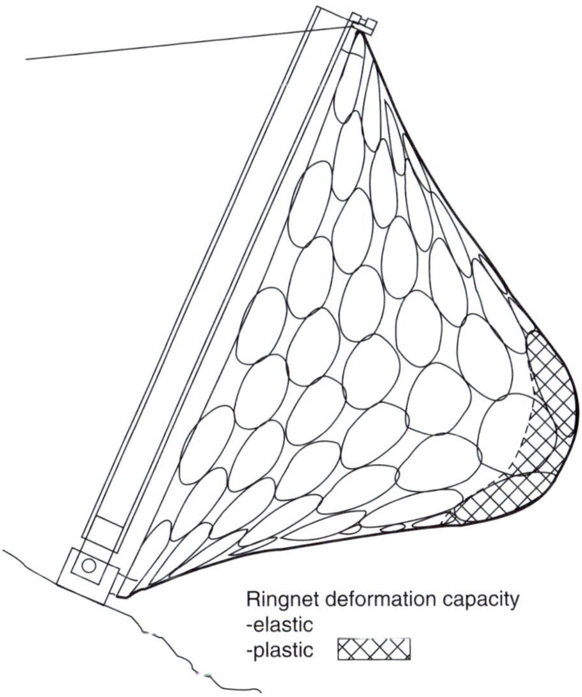

(b)

12.1 Continued

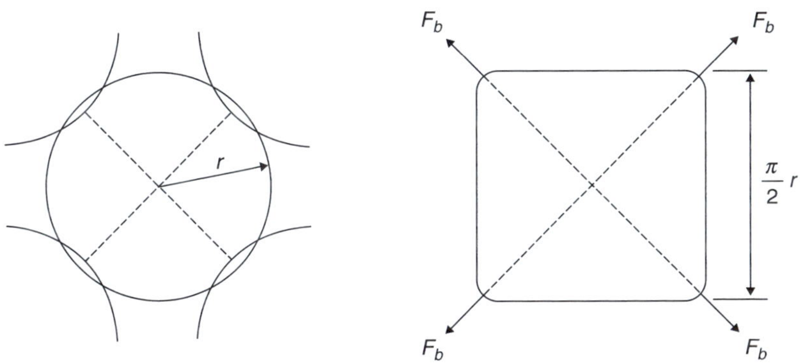

(a) (b)

12.2 Large deformation mechanism of a ring simultaneously pulled along two diagonal directions: (a) the original circular shape and (b) the final square shape.

rings. Therefore, the total bending energy dissipated in this shape change process is

$$(W_p^0)_b = 2\pi r M_p\left(\frac{1}{r}-0\right)+4M_p\frac{\pi}{2}=4\pi M_p \quad [12.1]$$

where the superscript 0 for the plastic dissipation W_p denotes a single ring whilst the subscript b denotes bending deformation; the two terms in the expression represent the energy required to straighten the circular arcs and that dissipated by the plastic hinges at the four corners, respectively and M_p is the fully plastic bending moment of the ring's cross-section. Obviously, Eq. [12.1] takes account of bending deformation only and disregards the energy related to the tensile deformation of the ring, which will be discussed in Section 12.1.4.

Consequently, if a rolling rock carries an initial kinetic energy K_o, then the number of rings being fully straightened can be approximately determined as

$$n_d = \frac{K_o}{(W_p^0)_b} \qquad [12.2]$$

Those n_d severely deformed rings are distributed in a circular region with a total area of $2 \times (r\pi/2)^2 \times n_d = 4.93r^2 \times n_d$, where $(\pi/2)r$ is the side length of the square frame, see Fig. 12.2(b). Hence, the radius of this region can be estimated as

$$s = \sqrt{\frac{4.93r^2 \times n_d}{\pi}} = \sqrt{\frac{\pi K_o}{2(W_p^0)_b}}\,r = 1.25\sqrt{\frac{K_o}{(W_p^0)_b}}\,r \qquad [12.3]$$

By comparing the horizontal sizes of a circular ring and a square frame, it is found that, before deformation, the radius of the above region was

$$s_o = \frac{\sqrt{2}r}{(\pi/2)r}s = \frac{\sqrt{2}}{\pi/2}s = 0.90s \qquad [12.4]$$

Besides, it is easily seen that, when a circular ring becomes square, the length along a diagonal direction changes by (refer to Fig. 12.2)

$$\delta_{dia} = \sqrt{2}\frac{\pi}{2}r - 2r = 0.22r \qquad [12.5]$$

Figure 12.3 sketches the profile of a protective net, viewed from the top, between two supporting poles $2L$ apart. The horizontal dashed line denotes the original profile of the net and the solid curve denotes the final profile of the net after it was impinged by a rolling rock. The geometry of undeformed profile gives

$$L = b + s_o = b + 0.90s \qquad [12.6]$$

while the deformed profile shown in Fig. 12.3 gives

$$b\cos\alpha + R\sin\alpha = L \qquad [12.7]$$

where $R = s/\alpha$ is the radius of the curved portion of the profile and α denotes an angle given in Fig. 12.3. Thus, a combination of Eqs. [12.6] and [12.7] gives

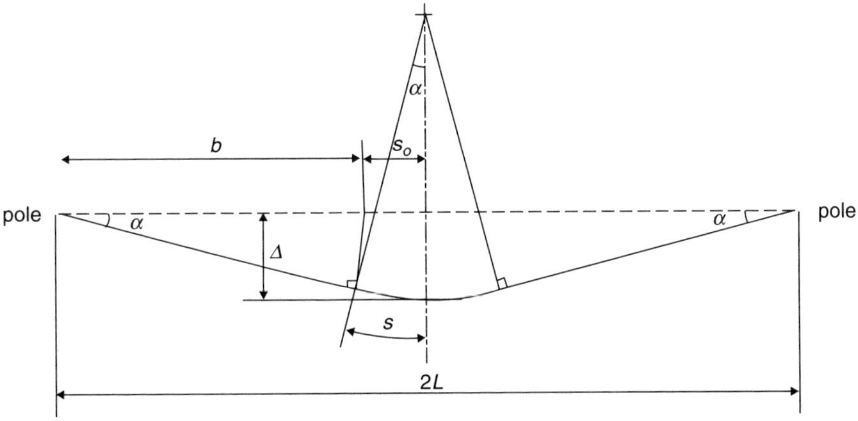

12.3 A sketch of the profile of a protective net, viewed from the top, between two supporting poles 2*L* apart.

$$(L - 0.90s)\cos\alpha + s\frac{\sin\alpha}{\alpha} = L \qquad [12.8]$$

When $\alpha^2 \ll 1$, this equation leads to

$$\alpha^2 \approx \frac{s}{5L - 2.8s} \qquad [12.9]$$

where s is calculated from Eq. [12.3].

Finally, as seen from Fig. 12.3, the maximum global deflection of the net caused by the rolling rock is

$$\Delta_{max} = b\sin\alpha + R(1 - \cos\alpha) \approx (L - 0.40s)\alpha \qquad [12.10]$$

12.1.3 Tensile deformation

In Section 12.1.2, only bending deformation of the rings is taken into account; in other words, the ring is assumed to be inextensible. This may be appropriate when the total applied force to the ring net is not large. However, when a ring has become a square frame after bending, its further deformation along its diagonal direction can only be accommodated by the tensile deformation of the frame.

To estimate the energy-absorption capacity of a square frame under its tensile deformation, first consider the tensile strain that has occurred during the bending deformation stage. It is evident that the most severe bending deformation during that stage occurred around the corner hinges. As illustrated in Section 2.2.2, a plastic hinge has an effective length $\lambda = (2 \sim 5)h$,

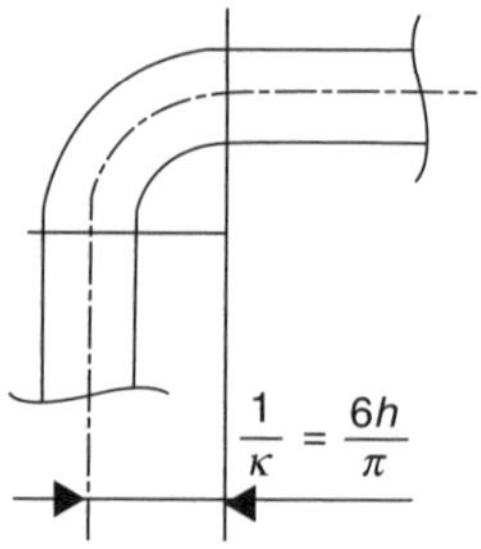

12.4 A region around a plastic hinge at a corner in the final deformation stage of a ring which has been pulled along two diagonal directions.

with h being the thickness of the element in bending. As seen in Fig. 12.2, in the bending deformation stage, the total rotation angle at a corner hinge is $(\pi/2)$. If λ is taken as $3h$, then Fig. 12.4 indicates that the final curvature of the central axis of the segment at a corner hinge is equal to $\kappa = (\pi/2)/(3h) = (\pi/6h)$. By noting that the extreme fibre of the circular ring has an original curvature of $1/r$, the maximum bending strain occurring at the extreme fibre of the segment is

$$(\varepsilon_b)_{\max} = \left(\kappa - \frac{1}{r}\right)\frac{h}{2} = \frac{\pi}{12} - \frac{h/2}{r} \approx 0.26 - \frac{h}{2r} \qquad [12.11]$$

Therefore, as the rigid-plastic ring (frame) has a fully plastic tensile force $N_p = YA$, with A being the cross-sectional area of the ring ($A = \pi c^2$ if the cross-section is a circle of radius c), the total energy dissipation due to tensile deformation at the four corner hinges until a tensile failure occurs at one of them will be

$$(W_p^0)_m = 2\pi rYA(\varepsilon_f - 0.26 + h/2r) = 2\pi^2 rc^2 Y(\varepsilon_f - 0.26 + c/r) \qquad [12.12]$$

where the subscript m denotes the tensile (i.e. membrane) deformation mode, ε_f is the maximum tensile strain at which the material fails by tensile tearing. Typically mild steel has $\varepsilon_f \approx 0.3$, so Eq. [12.9] indicates that the energy dissipated in the tensile deformation stage may not be very large because the rings may soon experience tensile failure.

12.1.4 Force magnitude

Assume that each of the original circular rings has radius r and circular cross-section of radius c. In the bending deformation stage (Fig. 12.2(b)), the mean force applied at each corner can be found from Eqs [12.1] and [12.5] as

$$\overline{F}_b = \frac{(W_p^0)_b}{2\delta_{dia}} = \frac{4\pi M_p}{2\times 0.22r} = 38.1\frac{Yc^3}{r} \quad [12.13]$$

While in the tensile deformation stage, the mean force applied at each corner is (Fig. 12.5)

$$F_m = \sqrt{2}N_p = \sqrt{2}\pi Yc^2 \quad [12.14]$$

Hence, the ratio between these two forces is

$$\frac{F_m}{F_b} = 0.117\frac{r}{c} \quad [12.15]$$

It implies that if $r/c < 8.5$, the force magnitude in the tensile deformation mode is roughly the same as or even smaller than the mean force in the bending deformation mode, so that a tensile failure may occur before the end of full bending deformation (i.e. before a ring becomes a square frame), which is not desirable. Therefore, it is recommended that the ring system should be designed to have $r/c > 8.5$ in order to ensure that most of the energy is dissipated by bending deformation.

12.1.5 A numerical example

Suppose a protective net is made of steel rings of radius r = 75 mm; the cross-section of the rings is a circle of radius c = 2.5 mm and the yield stress of the steel is Y = 240 MPa. Then, from Eq. [12.1], the energy absorption capacity by bending of a single ring is

$$(W_p^0)_b = 4\pi M_p = 4\pi(4Yc^3/3) = (16/3)\pi Yc^3 = 62.8\,\text{N m} = 62.8\,\text{J}$$

Suppose the net is supported by poles $2L$ = 3.2 m apart. Also, assume that a piece of rock of mass 100 kg slides down a slope at a speed of 7 m/s (about 25 km/h), then its initial kinetic energy is $K_o = 100 \times (7)^2/2 = 2450$ J. Based on Eqs [12.2] and [12.3], the collision of the rock on the net will severely

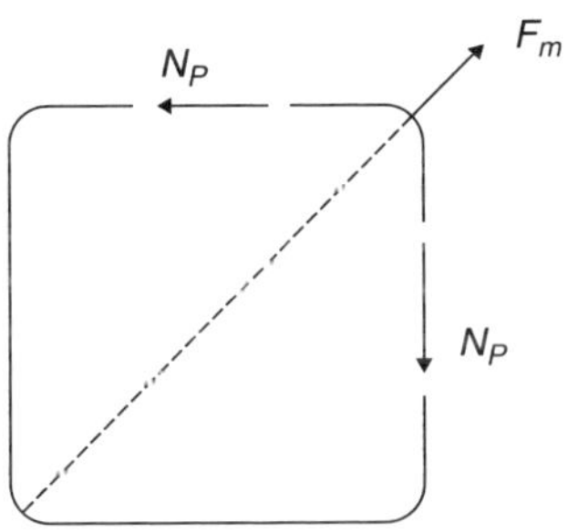

12.5 Forces applied at a corner.

deform (straighten about 39 rings) and the radius of the deformed region is estimated as $s = 585\,\text{mm}$.

By adopting this value of s, Eq. [12.9] gives $\alpha = 0.303 = 17.4°$, and Eq. [12.10] gives the final maximum deflection of the net as $\Delta_{max} = 414\,\text{mm}$. With these parameters, the final deformed profile of the net can easily be drawn.

It is known from Eq. [12.13] that the mean force in the bending deformation stage is

$$\bar{F}_b = 38.1\frac{Yc^3}{r} = 1.905\,\text{kN}$$

whilst the force in the tensile deformation stage is

$$F_m = \sqrt{2}\pi Yc^2 = 6.664\,\text{kN}$$

which is 3.5 times the mean force in the bending stage, so that the ring has almost become a square frame before the tensile plastic deformation begins. In fact, Eq. [12.12] gives

$$(W_p^0)_m = 2\pi^2 rc^2 Y(\varepsilon_f - 0.26 + c/r) = 162.8\,\text{J}$$

which is 2.6 times the bending energy dissipated in a single ring. This implies that the ring still has sufficient energy-absorption capacity even after it has been bent into a square frame.

The relationship between the applied force and the diagonal displacement is sketched in Fig. 12.6.

12.1.6 Comparison with a circular ring pulled along a diameter

As discussed in Section 4.2, similar behaviour has been observed in the large plastic deformation of a circular ring pulled along a diameter. The

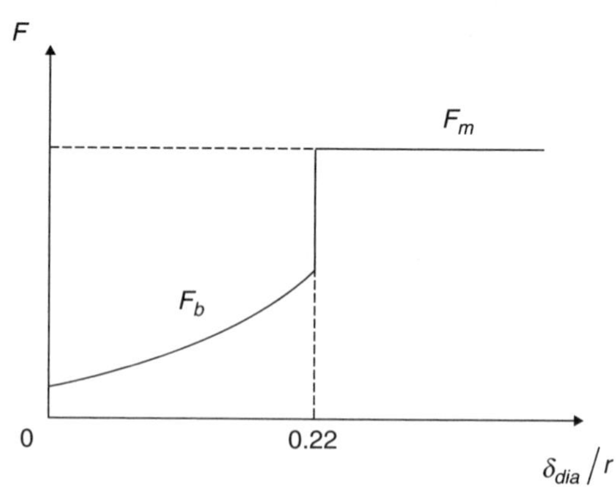

12.6 The applied force as a function of the diagonal displacement.

bending deformation mode of a rigid-plastic ring pulled along a diameter consists of four rigid circular arcs and two straight segments, connected by six plastic hinges (Fig. 4.3). The positions of these plastic hinges move with the increase in force F. The force–displacement relationship is given by Eqs [4.7] and [4.8].

In this case, force F increases with displacement δ. In fact, as $\theta \to \pi/2$, $F \to \infty$ and $\delta/D \to (\pi/2)-1 = 0.571$. Therefore, before reaching this limit case, the bending deformation mode is replaced by a plastic tension mode. In other words, the mode transition for a ring being pulled along a diameter is very similar to what happens in a ring net, analysed in Sections 12.1.2–12.1.4. Obviously, the load–displacement relationship for the ring net can be established by referring to the simple analysis given in Section 4.2 for a ring pulled along a diameter which has generally similar features to those sketched in Fig. 12.6.

12.2 Packaging using plastic foams

12.2.1 Protection of fragile objects during transportation

The packaging of fragile and valuable objects, such as electronic products and ceramic goods, during their transportation is aimed at protecting the objects from a potential drop on a hard surface, in which a large impact force is created by the object's own weight. Low density plastic foams are widely used for this purpose because of their excellent energy-absorption capacity as well as their light weight. In those situations, impact resistance and energy absorption become the main design criteria over static or low frequency dynamic loading (Soroka, 1999). Here we are interested in impact velocities in the range of 1 ~ 20 m/s, which means that the strain-rates in foams of thickness from 20–200 mm range from 55′ to 1000 s^{-1}. The subject of ballistic impact is not considered here, and as the foams are of low density the propagation of stress waves is not of importance in their dynamic response.

In the packaging of fragile and valuable objects, most of the foams used are closed-cell ones, because the trapped gas within the cells provides pneumatic cushioning, which happens to be of the correct magnitude for many applications. Open-cell foams, as discussed in Chapter 10, do not have this energy absorption mechanism, although there may be pneumatic losses as the air passes through restricted holes in the cell faces (Mills, 1994).

Some packaging applications involve simple blocks of foam, which are fixed to cardboard by suitable adhesives and then assembled into the packaging, as sketched in Fig. 12.7. In this case extruded foam slab stock is the lowest cost manufacturing route and a large number of polymers can be used. In the present section, we will focus on this simple geometry. As seen

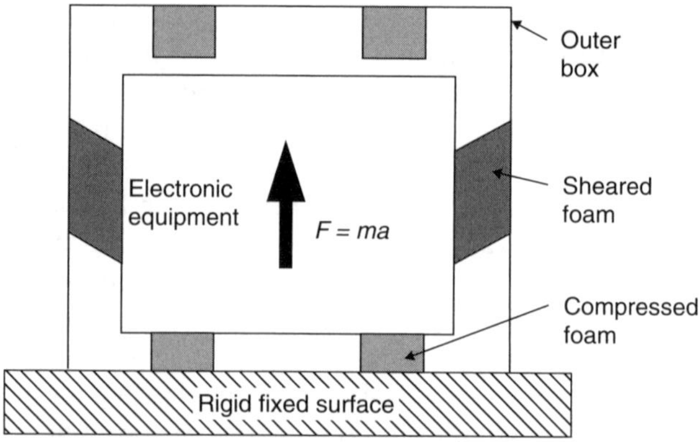

12.7 Schematic illustration of simple foam blocks used for packaging, which will deform by compression and shear when the box is dropped onto a rigid surface.

from Fig. 12.7, when the packaged object drops onto a rigid surface, the major deformation modes will be compression and shear.

12.2.2 Packaging design based on cushion curves

The widely accepted method of designing the packaging for transportation of delicate objects uses ***cushion curves***. Figure 12.8 shows a typical set of the cushion curves obtained by dropping a rectangular object of mass M from height H to foam blocks of various thicknesses T. The peak acceleration G of the falling mass measured in g (the gravitational acceleration, $= 9.81\,\mathrm{m/s^2}$) is recorded. The horizontal coordinate of Fig. 12.8 is the static stress $\sigma_s = Mg/A$, which is the compressive stress applied to the foam block when the mass M is resting on it, with A being the contact area between the mass and the foam. Each curve in Fig. 12.8 represents the value of peak acceleration G against σ_s obtained from the drop tests on the foam of a particular thickness T, while the whole graph is constructed under a particular drop height H.

The sample dimensions and the magnitudes of drop heights, etc. are specified in relevant test standards, such as BS 4443 and AST M D 1596. In general, the height of the test block must not exceed twice its width or else it might buckle during the test. The drop heights, H, of objects depend on the type of transportation or handling, while some typical values are given in Table 12.1, where the values of H in the second column from the right are taken from Brown (1959) with imperial units being converted into metric.

Table 12.1 Typical drop heights

Type of handling	Mass M (kg)	Drop height H (m)	H (m) Eq. [12.16]
1 man throwing	0–9	1.05	1.80–1.18
1 man carrying	10–22	0.90	1.16–0.93
2 men carrying	23–110	0.75	0.92–0.48
Light handling equipment	111–225	0.60	0.48–0.28
Medium handling equipment	225–450	0.45	0.28–0.09
Heavy handling equipment	>450	0.30	<0.09

Table 12.2 Typical fragility factors

Classification	Fragility factor G	Objects
Highly fragile	15–25 g	Precision instruments with sensitive mechanical bearings, hard disc drives, gyroscopic instruments
Very fragile	20–40 g	Electro-mechanical measuring instruments
Fragile	40–60 g	Electro-mechanical equipment, e.g. computer monitors, electric typewriters, etc.
Moderately fragile	60–80 g	Audio and television equipment, floppy disc drives, optical projectors
Fairly robust	80–100 g	Household appliances and furniture, e.g. washing machines, refrigerators, cookers, etc.
Robust	100–120 g	Radiators, sewing machines, machine tools

Alternatively, the US Defence guide suggests that the drop height H can be chosen as

$$H = 1.8 - 0.28 \times \ln(Mg) \qquad [12.16]$$

As shown in the rightmost column, the values of H chosen from Eq. [12.16] are lower than those given by Brown (1959), especially when M is large.

Another factor involved in the design of packaging is the choice of an appropriate value of G for a given object, called the ***fragility factor***. The value of this factor depends on the type of object, as listed in Table 12.2 (refer to Paine, 1991 and Mills, 1994).

In the design of packaging, the weight and type of object are known, so G and H can be selected based on Tables 12.2 and 12.1, respectively. If the packaging material (plastic foam) is selected, then the supporting area A and the foam thickness T can be determined by the cushion curves (see Fig. 12.8) of this foam.

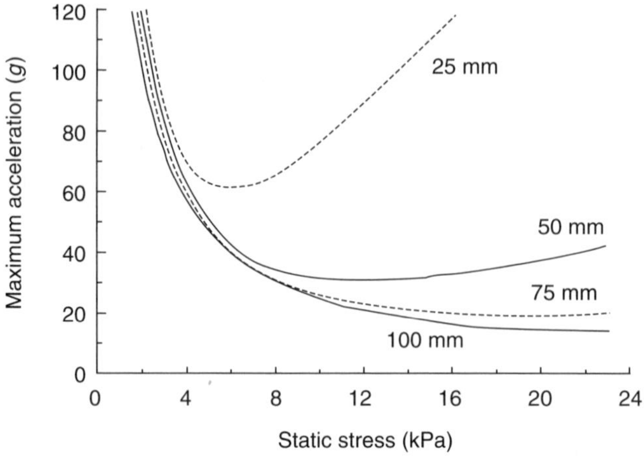

12.8 Typical cushion curves obtained by dropping a rectangular object of mass *M* from height *H* on to foam blocks of various thickness *T*.

12.2.3 How to construct cushion curves from the stress–strain curve of a foam

In fact, if the impact stress–strain behaviour of the plastic foam under concern follows a single curve (called the ***master curve***, i.e. the behaviour is independent of strain-rate), it is not necessary to conduct numerous drop tests for the same cushion material of various thicknesses. That is, as long as one cushion curve has been obtained from drop tests on the foam of a particular thickness, the cushion curves for other thicknesses can be constructed as illustrated below.

For a given drop height the kinetic energy of the object at the moment of impact is MgH. When the object has been arrested all this energy must be absorbed by the foam block of initial volume AT. Therefore, the energy density U input into the foam is given by

$$U = \frac{MgH}{AT} = \int_0^{\sigma_m} \sigma \mathrm{d}\varepsilon \qquad [12.17]$$

The integral represents the area under the stress–strain curve (Fig. 12.9(a)) up to the maximum stress σ_m, resulting in a function $U(\sigma_m)$ of the maximum stress. On the other hand, $\sigma_s = Mg/A$ represents a static pressure produced by the object on the supported foam. Thus, Eq. [12.17] results in

$$U(\sigma_m) = \frac{H}{T}\sigma_s \qquad [12.18]$$

It is noted that the static stress σ_s applies to the foam when the object possesses an acceleration of 1 g (i.e. its own weight), while the maximum acceleration during the impact response, G, occurs when the compressive stress reaches its maximum value σ_m. Therefore, G can be regarded as the ratio of the accelerations in these two cases and it should be given by

$$G = \frac{\sigma_m}{\sigma_s} = \frac{H}{T}\frac{\sigma_m}{U(\sigma_m)} \qquad [12.19]$$

Therefore, it is possible to calculate G as a function of $\sigma_s = Mg/A$ based on the master stress–strain curve of the foam and Eq. [12.19], which only involves the ratio H/T.

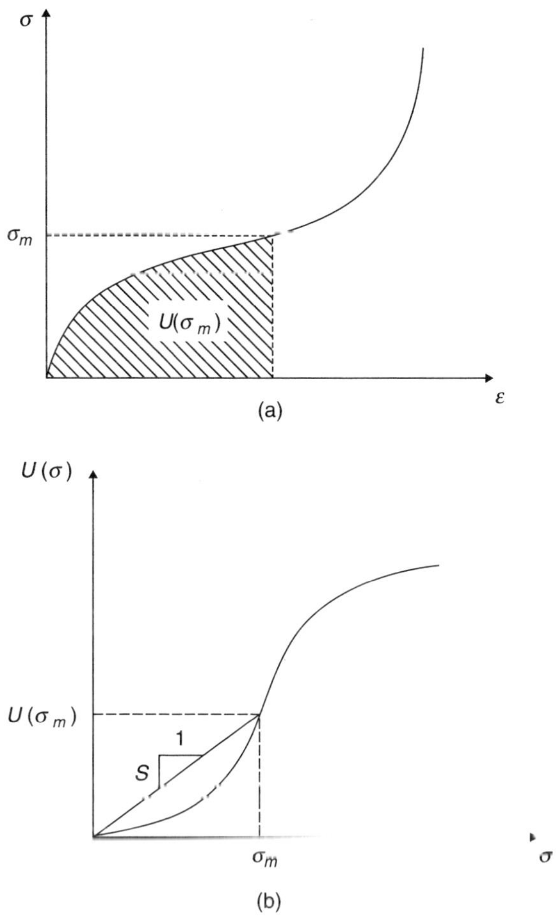

12.9 Behaviour of foam used for packaging: (a) the stress–strain curve and (b) the energy density U absorbed till the maximum stress σ_m is reached

The above discussion can be related to the energy-absorption capacity of foams illustrated in Section 10.2. In general, an energy-absorption diagram can be constructed from the stress–strain curve of the foam, expressing the energy absorption of the foam as a function of its normalised stress, so that Fig. 12.9(b) shows such a curve calculated from the stress–strain curve given in Fig. 12.9(a), in which the ordinary coordinates are adopted (rather than the double logarithmic coordinates adopted in Chapter 10). It is evident that in this diagram $S \equiv U(\sigma_m)/\sigma_m$ represents the slope of a secant connecting the origin with a point in the curve shown in Fig. 12.9(b). From Eq. [12.19], we obtain

$$G = \frac{H}{T} \times \frac{1}{S} \qquad [12.20]$$

Hence, at any specified value of σ_m in Fig. 12.9(b), by taking the slope of the secant at the corresponding point, we are able to calculate the value of G from Eq. [12.20], provided H and T are given. This explains how to construct a whole series of cushion curves for various values of H and T from the single energy-absorption diagram of the selected foam shown in Fig. 12.9(b).

12.2.4 Discussion

(1) If a safety factor SF is needed in a design (e.g. $SF = 1.10$ is suggested for some applications), then the thickness of the foam, T, can be calculated from

$$T = (SF) \times \frac{H}{GS} \qquad [12.21]$$

or chosen from the cushion curves of a selected foam in a figure like Fig. 12.8 with the given static stress $\sigma_s = M/A$ and given value of $G/(SF)$. In fact, it is easy to see that in order to minimise the material cost of the foam for packaging, the static stress $\sigma_s = M/A$ should be taken to be as large as possible; that is, a larger supporting area A for an object being packaged is preferable.

(2) If a plastic foam possesses a nearly constant yield stress, Y, and the elastic energy is neglected during its large plastic deformation, then $\sigma_m = Y$ and $U(\sigma_m) \approx Y\varepsilon$ holds until the densification strain ε_D. Therefore, the maximum value of $S \equiv U(\sigma_m)/\sigma_m$ will be $S_{\max} = \varepsilon_D$. Consequently, the minimum thickness of the foam adopted in the packaging can be calculated by

$$T_{\min} = (SF) \times \frac{H}{G\varepsilon_D} \qquad [12.22]$$

where H and G can be selected from Tables 12.1 and 12.2, respectively.

(3) The above method of constructing cushion curves for foams is applicable if and only if the foam possesses a master curve, that is, the foam exhibits no rate-dependence in its stress–strain behaviour. As the energy-absorption capacity of polymer foams usually increases with the strain-rate (see Section 10.2), a packaging design based on the quasi-static properties of foams would be on the conservative side, or would be regarded as leaving a margin for safety.

12.2.5 A numerical example

As a particular case, let us consider the packaging of a DVD machine of mass 5 kg. Its width and height are $B = 440$ mm and $H = 80$ mm. The packaging configuration is sketched in Fig. 12.10. Suppose that a polystyrene foam of relative density $\rho^*/\rho_s = 0.1$ is selected as the candidate packaging material. As given in Section 8.6 of the book by Gibson and Ashby (1997), Figs. 12.11(a) and (b) depict the stress–strain curve and the energy-absorption diagram of this foam at a low strain-rate of $4 \times 10^{-3}\ \mathrm{s}^{-1}$.

Equation [12.22] can be applied to obtain the minimum thickness of the packaging form for the DVD machine. Now, we can choose $SF = 1.10$, fragility factor $G = 60g$ for audio equipment (see Table 12.2) and $H = 1.35$ m for a typical one-man-throwing drop height (see Table 12.1 and Eq. [12.16]). Finally, assuming the densification strain $\varepsilon_D = 0.6$ which is taken from Fig. 12.11(a), we obtain from Eq. [12.22] that

$$T_{\min} = 4.2\,\mathrm{mm}$$

However, this calculation is based on the assumption that the whole bottom surface of the DVD machine is supported by the packaging foam, while the actual contact length between the machine and the foam along the width direction is only $2a_1$. Thus, the minimum foam thickness should be determined by

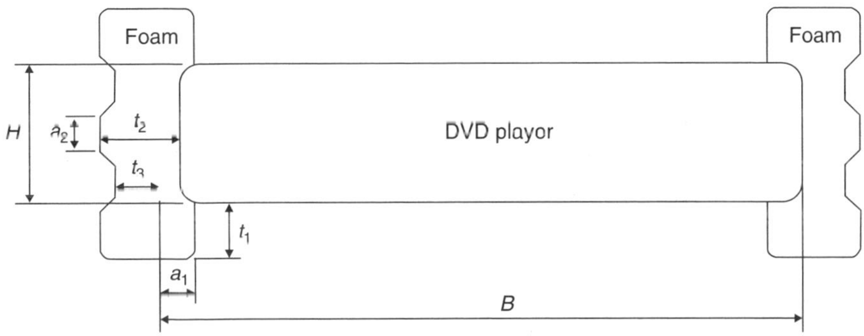

12.10 A packaging configuration of a DVD machine of mass 5 kg.

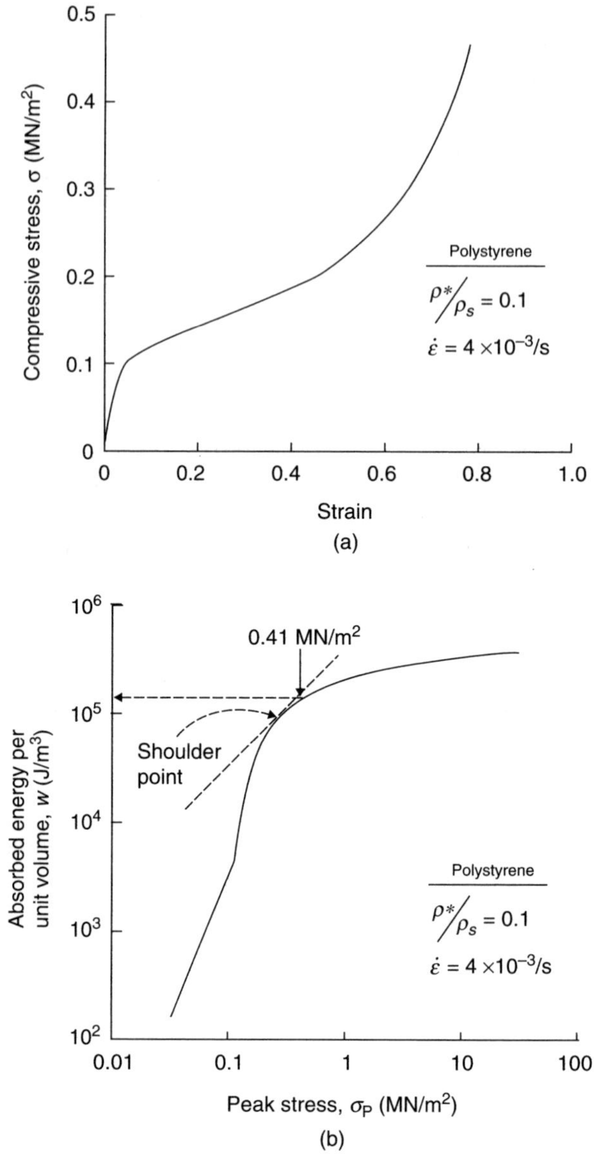

12.11 A polystyrene foam of relative density $\rho^*/\rho_s = 0.1$: (a) the stress–strain curve at a low strain rate of $4 \times 10^{-3}s^{-1}$; (b) the energy-absorption diagram. Adopted from Gibson and Ashby (1997) (reproduced with kind permission of Cambridge University Press).

$$t_{1\min} = \frac{B}{2a_1} T_{\min}$$

If a_1 is selected as 20 mm, then the foam thickness should be

$$t_{1\min} = 46\,\text{mm}$$

The other thicknesses, i.e. t_2 and t_3, shown in Fig. 12.10 can be similarly determined.

12.3 Design of vehicle interior trim

12.3.1 Introduction

To reduce the possible injury to the occupants during a vehicle's collision, the design of vehicle interior trim, e.g. foam padding, plastic ribbed trim, head impact airbag, etc., should meet the head impact requirement, e.g. the new head impact safety standard in the USA, an amendment to FMVSS 201, which came into effect in the 1999 model year with a phase-in over five years.

In FMVSS 201 head impact tests, the recorded Head Injury Criterion (HIC) value depends on many factors, including initial impact velocity, properties of vehicle structure and interior trim, impact location on the free motion headform (FMH), launch angle and geometrical configuration around point of impact. Gabler *et al.*, (1991) experimentally studied many of the above factors and concluded that head impact injury potential is a strong function of vehicle upper interior design. Using the finite element method, Barbat and Prasad (1995) analysed foam padding as a countermeasure and obtained data on the relation between HIC response, padding thickness and foam properties. More recently, Deb *et al.* (1997) evaluated the effectiveness of foam and stiffened plastic trim padding and found that the addition of plastic stiffeners under the trim may be more effective.

To provide insight into the design of vehicle interior trim to meet the FMVSS 201 head impact requirements, Zhou *et al.* (1998) made an analytical study of system variables. The parameters included in the model are initial impact velocity, shape of acceleration pulse, peak acceleration, time at rebound, stopping distance and HIC. The following contents are basically adopted from their analytical approach.

12.3.2 Analytical model

The HIC is defined in Section 1.1.3 and given by Eq. [1.2]. Since the FMH is used in FMVSS 201 tests, a linear regression formula is provided in FMVSS 201 to adjust its HIC to account for the difference between this free-flight headform and a Hybrid III dummy's head. This adjusted HIC is called HIC(d) and is defined by

$$HIC(d) = 0.75446[HIC(FMH)] + 166.4 \quad [12.23]$$

The US federal rule requires that HIC(d), for a 24.1 km/h (15 mph) head impact, must be less than 1000.

Among many different types of hypothetical acceleration pulses, the haversine shape is a good representation of most actual test pulses, and it can be expressed as a function of time t by

$$a_g(t) = \frac{a_p}{2}\left(1 - \cos\frac{2\pi t}{T}\right) \quad [12.24]$$

where a_p is the peak acceleration in g and T is the impact duration. However, unlike the pulse described by Eq. [12.24], actual test pulses are generally not symmetric. A more realistic model, referred to as an ***asymmetric haversine pulse***, combines opposite halves of two separate haversine waves. As an example, an asymmetric haversine pulse is shown as the dashed line in Fig. 1.5 together with a measured test pulse. HIC values calculated from the test pulse and from this particular haversine function are very close because HIC is determined by the central part of the pulse, which is the best-fit portion of the two curves. The asymmetric haversine pulse is expressed as

$$a_g(t) = \begin{cases} \dfrac{a_p}{2}\left(1 - \cos\dfrac{2\pi t}{T_L}\right) & 0 \le t \le \dfrac{T_L}{2} \\ \dfrac{a_p}{2}\left(1 - \cos\dfrac{2\pi\left(t + \dfrac{T_R}{2} - \dfrac{T_L}{2}\right)}{T_R}\right) & \dfrac{T_L}{2} \le t \le \dfrac{T_L + T_R}{2} = T \end{cases}$$

[12.25]

where the subscripts L and R indicate the left half and the right half, respectively. Peak acceleration occurs at $t_p = T_L/2$. The parameter $\gamma \equiv T_R/T_L$ is introduced as a measure of asymmetry of the pulse, and $\gamma < 1$ indicates a late peak pulse, while $\gamma > 1$ suggests an early peak pulse. By adjusting the three parameters a_p, T_L and T_R (or a_p, T_L and γ) the asymmetric haversine function defined in Eq. [12.25] can cover a wide range of acceleration pulse shapes in head impact tests.

With the explicit expression of the asymmetric haversine pulse [12.25], HIC as defined in Eq. [1.2] can be calculated mathematically. Skipping the lengthy derivation, the final expression of HIC is

$$HIC = 0.1515(1 + \gamma)a_p^{2.5}T_L \quad [12.26]$$

Neglecting the headform rotation, it is straightforward to calculate velocity and displacement profiles of the headform by integrating the acceleration pulse given by Eq. [12.25].

The next step is to determine the time at which the displacement reaches its maximum, i.e. the time at rebound. From high speed films of head impact tests, it has been observed that, as the headform contacts the target surface, the rubber skin of the headform is compressed, the trim system or countermeasure is crushed and the structure (pillar, rail or roof) is deflected. Meanwhile, the headform slows down, comes to rest and eventually rebounds. The maximum displacement, referred to as the ***stopping distance***, includes deformation of the headform skin, the deformation of the interior trim and the global elastic deflection of the pillar. Both the head impact tests and finite element analyses have shown that the rebound occurs between the time at peak acceleration and the time at the end of impact when the acceleration and the impact force are nearly zero. So the headform impact with a surface with energy-absorbing countermeasure (foam, plastic ribbed trim, or others) is neither purely elastic nor purely viscous but viscous-elastic or viscous-plastic. This is because the impact involves a combined compression of rubber skin, interior trim and sheet metal. These materials have quite different stress–strain relations, strain-rate dependence and unloading paths. For a further study, the parameter ζ is introduced to represent a relative time for the rebound

$$\zeta \equiv \frac{t_r - t_p}{T - t_p} \tag{12.27}$$

where t_r is the time at rebound (zero velocity), t_p is the time at peak acceleration and $T - t_p$ is the time duration from the peak acceleration to the zero acceleration at the end of the acceleration pulse. Physically, $\zeta = 0$ indicates a purely elastic impact for which the maximum displacement occurs at the same time as the peak acceleration ($t_r = t_p$); $\zeta = 1$ represents a purely viscous impact with no rebound for which the maximum displacement occurs at the end of impact ($t_r = T$) and $\zeta = 0.5$ characterises a viscous-elastic countermeasure for which the rebound occurs in the middle.

The two parameters, ζ and γ, introduced here, reflect the properties of an interior trim. To find out their typical ranges in head impact tests, Zhou *et al.* (1998) examined dozens of finite element cases and actual tests involving different types of countermeasures. They found that ζ ranges from 0.2 to 0.9 and γ ranges from 0.6 to 1.1. Hence, $\zeta = 0.5$ and $\gamma = 0.9$ may be representative values of most prevailing countermeasures.

The headform velocity reaches zero when $t = t_r$. Applying this condition in the velocity and displacement profiles and for $\zeta = 0.5$ and $\gamma = 0.9$, the initial impact velocity and the stopping distance are found to be

$$V_o = 0.4341 g a_p T_L \qquad S_T = S_{\max} = 0.457 V_o T_L \qquad [12.28]$$

respectively, where g is the gravitational acceleration.

Then, based on Eqs [12.26] and [12.23], the HIC and HIC(d) are found as

$$HIC = \frac{0.7163 V_o^4}{g^{2.5} S_T^{1.5}} \qquad HIC(d) = \frac{0.5404 V_o^4}{g^{2.5} S_T^{1.5}} + 166.4 \qquad [12.29]$$

respectively.

12.3.3 Discussion

Effects of ζ and γ

In the design of interior trim to meet the head impact requirements, the available stopping distance is usually dictated by how much interior space can be sacrificed. Adding the deformation of the headform rubber skin (about 5 mm) and the deflection of the pillar (about 10 mm), 30 mm may be a reasonable value for the total stopping distance. In this limited space, countermeasure design will play a major role. As shown above, ζ and γ are the parameters representing properties of countermeasure design which directly affect efficiency of energy absorption and effectiveness of HIC reduction. Equation [12.29] relates HIC(d) to the initial impact velocity and stopping distance for $\zeta = 0.5$ and $\gamma = 0.9$. Following a similar approach, the HIC(d) values for other ζ and γ values are shown in Table 12.3 for impact velocity $V_o = 24.1$ km/h (15 mph) and stopping distance $S_T = 30$ mm.
The following observations can be drawn from Table 12.3:

- Very elastic padding must be avoided since it results in very high HIC(d) values;
- In general, more viscous padding is better, but not much improvement can be achieved in reducing HIC(d) just by making the padding very viscous (e.g. ζ approaches 1); besides, it is very difficult to design a very viscous padding in a very limited space;

Table 12.3 HIC(d) for different ζ and γ values given impact velocity of 24.1 km/h (15 mph) and stopping distance of 30 mm

		Elastic $\zeta = 0$	$\zeta = 0.5$	Viscous $\zeta = 1$
Late peak	$\gamma = 0.5$	1602	957	880
	$\gamma = 0.9$	2034	864	775
Early peak	$\gamma = 1.5$	2623	774	686

- Early peak acceleration pulses can help to reduce HIC(d) values for most prevailing countermeasures ($\zeta > 0.5$);
- In summary, a countermeasure that can achieve early peak acceleration pulse and which has a more viscous nature is preferred.

HIC(d) dependence on stopping distance and impact velocity

The HIC(d) expression given in Eq. [12.29] can be used to guide the interior trim design and is graphically shown in Fig. 12.12(a) and (b). For the

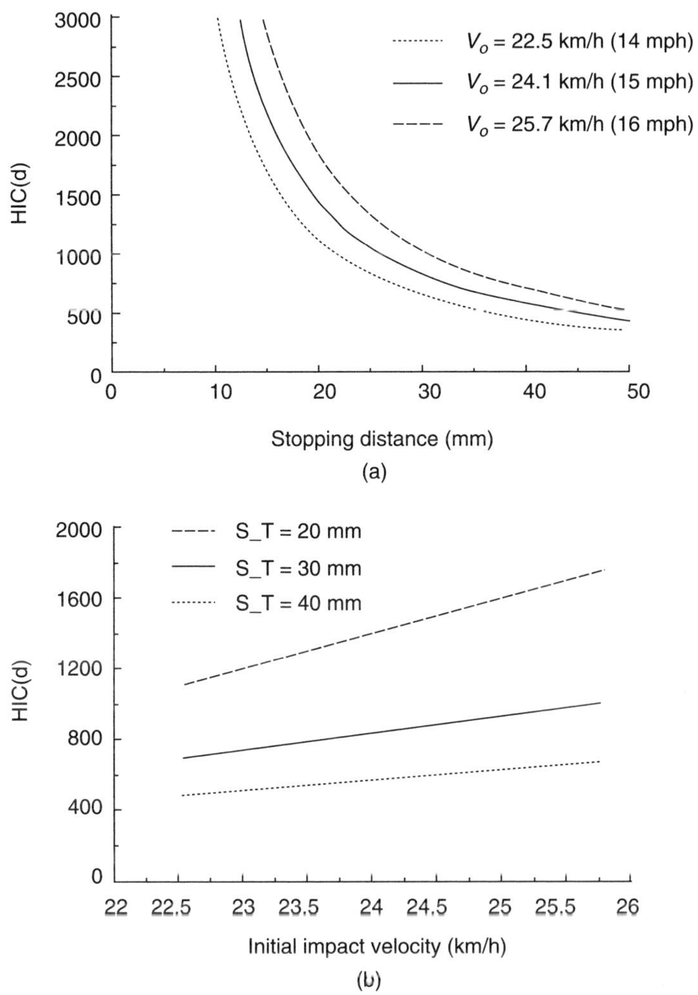

12.12 The dependence of HIC(d) on (a) the stopping distance S_T and (b) the initial impact velocity V_o (reproduced with kind permission of American Society of Mechanical Engineers).

three impact velocities, $V_o = 22.5$, 24.1 and 25.7 km/h (14, 15 and 16 mph), the dependence of HIC(d) on the stopping distance S_T is plotted in Fig. 12.12(a). Three regions are identified along this power-law-decreasing trend. For small stopping distance, say less than 20 mm, HIC(d) is very sensitive to the stopping distance. On the other hand, for large stopping distance, say greater than 40 mm, HIC(d) is relatively insensitive. Thus, the desired head impact protection countermeasure should result in a stopping distance between 20 and 40 mm.

Impact velocity in tests may vary by 0.3–0.5 km/h. A family of HIC(d) versus impact velocity curves is plotted in Fig. 12.12(b) for stopping distances of 20, 30 and 40 mm to show the HIC(d) sensitivity to the velocity. For stopping distances from 30–40 mm, approximately every 0.1 km/h increase in the impact velocity results in an increase in HIC(d) of 8–11. Hence, a typical variation in velocity, say 0.4 km/h, will result in a variation in HIC(d) of no more than 50.

12.3.4 Validation of the analytical model

Zhou *et al.* (1998) also used both test results and finite element analysis (FEA) results to validate their analytical model. A series of head impact tests were conducted with a variety of padding designs using different materials and thicknesses. The average discrepancy in HIC(d) between the tests and the analytical model is 13.6 % and the maximum difference is 33.4 %.

In their FEA, three impact velocities were chosen at 22.5, 24.1 and 25.7 km/h (14, 15 and 16 mph) and contact locations on the headform were changed in a 40 mm range within the required contact window on the headform forehead. The average discrepancy in HIC(d) between the FEA results and the analytical model in HIC(d) is only 7.6 % and the maximum is 16.5 %. The variation is easier to control in the FEAs than in the tests so, as expected, the discrepancy is much lower.

12.3.5 Concluding remarks

By assuming an asymmetric haversine pulse for the acceleration of the headform in a head impact test, general equations have been derived by Zhou *et al.* (1998) showing relationships between system variables including HIC(d), impact velocity, stopping distance, peak acceleration, impact duration, time at peak acceleration and time at rebound. The analytical model has been validated using the results from FEA and head impact tests. For current types of interior trim designs, the asymmetric haversine shows a good representation in modelling the headform response. Its generality and flexibility make the model useful in relating the physical parameters of the impact test to the resulting acceleration pulse and HIC value.

As a result of this analysis, a preferred design range of stopping distance is identified and HIC sensitivity to impact velocity is revealed. The analytical model and its results may be utilised to guide the design of vehicle interior trim to meet FMVSS 201 head impact requirements.

In particular, parameter ζ is identified to be the key parameter for assessing the effects to a purely elastic impact, a viscous-elastic-plastic impact, or a purely viscous impact achieved by different energy-absorbing countermeasure designs. Although a purely viscous impact ($\zeta = 1$) is ideal for energy-absorbing, HIC sensitivity to ζ is greatly reduced when $\zeta > 0.5$ and such a design is not efficient in terms of the benefits it can achieve. On the other hand, a small ζ energy-absorbing design must be avoided because its more elastic impact results in high HIC values. Another important discovery is that for a more viscous impact ($\zeta > 0.5$), early peak ($\gamma > 1$) can help to reduce HIC. Because of its rib buckling mechanism, plastic ribbed trim is a countermeasure that may result in an earlier peak pulse compared to foam padding.

The power-law decreasing relationship between HIC(d) and stopping distance (see Fig. 12.12(a)), derived from the above model, shows that a preferred design range of stopping distance is approximately 20–40 mm. HIC(d) is extremely sensitive to stopping distances below 20 mm, while stopping distances greater than 40 mm do not significantly reduce HIC(d). This analytical study also shows that HIC(d) depends on initial impact velocity to the 4th power for a given stopping distance. For stopping distances in the range 30–40 mm, approximately every 0.1 km/h increase in initial impact velocity will result in an increase in HIC(d) of 8–11.

12.4 Corrugated guardrail beams

12.4.1 Introduction

Around the world the corrugated steel W-beam guardrail system along the roadside is the most popular energy-absorbing system intended to dissipate vehicles' kinetic energy in collision events and to reduce the damage to both the car and the occupants. Its dynamic behaviour is of great importance for road safety.

Owing to its significance in saving lives during car accidents, many full-scale evaluations of the performance of W-beam guardrail systems have been conducted in the USA (Bank *et al.*, 1998a) and in Japan (Ando *et al.*, 1995). However, to conduct full-scale impact testing on a guardrail system is very time-consuming and costly, so that an attempt has been made to simplify the investigation by examining downscaled prototypes of guardrail beams for better controlled input and a more complete record (Bank *et al.*, 1998b).

Based on the studies conducted in the Hong Kong University of Science and Technology (HKUST) (refer to Hui *et al.*, 2003, and Yu *et al.*, 2003), this section will briefly summarise the findings from an experimental study of downscaled W-beams and W-beam guardrail systems with a scaling factor $\beta = 1/3.75$. The experiments carried out at HKUST were divided into two main categories: the first set of tests focused on the static and dynamic behaviour of the W-beam itself, while the other set had two supporting posts attached to the W-beam sample to form a prototype of one segment of the guardrail system during impact tests.

12.4.2 Experiments

Downscaled testing

Tests were conducted on a geometrically similar but β-downscaled prototype of the W-shaped guardrail beam conventionally used in Hong Kong (Fig. 12.13) to acquire the response characteristics. The downscaled W-beam samples were tested both statically and dynamically in the regime of large plastic deformation, for the W-beam itself as well as for a guardrail system consisting of both the W-beam and the supporting posts. Other than the sample size, other geometrical features, like the supported span of the W-beam and the dimensions of the supporting posts, were also scaled down by factor β. The detailed dimensions of the downscaled prototype are shown in Table 12.4. Apart from the scaling factor, the material of the specimens was chosen to have properties comparable to those of the real beams.

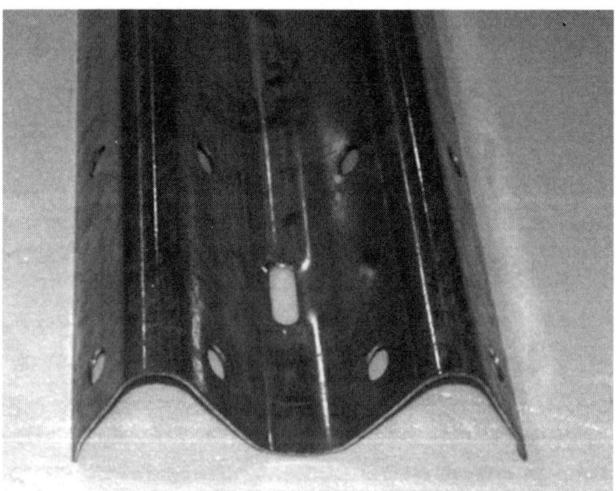

12.13 A W-shaped guardrail beam conventionally used in Hong Kong (reproduced with kind permission of Korean Society of Automotive Engineers).

Table 12.4 Comparisons of dimensions of a conventional guardrail system and the experimental setup

	W-beam		Circular hollow post		Rectangular hollow post		
	Width	Thickness	Diameter	Thickness	Width	Height	Thickness
Original system	310–317 mm	3 mm	115 mm	4 mm	50 mm	100 mm	5 mm
Scaled-down system	81–84 mm	0.8 mm	30 mm	0.8 mm	13 mm	25 mm	1 mm

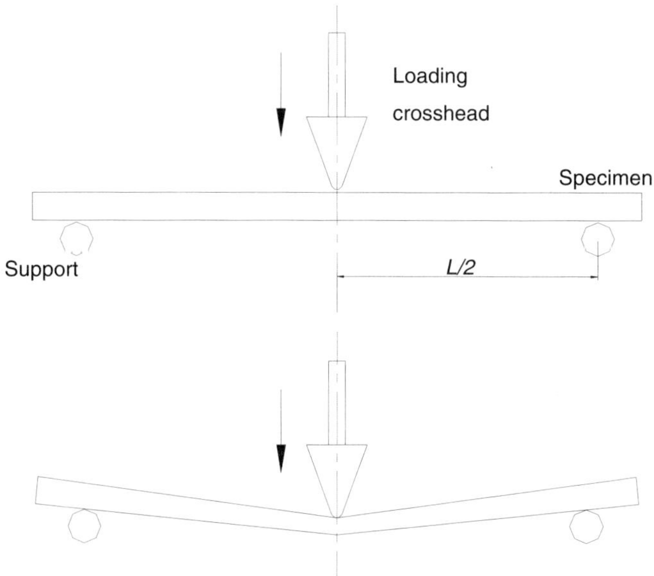

12.14 Schematic illustration of an experimental configuration (reproduced with kind permission of Korean Society of Automotive Engineers).

Experimental setup

For the beam tests, the β-downscaled W-beams of 600 mm in length were symmetrically placed on the supports with a span L = 535 mm, and transversely loaded statically or dynamically at the mid-span by a rigid wedge-head perpendicular to the beam axis (Fig. 12.14). There were three supporting conditions: (a) simply roller-supported (RS), (b) simply box-supported (BS) and (c) axially constrained roller-supported (AR). The main difference between the roller support and the box support was that on the roller support, the cross-section of the beam was transversely constrained, while on the box support, the edges of the cross-section were free to move or rotate. In the quasi-static tests, the samples were loaded with a crosshead speed of 5 mm/min and the load was removed at every 10 mm-interval until a final transverse displacement of 120 mm. In the impact tests, the samples were subjected to impact by a wedge-headed drop weight assembly of 12.92 kg with three different impact velocities.

For the system tests, each downscaled W-beam was mounted on two supporting posts for impact loading. Two types of posts were tested: circular hollow tubes and rectangular hollow tubes. The same drop weight assembly of 12.92 kg was used and the supporting distance between two posts remained as $L = 535$ mm. The two supporting posts were clamped at the roots at a distance $H = 160$ mm from the clamped end to the bolt connection to the W-beam sample.

Test results

The load versus transverse displacement curves of the W-beam samples under quasi-static three-point bending with different supporting conditions are shown in Fig. 12.15. In general, the load-carrying capacities of the simply supported samples were very similar. The load rose very quickly to a peak

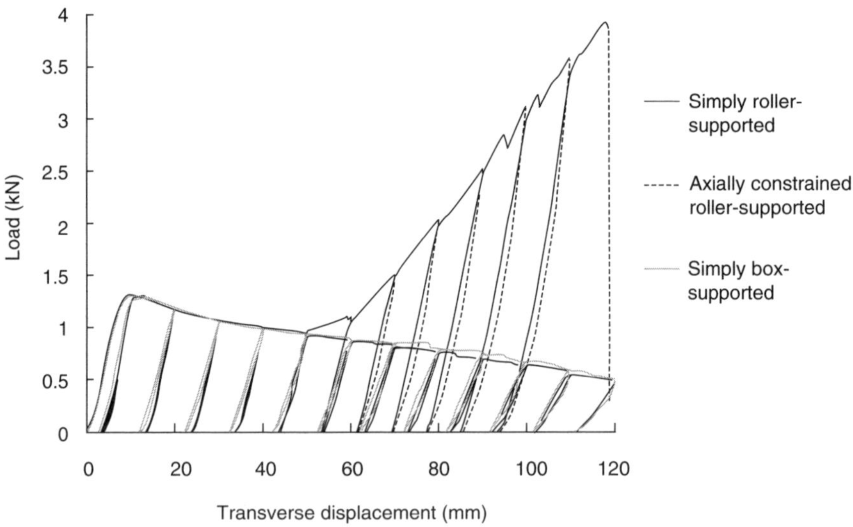

12.15 Load–transverse displacement curves for quasi-static three-point bending of downscaled W-beams with different supporting end conditions (reproduced with kind permission of Korean Society of Automotive Engineers).

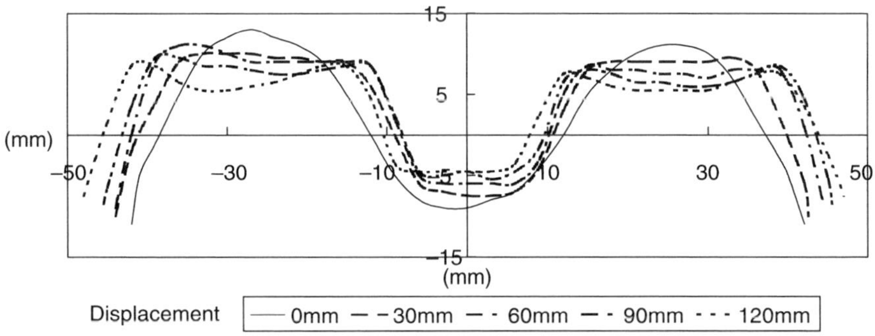

12.16 Local cross-sectional distortion at different transverse displacements for downscaled simply roller-supported W-beams under quasi-static three-point bending (reproduced with kind permission of Korean Society of Automotive Engineers).

then decreased gradually. Along with the global flexural deformation, there was a local cross-sectional distortion right under the loading wedge, as recorded in Fig. 12.16 for a simply roller-supported sample. The material's strain-hardening contributed to the rise in load-carrying capacity until the beam reached its limit instability. At that point serious distortion of the local cross-section overwhelmed the material's hardening, resulting in a decrease in the load-carrying capacity. The deviation of the performance of the axially constrained beams from that of the simply supported ones was due to the effect of the axial constraints in strengthening the beam by the induced tensile force and compensating for the structural softening effect, resulting in the load rising again.

Except that the increase in width at the local cross-section was more severe for beams with axial constraints, leading to larger transverse displacement, the local cross-sectional distortions for beams under different supporting conditions were very similar. That is, flattening of the top portions was followed by a further collapse of the whole cross-section with bulges formed at the top, as shown in Fig. 12.16. From the viewpoint of solid mechanics, during the bending process of a beam of thin-walled cross-section, some cross-sectional distortions must occur and accompany the flexural deformation to minimise the total deformation energy of the beam. A similar phenomenon was previously observed and analysed for four-point bending of beams of angle-section (Yu and Teh, 1997).

In impact tests of the downscaled W-beams, the initial peak loads for all impact velocities were similar and were approximately 1.5 times the value of the static initial peak load. Another sudden jump in load was observed when the drop weight reached its maximum transverse displacement and just started to rebound. The higher the initial impact velocity, the higher is this abrupt peak; for lower velocity impacts, this sharp increase was absent.

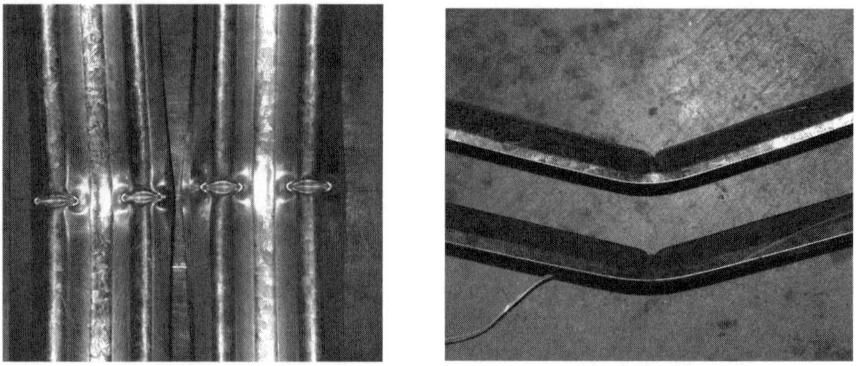

(a) (b)

12.17 Comparison of the local cross-sectional distortions of static- and impact-tested downscaled W-beams with a similar final transverse displacement: (a) bird's eye view; (b) side view (reproduced with kind permission of Korean Society of Automotive Engineers).

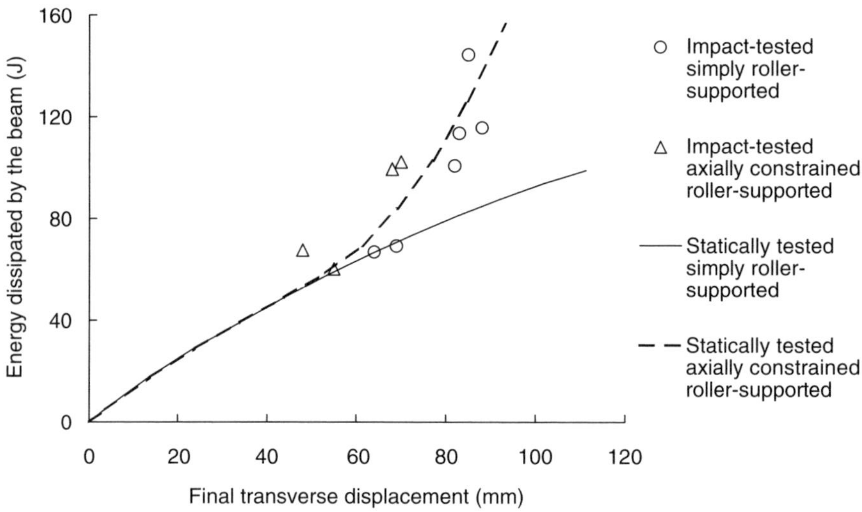

12.18 Correlation between the energy dissipation by the beam and the final transverse displacement for the static- and impact-tested beam samples (reproduced with kind permission of Korean Society of Automotive Engineers).

The local cross-sectional distortion mechanism which occurred in impact tests was very similar to those observed in the static three-point bending, as shown in Fig. 12.17.

Figure 12.18 presents the correlation between the energy dissipated by the beam and its final transverse displacement. It indicates that, to achieve the same final displacement, more input energy is required when the W-beam is subjected to an impact loading compared with a quasi-static loading, regardless of the type of end support.

In impact tests on the downscaled W-beam guardrail system, the initial peak loads for different supporting posts were similar to those in the beam-only tests. A similar 'abrupt-peak-load' phenomenon also occurred at the maximum transverse displacement but a higher impact velocity was required for the occurrence compared with that in the beam-only testing. The energy dissipation partitioning between the downscaled W-beam and the supporting posts can be estimated by considering the final deformation of the W-beam. Based on the final deformation of the beam in the system test with a comparison to that measured in the beam-only test, the corresponding energy dissipation ratios α_b and α_p are shown in Fig. 12.19, where

$$\alpha_b = (\text{energy dissipated by the beam})/(\text{total energy dissipated by the system})$$

$$\alpha_p = (\text{energy dissipated by the posts})/(\text{total energy dissipated by the system})$$

It is found that for low impact energy, the W-beam dissipated most of the input energy, while for high impact energy, the supporting posts shared more energy, and eventually dissipated a larger portion than the W-beam.

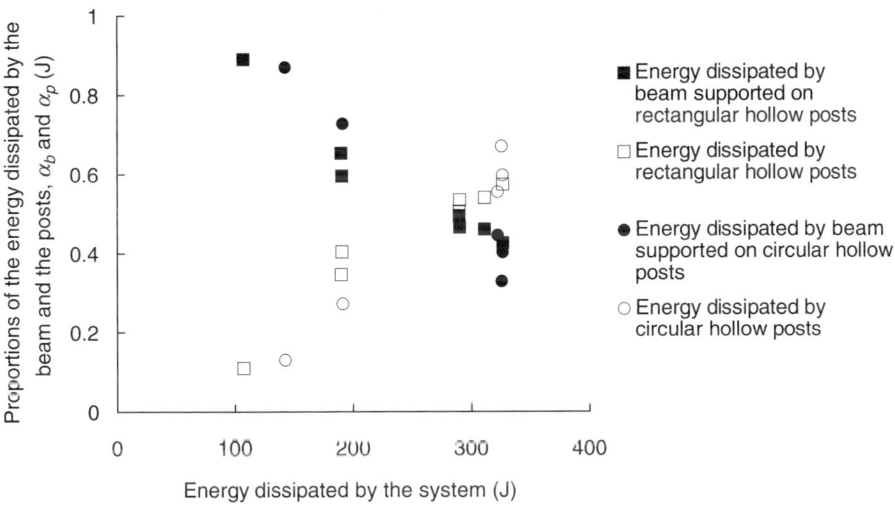

12.19 Proportions of the energies dissipated by beam and posts under impact loading vary with the total energy dissipation in the beam-post system (reproduced with kind permission of Korean Society of Automotive Engineers).

12.4.3 A mass-spring model

Experimental study of the guardrail beams is always limited to certain loading conditions (i.e. the mass and initial velocity of the drop weight). It is very difficult to assess the variety of real collisions which can take place between vehicles and roadside parapets. Hence, a theoretical model should be developed to reveal the fundamental issues of experimental observations and to apply them in real situations.

In general, the dynamic response of a structure is governed by two major factors: one is the inertia which is characterised by the mass or effective mass of the structure; the other is the resistance force which is dictated by the stiffness of the material and structure and is able to be formulated as a function of deformation. Hence, a single-DoF (Degree of Freedom) or multi-DoF mass-spring system can be employed as the simplest model for analysing the dynamic behaviour of structures without losing the essential physical significance. For example, Symonds and Frye (1988) used a single DoF mass-spring model to study the validity of rigid, perfectly plastic idealisation. The applicability of simple mass-spring models in predicting dynamic deformation of elastic-plastic structures under impact was examined by Wu and Yu (2001).

It is noted that a mass-spring model transforms a continuum structure into a finite DoF system, resulting in great simplification. In this sense, the modal approximate technique (MAT) proposed by Martin and Symonds (1966), which transforms a beam or plate's dynamic response into a single DoF problem by properly assuming the deformation pattern (i.e. the mode) of the structure, can also be regarded as a mass-spring model. Although the conventional MAT is based on the rigid-plastic material idealisation, it can also be extended to consider elastic effect, i.e. be developed to an elastic-plastic modal approximate technique, as proposed by H.H. Ruan *et al.* (2003). Hence, the mass-spring model based on MAT can incorporate both elastic and plastic behaviour of a structure.

For the present problem, the deformation profile shown in Fig. 12.17(b) indicates a single-plastic-hinge mechanism, which actually is the primary plastic mode of a simply supported beam under impact at its mid-point. Theoretically speaking, the validity of this modal solution is related to the large mass ratio of the drop weight (or vehicles) to the W-beams. This ensures that most of the input energy will be dissipated in the modal phase (refer to Stronge and Yu, 1993). Bu adopting the small deflection assumption the governing equation of such a deformation pattern (Fig. 12.20) is found as

$$\frac{1}{3}\rho\left(\frac{L}{2}\right)^2\ddot{W}+\frac{G}{2}\left(\frac{L}{2}\right)\ddot{W}=M_B \qquad [12.30]$$

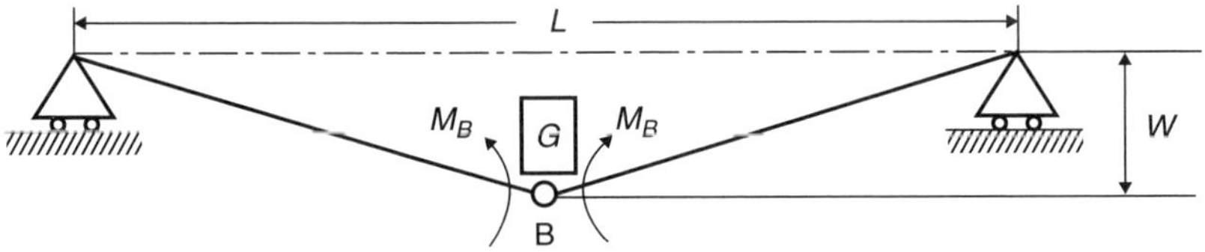

12.20 Single-hinge deformation mechanism for a simply-supported beam subjected to a striking mass *G* at the mid-span.

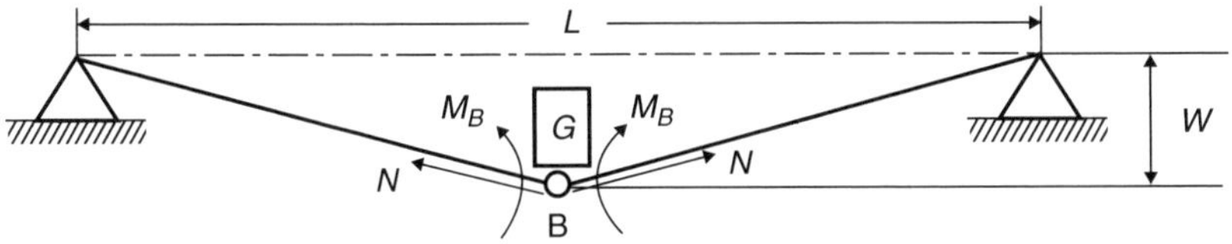

12.21 Single-hinge deformation mechanism for an axially constrained simply-supported beam subjected to a striking mass *G* at the mid-span.

where ρ is the mass per unit length of the beam, L the span between the supports, W the displacement of the mid-point, G the mass of the drop weight assembly and M_B the bending moment at mid-section B.

Letting $m = \rho L/3$, $F = M_B/L$, Eq. [12.30] is recast as

$$(m+G)\ddot{W} = F \qquad [12.31]$$

Similarly, for the beam with axial constraint at the supports, assuming that the deformation mechanism is the same as that without axial constraint (Fig. 12.21), the equation of motion is

$$\frac{1}{3}\rho\left(\frac{L}{2}\right)^2 \ddot{W} + \frac{G}{2}\left(\frac{L}{2}\right)\ddot{W} = M_B + NW \qquad [12.32]$$

where N is the axial force induced by the axial constraints at the supports.

Letting $F' = F + NW/L$, Eq. [12.32] is rewritten as

$$(m+G)\ddot{W} = F' \qquad [12.33]$$

Equations [12.31] and [12.33] represent the mass-spring models sketched in Figs 12.22(a) and (b), respectively. Herein, both blocks on top of the springs have mass m, which can be regarded as the equivalent mass of the beam. The resistance forces of the non-linear springs shown in Fig. 12.22(a) and (b) are given by F and F', respectively, which are both functions of the mid-point displacement W.

If the load-carrying capacity of a beam is assumed not to be sensitive to the deformation rate, then the static load–displacement curve can be employed as the mechanical property of the non-linear spring for the mass-

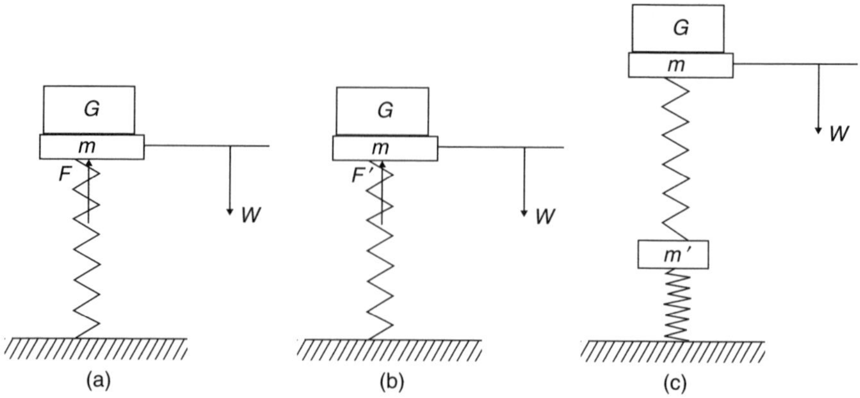

12.22 Mass-spring models of (a) a simply-supported W-beam, (b) an axially constrained simply-supported W-beam and (c) a guardrail system, under impact by a striking mass *G*.

spring system. However, since the material used in both the real W-beam system and the downscaled model, steel, is a strain-rate sensitive material, the resistant force resulting from an impact loading on a downscaled W-beam must be higher than its quasi-static counterpart. Secondly, the cross-sectional distortion of the W-beam during dynamic response is smaller than that in static deformation, leading to an increase in the resistance of the beam when the same mid-point deflection is attained. Moreover, the lateral inertia related to cross-sectional distortion (as shown in Fig. 12.17(a)), which cannot be represented by the mass block in the mass-spring system, also results in an increase in the resistance of the W-beam. Therefore, there is a need to account for the dynamic effect on the spring's behaviour. Herein, based on the experimental results, a ***dynamic enhancement factor*** is introduced to modify the static load–displacement characteristics of the W-beams to account for the various effects caused by dynamic loading. As the combination of the strain-rate effect of material, the inertia effect of the structure and the dynamic effect on the flexural rigidity of the beam related to the local cross-sectional distortion, this dynamic enhancement factor η is defined as

$$\eta = \frac{E^d|_{\delta_b}}{E^s|_{\delta_b}} \qquad [12.34]$$

where E denotes the energy dissipation, the superscripts d and s pertain to the dynamic and static loading, respectively, and δ_b is the maximum mid-point displacement.

For the beam–post system, it is observed that the post's deformation is mainly concentrated at the region close to its clamped end. Hence, it is reasonable to assume that the post deforms in the primary plastic mode of a tubular cantilever beam, in which a stationary hinge is formed at the root. Thus, a two-DoF mass spring system as shown in Fig. 12.22(c) can be constructed and formulated.

When a collision occurs, a local contact deformation must happen in the contact area. A more complicated model may account for the local deformation by employing contact springs (refer to Section 7.1.3) as proposed by Wu and Yu (2001) and Ruan and Yu (2003). For the present problem, since the mass of the drop weight or real vehicle is much larger than that of the guardrail beams, the energy dissipated in the local deformation will be merely a minor portion of the input energy according to Wu and Yu (2001) and Ruan and Yu (2003). Hence, the local deformation may be neglected. The stick assumption, which assumed that the striker is adhered to the beam immediately after impact, can also be employed, owing to the negligible local energy dissipation. Accordingly, based on the conservation of momentum, the initial condition of the mass-spring system is obtained as

$$\dot{W}\big|_{t=0} = \frac{GV_o}{m+G}, \quad W\big|_{t=0} = 0 \qquad [12.35]$$

Employing the above model for the downscaled W-beams, Fig. 12.23 depicts the numerical predictions together with the corresponding experimental results.

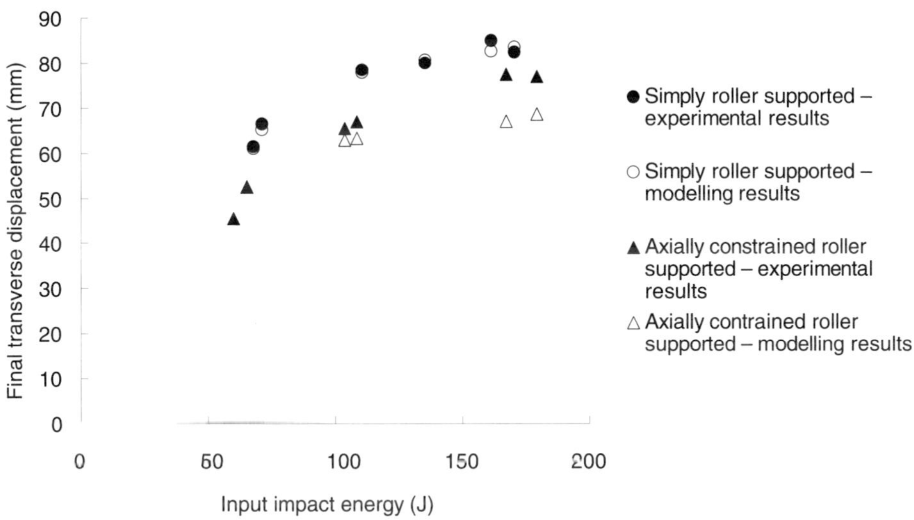

12.23 Experimental results (solid circles and solid triangles) and numerical predictions (hollow circles and hollow triangles) for downscaled W-beam tests with different end supporting conditions.

12.4.4 Concluding remarks

Quasi-static three-point bending experiments were first conducted on the scaled-down W-beams to obtain the load–displacement characteristics and the deformation mechanism. Three factors are found to affect the load-carrying capacity of the beam: (i) the material's strain-hardening, (ii) the structural softening caused by the local cross-sectional distortion and (iii) the tensile force due to the axial constraints. When the beam samples were impact tested, either alone or with two supporting posts, the deformation mechanism of the beam was generally similar to that in quasi-static tests. The energy dissipation partitioning between the beam and the supporting posts changed with different input energy. Finally, simple mass-spring models are proposed to predict the dynamic behaviour of the guardrail or the guardrail–post system subjected to impact loading. The predictions of the model on the final displacement of the guardrail system show a good agreement with the experimental measurements.

References

Abramowicz, W. (1983) The effective crushing distance in axially compressed thin-walled metal columns. *Int. J. Impact Engng*. **1**, 309–17.

Abramowicz, W. and Jones, N. (1984a) Dynamic axial crushing of square tubes. *Int. J. Impact Engng*. **2**, 179–208.

Abramowicz, W. and Jones, N. (1984b) Dynamic axial crushing of circular tubes. *Int. J. Impact Engng*. **2**, 263–81.

Abramowicz, W. and Jones, N. (1986) Dynamic progressive buckling of circular and square tubes. *Int. J. Impact Engng*. **4**, 243–70.

Abrate, S. (1998) *Impact on Composite Structures*. Cambridge University Press, Cambridge, UK.

Alexander, J.M. (1960) An approximate analysis of the collapse of thin cylindrical shells under axial load. *Quart. J. Mech. App. Math.* **13**, 10–15.

Al-Hassani, S.T.S., Johnson, W. and Lowe, W.T. (1972) Characteristics of inversion tubes under axial loading. *J. Mech. Engng. Sci.* **14**, 370–81.

Ando, K., Fukuya, T., Kaji, S. and Seo, T. (1995) Development of guardrails for high speed collisions. *Transportation Research Record* **1500**, 52–8.

Andrews, K.R.F., England, G.L. and Ghani, E. (1983) Classification of the axial collapse of circular tubes under quasi-static loading. *Int. J. Mech. Sci.* **25**, 687–96.

Ashby, M.F., Evans, A., Fleck, N.A., Gibson, L.J., Hutchinson, J.W. and Wadley, H.N.G. (2000) *Metal Foams: A Design Guide*. Butterworth-Heinemann, Boston MA.

Atkins, A.G. (1987) On the number of cracks in the axial splitting of ductile metal tubes. *Int. J. Mech. Sci.* **29**, 115–21.

Atkins, A.G. (1988) Scaling in combined plastic flow and fracture. *Int. J. Mech. Sci.* **30**, 173–91.

Atkins, A.G. (1989) Tearing of thin metal sheets, in *Structural Failure* (eds T. Wierzbicki and N. Jones). John Wiley & Sons, New York, 107–32.

Aya, N. and Takahashi, K. (1974) Energy absorption characteristics of vehicle body structures. *Trans. Society Auto. Engrs. Japan*, No. 7, May.

Baker, W.E., Togami, T.C. and Weydert, J.C. (1998) Static and dynamic properties of high-density metal honeycombs. *Int. J. Impact Engng.* **21**, 149–63.

Bank, L.C., Gentry T.R. and Yin J. (1998a) Pendulum impact tests on steel W-beam guardrails. *J. of Transportation Engng.* July/August, 319–25.

Bank, L.C., Hargarve, M., Svenson, A. and Tcibci, A. (1998b) Impact performance of pultruded beams for highway safety applications. *Composite Structures* **42**, 231–7.

Barbat, S.D. and Prasad, P. (1995) Finite element modeling of structural foam and head impact interaction with vehicle upper interior. SAE-950885, SAE International Congress & Exposition, Detroit, USA.

Belingardi, G., Chiara, A. and Vadori, R. (1992) Experimental verification of the axial crushing behaviour of thin-walled columns with different sections. *Proceedings of the 3rd International Conference of Innovation and Reliability in Automotive Design and Testing*, April, **1**, 59–68.

Berry, J.P. (1984) Energy absorption and failure mechanisms of axially crushed GRP tubes. PhD thesis, University of Liverpool.

Birch, R.S. and Jones, N. (1990) Dynamic and static axial crushing of axially stiffened cylindrical shells. *Thin-Walled Structures* **9**, 29–60.

Bodner, S.R. and Speirs, W.G. (1963) Dynamic plasticity experiments on aluminium cantilever beams at elevated temperature. *J. Mech. Phys. Solids* **11**, 65–77.

Bodner, S.R. and Symonds, P.S. (1960) Plastic deformation in impact and impulsive loading of beams, in *Plasticity, Proceedings of the Second Symposium of Naval Structural Mechanics* (eds E.H. Lee and P.S. Symonds). Pergamon Press, Oxford, 488–500.

Bodner, S.R. and Symonds, P.S. (1962) Experimental and theoretical investigation of plastic deformation of cantilever beams subjected to impulsive loading. *ASME J. Appl. Mech.* **29**, 719–28.

Booth, E., Collier, D. and Miles, J. (1983) Impact scalability of plated steel structures, in *Structural Crashworthiness* (eds N. Jones and T. Wierzbicki). Butterworths, London, 136–75.

Bridgman, P.W. (1922) *Dimensional Analysis*. Yale University Press, New Haven, CT.

Brown, J.C. and Tidbury, G.H. (1983) An investigation of the collapse of thin-walled rectangular beams in bi-axial bending. *Int. J. Mech. Sci.* **25**, 733–46.

Brown, K. (1959) *Package Design Engineering*. Wiley, New York.

Buckingham, E. (1914). On physically similar systems. *Phys. Rev. London*, **4**, 345.

Burton, R.H. and Craig, J.M. (1963) An investigation into the energy absorbing properties of metal tubes loaded in the transverse direction. BSc (Eng) Report, University of Bristol, UK.

Calladine, C.R. (1968) Simple ideas in the large-deflection plastic theory of plates and slabs, in *Engineering Plasticity* (eds J. Heyman and F.A. Leckie). Cambridge University Press, Cambridge, UK, 93–127.

Calladine, C.R. (1983a) An investigation of impact scaling theory, in *Structural Crashworthiness* (eds N. Jones and T. Wierzbicki). Butterworths, London, 169–74.

Calladine, C.R. (1983b) *Theory of Shell Structures*. Cambridge University Press.

Calladine, C.R. (1986) Analysis of large plastic deformation in shell structures, in *Inelastic Behaviour of Plates and Shells. IUTAM Symposium* (eds L. Bevilacqua, R. Feijoo and R. Valid). Springer, Berlin/Heidelberg, 69–101.

Calladine, C.R. (1990) The teaching of some aspects of the theory of inelastic collisions. *Int. J. Mech. Engng. Educ.* **18**, 301–10.

Calladine, C.R. (1993) Some problems in propagating plasticity, in *Plasticity and Impact Mechanics* (ed. N.K. Gupta). Wiley Eastern Ltd, New Delhi, 1–12.

Calladine, C.R. and English, R.W. (1984) Strain-rate and inertia effects in the collapse of two types of energy-absorbing structure. *Int. J. Mech. Sci.* **26**, 689–701.

Carney III, J.F. (1993) Motorway impact attenuation devices: past, present and future, in *Structural Crashworthiness and Failure* (eds N. Jones and T. Wierzbicki). Elsevier, London & New York, 423–66.

Carney III, J.F., Austin, C.D. and Reid, S.R. (1982) Energy dissipation characteristics of steel tube clusters. *Twenty-Third Structures, Structural Dynamics and Materials Conference,* AIAA/ASME/ASCE/AHS, May 10–12, 1982, New Orleans, LA. *AIAA* 82-0759-CP.

Carruthers, J.J., Kettle, A.P. and Robinson, A.M. (1998) Energy absorption capability and crashworthiness of composite material structures: a review. *Appl. Mech. Rev.* **51**, 635–49.

Cavanaugh, J.M. (1993) The biomechanics of thoracic trauma. Chapter 15 in *Accidental Injury – Biomechanics and Prevention* (eds A.M. Nahum and J.W. Melvin). Springer-Verlag, New York.

Chang, F.S, Song, Y., Lu, D.X. and de Silva, C.N. (1998) Unified constitutive equations of foam materials. *ASME J. Engng. Mat. & Tech.* **120**, 212–17.

Chater, E. and Hutchinson, J.W. (1984) On the propagation of bulges and buckles. *J. Appl. Mech.* **51**, 269–77.

Chen, W., Wierzbicki, T., Breuer, O. and Kristiansen, K. (2001) Torsional crushing of foam-filled thin-walled square columns. *Int. J. Mech. Sci.* **43**, 2297–317.

Chen, W.F. and Atsuta, T. (1976) *Theory of Beam-Columns*. McGraw-Hill, New York.

Chou, C.C., Howell, R.J. and Chang, B.Y. (1988) A review and evaluation of various HIC algorithms. SAE Technical Paper Series 880656.

Chung, J. and Waas, A.M. (2002a) Compressive response of honeycombs under in-plane uniaxial static and dynamic loading, part 1: experiments. *AIAA Journal* **40**, 966–73.

Chung, J. and Waas, A.M. (2002b) Compressive response of honeycombs under in-plane uniaxial static and dynamic loading, part 2: simulations. *AIAA Journal* **40**, 974–80.

Cimpoeru, S.J. and Murray, N.W. (1993) The large deflection pure bending properties of a square thin-walled tube. *Int. J. Mech. Sci.* **35**, 247–56.

Colokoglu, A. and Reddy, T.Y. (1996) Strain rate and inertia effects in free external inversion of tubes. *Int. J. Crashworthiness* **1**, 93–106.

Corbett, G.G., Reid, S.R. and Al-Hassani, S.T.S. (1991) Static and dynamic penetration of steel tubes by hemispherically nosed punches. *Int. J. Impact Engng.* **9**, 165–90.

Cotterell, B. and Reddel, J.K. (1977) The essential work of plane stress ductile fracture. *Int. J. Fracture* **13**, 267–77.

Cowper, G.R. and Symonds, P.S. (1957) Strain-hardening and strain-rate effects in the impact loading of cantilever beams. Brown Univ. Dept. Appl. Math. T.R. 28.

Croll, J.G.A. (1985) Analysis of buckle propagation in marine pipelines. *J. Constr. Steel Res.* **5**, 103–22.

Crozier, W.D. and Hume, W. (1957) High-velocity, light-gas gun. *J. Appl. Phys.* **28**, 892–4.

de Oliveira, J.G. and Wierzbicki, T. (1982) Crushing analysis of rotationally symmetric plastic shells. *J. Strain Analysis* **17**, 229–36.

de Oliveira, J., Wierzbicki, T. and Abramowicz, W. (1982) Plastic behaviour of tubular members under lateral concentrated loading. *Nor. Veritas Tech. Rep.* 82-0708.

de Runtz, J.A. and Hodge, P.G. (1963) Crushing of a tube between rigid plates. *ASME J. Appl. Mech.* **30**, 391.

Deb, A., Calso, S. and Saha, N. (1997) Effectiveness of countermeasures in upper interior head impact. SAE-970391, SAE International Congress & Exposition, Detroit, USA.

Deshpande, V.S., Fleck, N.A. (2000) High strain rate compressive behaviour of aluminium alloy foams. *Int. J. Impact Engng*. **24**, 277–98.

Dewalt, W.J. and Herbein, W.B. (1972) *Energy absorption by compression of aluminium tubes*. Report No. 12-72-23, Alcoa Research Laboratories, Alcoa Center, Pennsylvania.

Duffey, T.A. (1971). Scaling laws for fuel capsules subjected to blast, impact and thermal loading. SAE Paper 719107.

Fairfull, A.H. and Hull, D. (1987) Effects of specimen dimensions on the specific energy absorption of fibre composite tubes, in *Proceedings of the Sixth International Conference on Composite Materials*, eds F.L. Matthews, N.C.R. Buskell, J.M. Hodgkinson and J. Morton, Elsevier Applied Science, London, 3.36–3.45.

Fan, H., Wang, B. and Lu, G. (2002) On the tearing energy of thin sheet. *Int. J. Mech. Sci*. **44**, 407–21.

Farley, G.L. (1983) Energy absorption of composite materials. *J. Composite Mat.* **17**, 267–79.

Farley, G.L. (1986) Effect of specimen geometry on the energy absorption capability of composite materials. *J. Composite Mat.* **20**, 390–400.

Farley, G.L. and Jones, R.M. (1992) Crushing characteristics of composite tubes with 'near elliptical' cross sections. *J Composite Mat.* **26**, 1741–51.

Fok, W.C., Lu, G. and Seah, L.K. (1993) A simplified approach to buckling of plain C channels under pure bending. *IMechE Proc. Instn Mech. Engrs. Part C* **207**, 255–62.

Forrestal, M.J. and Sagartz, M.J. (1978) Elastic-plastic response of 304 stainless steel beams to impulsive load. *ASME J. Appl. Mech.* **45**, 685–7.

Forrestal, M.J. and Wesenberg, D.L. (1977) Elastic-plastic response of simply supported 1018 steel beams to impulsive load. *ASME J. Appl. Mech*. **44**, 779–80.

Frisch-Fay, R. (1962) *Flexible Bars*. Butterworths, London.

Gabler, H.C., Willke, D.T. and Wagner, J.J. (1991) Upper interior head impacts: the safety performance of passenger vehicles. *The 13th International Technical Conference on Experimental Safety Vehicles*, Paris, France.

Gao, Z.Y., Yu, T.X. and Lu, G. (2003a) A study on type II structures, part I: a one-dimensional model. Submitted to *Int. J. Impact Engng*.

Gao, Z.Y., Yu, T.X. and Lu, G. (2003b) A study on type II structures, part II: behaviour of a chain of pre-bent plates. Submitted to *Int. J. Impact Engng*.

Gibbings, J.C. (1982) A logic of dimensional analysis. *J. Phys. A* **15**, 1991–2002.

Gibson, L.J. and Ashby, M.F. (1997) *Cellular Solids, Structure and Properties.* Second edition. Cambridge University Press, Cambridge, UK.

Gilchrist, A. and Mills, N.J. (2001) Impact deformation of rigid polymeric foams: experiments and FEA modeling. *Int. J. Impact Engng.* **25**, 267.

Gill, S.S. (1976) Large deflection rigid plastic analysis of a built-in semi-circular arch. *Int. J. Mech. Engng. Educ.* **4**, 339–55.

Goldfinch, A.C. (1986) Plate tearing energies. Part II Project Report, Engineering Department, Cambridge University.

Goldsmith, W. and Sackman, J.L. (1992) An experimental study of energy absorption in impact on sandwich plates. *Int. J. Impact Engng*. **12**, 241–62.

Gould, P.L. (1999) *Analysis of Shell and Plates*. Prentice Hall, Englewood Cliffs, NJ.

Grzebieta, R.H. (1990) An alternative method for determining the behaviour of round stocky tubes subjected to axial crush loads. *Thin-Walled Structures* **9**, 66–89.

Grzebieta, R.H. and Murray, N.W. (1985) The static behaviour of struts with initial kinks at their centre point. *Int. J. Impact Engng.* **3**, 155–65.

Grzebieta, R.H. and Murray, N.W. (1986) Energy absorption of an initially imperfect strut subjected to an impact load. *Int. J. Impact Engng.* **4**, 147–59.

Guillow, S.R., Lu, G. and Grzebieta, R.H. (2001) Quasi-static axial compression of thin-walled circular aluminium tubes. *Int. J. Mech. Sci.* **43**, 2103–23.

Guist, L.R. and Marble, D.P. (1966) Prediction of the inversion of a circular tube. Technical Note TN-D-3622. Washington: National Aeronautics and Space Administration.

Gupta, N.K. and Ray, P. (1998) Collapse of thin-walled empty and filled square tubes under lateral loading between rigid plates. *Int. J. Crashworthiness* **3**, 265–85.

Gupta, N.K. and Sinha, S.K. (1990a) Lateral compression of crossed layers of square-section tubes. *Int. J. Mech. Sci.* **32**, 565–80.

Gupta, N.K. and Sinha, S.K. (1990b) Transverse collapse of thin-walled square tubes in opposed loadings. *Thin-Walled Structures* **10**, 247–62.

Gupta, N.K., Velmurugan, R. and Gupta, S.K. (1997) An analysis of axial crushing of composite tubes. *J. Composite Mat.* **31**, 1262–86.

Gupta, N.K. and Velmurugan, R. (1999) Axial compression of empty and foam filled composite conical shells. *J. Composite Mat.* **33**, 567–91.

Hackney, J.R., Monk, M.W., Hollowwell, W.T., Sullivan, L.K. and Willke, D.T. (1984) Results of the national highway safety administration's thoracic side impact protection research program. SAE Technical Paper Series 840886.

Hanefi, E.H. and Wierzbicki, T. (1996) Axial resistance and energy absorption of externally reinforced metal tubes. *Composites: Part B*, **27B**, 387–94.

Hanssen, A.G., Langseth, M. and Hopperstad, O.S. (1999) Static crushing of square aluminium extrusions with aluminium foam filler. *Int. J. Mech. Sci.* **4**, 967–93.

Harris, H.G. and Sabnis, G.M. (1999) *Structural Modeling and Experimental Techniques*. Second edition. CRC Press, Boca Raton, FL.

Harte, A.M., Fleck, N.A. and Ashby, M.F. (2000) Energy absorption of foam-filled circular tubes with braided composite walls. *Eur. J. Mech. A/Solids* **19**, 31–50.

Hill, R. (1950) *Mathematical Theory of Plasticity*. Oxford University Press, Oxford, UK.

Hodge, P.G. Jr. (1959) *Plastic Analysis of Structures*. McGraw Hill, London.

Hönig, A. and Stronge, W.J. (2000) Dynamic buckling of an imperfect elastic, viscoplastic plate. *Int. J. Impact Engng.* **24**, 907–23.

Hopkinson, B. (1914) A method of measuring the pressure produced in the detonation of high explosives or by the impact of bullets. *Roy. Soc. Phil. Trans.* **A213**, 437–56.

Hou, W.J., Yu, T.X. and Su, X.Y. (1995) Elastic effects in the dynamic response of plastic cantilever beams (in Chinese). *Acta Mechanica Solida Sinica*, **16**(1), 13–21.

Huang, X., Lu, G. and Yu, T.X. (2002a) Axial splitting of square tubes. *Thin-Walled Structures* **40**, 153–65, 2002.

Huang, X., Lu, G. and Yu, T.X. (2002b) On the axial splitting and curling of circular metal tubes. *Int. J. Mech. Sci.* **44**, 2369.

Huang, X. and Lu, G. (2003) Axisymmetric progressive crushing of circular tubes. *Int. J. Crashworthiness* **8**, 87–95.

Hui, S.K. and Yu, T.X. (2000) Large plastic deformation of W-beams used as guardrails on highways, in *Advances in Engineering Plasticity*, Part 2 (eds T.X. Yu, Q.P. Sun and J.K. Kim), special volume of *Key Engng. Mat.* **177–80**, 751–6.

Hui, T.Y.J., Ruan, H.H. and Yu, T.X. (2003) Dynamic characteristics of scaled-down W-beams under impact. *Int. J. Automotive Technology* **4**(1), 31–40.

Hull, D. (1983) Axial crushing of fibre reinforced composite tubes. Chapter 5 in *Structural Crashworthiness* (eds N. Jones and T. Wierzbicki). Butterworths, London.

Hull, D. (1991) A unified approach to progressive crushing of fibre-reinforced composite tubes. *Composites Sci. and Tech.* **40**, 377–421.

Hull, D. and Clyne, T.W. (1996) *An Introduction to Composite Materials*. Second edition. Cambridge University Press, Cambridge UK.

Johnson, K.L. (1985) *Contact Mechanics*. Cambridge University Press, Cambridge UK.

Johnson, W. (1972) *Impact Strength of Materials*. Edward Arnold, London.

Johnson, W. (1983) Structural damage in airship and rolling stock collisions. Chapter 15 in *Structural Crashworthiness* (eds N. Jones and T. Wierzbicki). Butterworths, London.

Johnson, W. (1990) The elements of crashworthiness: scope and actuality. *Proc. Instn. Mech. Engrs. Part D: J. Automobile Engng.* **204**, 255–73.

Johnson, W., Ghosh, S.K., Mamalis, A.G., Reddy, T.Y. and Reid, S.R. (1979) The quasi-static piercing of cylindrical tubes or shells. *Int. J. Mech. Sci.* **9**, 9–20.

Johnson, W. and Mamalis, A.G. (1978) *Crashworthiness of Vehicles*. Mechanical Engineering Publications Ltd, London.

Johnson, W., Mamalis, A.G. and Reid, S.R. (1982) Aspects of car design and human injury. Chapter 4 in *Human Body Dynamics, Impact, Occupational and Athletic Aspects* (ed. D.N. Ghista). Clarendon Press, Oxford, UK, 164–80.

Johnson, W. and Mellor, P.B. (1983) *Engineering Plasticity*. Ellis Horwood, Chichester, UK.

Johnson, W. and Reid, S.R. (1977) Metallic energy dissipating systems. *ASME Appl. Mech. Rev.* **31**, 277–88.

Johnson, W. and Reid, S.R. (1986) Update to 'Metallic energy dissipating systems'. *ASME Appl. Mech. Update*, 315–19.

Johnson, W., Reid, S.R. and Reddy, T.Y. (1977a) The compression of crossed layers of thin tubes. *Int. J. Mech Sci.* **19**, 423.

Johnson, W., Soden, P.D. and Al-Hassani, S.T.S. (1977b) Inextensional collapse of thin-walled tubes under axial compression. *J. Strain Analysis* **12**, 317–30.

Johnson, W. and Walton, A.C. (1983a) Protection of car occupants in frontal impact with heavy lorries: frontal structures. *Int. J. Impact Engng.* **1**, 111–24.

Johnson, W. and Walton, A.C. (1983b) An experimental investigation of the energy dissipation of a number of car bumpers under quasi-static loads. *Int. J. Impact Engng* **1**, 301–8.

Johnson, W. and Yu, T.X. (1981) Approximate calculations on the large plastic deformation of helical springs and their possible applications on a vehicle arresting system. *J. Strain Analysis* **16**, 111–21.

Johnson, W. and Yu, T.X. (1989) An outline of engineering dynamic elasticity and plasticity. Chapter 4 in *Plasticity and Modern Metal-Forming Technology* (ed. T.Z. Blazynski), Elsevier, New York, 73–114.

Jones, N. (1989a) *Structural Impact*. Cambridge University Press, Cambridge, UK.

Jones, N. (1989b) On the dynamic inelastic failure of beams, in *Structural Failure* (eds T. Wierzbicki and N. Jones). John Wiley, New York, 133–59.

Jones, N. and Birch, R.S. (1987) A study of plate tearing for ship collision and grounding damage. *J. Ship Research*, **31,** 253.

Jones, N. and Birch, R.S. (1990) Dynamic and static axial crushing of axially stiffened square tubes. *Proc. Instn. Mech. Engrs. Part C* **204**, 293–310.

Jones, N., Birch, S.E., Birch, R.S., Zhu, L. and Brown, M. (1992) An experimental study on the lateral impact of fully clamped mild steel pipes. *Proc. Instn. Mech. Engrs. Part E* **206**, 111–27.

Jones, N. and Shen, W.Q. (1992) A theoretical study of the lateral impact of fully clamped pipelines. *Proc. Instn. Mech. Engrs. Part E*, **206**, 129–46.

Kamalarasa, S. and Calladine, C.R. (1988) Buckle propagation in submarine pipelines. *Int. J. Mech. Sci.* **30**, 217–28.

Karagiozova, D., Alves, M. and Jones, N. (2000) Inertia effect in axisymmetrically deformed cylindrical shells under axial impact. *Int. J. Impact Engng.* **24**, 1083–115.

Karagiozova, D. and Jones, N. (1995a) Some observations on the dynamic elastic-plastic buckling of a structural model. *Int. J. Impact Engng.* **16**, 621–35.

Karagiozova, D. and Jones, N. (1995b) A note on the inertia and strain-rate effects in the Tam and Calladine model. *Int. J. Impact Engng.* **16**, 637–49.

Kardaras, C. and Lu, G. (2000) Finite element analysis of thin-walled tubes under point loads subjected to large plastic deformation, in *Advances in Engineering Plasticity*, Part 2 (eds T.X. Yu, Q.P. Sun and J.K. Kim), special volume of *Key Engng. Mat.* **177–80**, 733–8.

Keal, R. (1983) Post failure energy absorbing mechanisms of filament wound composite tubes. PhD thesis, University of Liverpool.

Kecman, D. (1983) Bending collapse of rectangular and square section tubes. *Int. J. Mech. Sci.* **25**, 623–36.

Kecman, D. and Suthurst, G.D. (1984) Theoretical determination of the maximum bending strength in the car body components. *Proc. IMechE*, **C181**(84), 53–61.

Kim, C.S., Lee, Y.R., Chung, T.E., Song, Y.J., King, S.Y. and Chirwa, E.C. (1997) A study on large deflection of thin-walled tubes under pure bending. *Int. J. Crashworthiness* **2**, 273–86.

Kim, T.H. and Reid, S.R. (2001a) Bending collapse of thin-walled rectangular section columns. *Computers and Structures* **79**, 1897–1911.

Kim, T.H. and Reid, S.R. (2001b) Multiaxial softening hinge model for tubular vehicle roll-over protective structures. *Int. J Mech. Sci.* **43**, 2147–70.

Kindervater, C.M. (1990) Energy absorption of composites as an aspect of aircraft structural crash-resistance, in *Developments in the Science and Technology of Composite Materials*, Elsevier Applied Science, 643–51.

Klintworth, J.W. and Stronge, W.J. (1988) Elasto-plastic yield limits and deformation laws for transversely crushed honeycombs. *Int. J. Mech. Sci.* **30**, 273–92.

Klintworth, J.W. and Stronge, W.J. (1989) Plane punch indentation of a ductile honeycomb. *Int. J. Mech. Sci.* **31**, 359–78.

Kolsky, H. (1953). *Stress Waves in Solids*. Clarendon Press, Oxford, UK.

Kyriakides, S., Yeh, M.K. and Roach, D. (1984) On the determination of the propagation pressure of long circular tubes. *ASME J. Pressure Vessel Tech.* **106**, 150.

Leckie, F.A. and Penny, R.K. (1968) Plastic instability of a spherical shell, in *Engineering Plasticity* (eds J. Heyman and F.A. Leckie). Cambridge University Press, Cambridge, UK, 401–11.

Lee, E.H. and Symonds, P.S. (1952) Large deformation of beams under transverse impact. *ASME J. Appl. Mech.* **19**, 308–14.

Lee, E.H. and Tupper, S.J. (1954) Analysis of plastic deformation in a steel cylinder striking a rigid target. *ASME J. Appl. Mech.* **21**, 63.

Lenard, J.G. (1978) On the inversion of a superplastic thin walled tube. *ASME J. Engng. Mat. and Tech.* **100**, 428–30.

Lim, B.B. (2001) *Transient response of pipe to shock loading.* MEng Thesis, Nanyang Technological University, Singapore.

Lowe, W.T., Al-Hassani, S.T.S. and Johnson, W. (1972) Impact behaviour of small scale model motor coaches. *Proc. Instn. Mech. Engng. Auto Div.* **186**(36/72), 409–19.

Lu, G. (1993a) The equivalent structure technique including axial force effect in the plastic large deformation analysis of structures. *Int. J. Mech. Engng. Educ.* **21**, 54–64.

Lu, G. (1993b) A study of the crushing of tubes by two indenters. *Int. J. Mech. Sci.* **35**, 267–78.

Lu, G. and Calladine, C.R. (1990) On the cutting of a plate by a wedge. *Int. J. Mech. Sci.* **32**, 293–314.

Lu, G., Fan, H. and Wang, B. (1998) An experimental method to determine ductile tearing energy under plane stress. *Metals and Materials* **4**, 432–5.

Lu, G., Ong, L.S., Wang, B. and Ng, H.W. (1994) An experimental study on tearing energy in splitting square metal tubes. *Int. J. Mech. Sci.* **36**, 1087–97.

Lu, G., Wang, B. and Zhang, T.G. (2001) Taylor impact tests for ductile-porous materials Part 1: theory. *Int. J. Impact Engng.* **25**, 981–91.

Lu, G. and Wang, X. (2002) Quasi-static piecing of square tubes. *Int. J. Mech. Sci.* **44**, 1101–15.

Macaulay, M.A. and Redwood, R.G. (1964) Small scale model railway coaches under impact. *The Engineer* 1041–6, 25 Dec.

Magee, C.L. and Thornton, P.H. (1978) Design consideration in energy absorption by structural collapse. Society of Automotive Engineers (SAE), Paper No. 780434.

Mai, Y-W. and Cotterell, B. (1984) The essential work of fracture for tearing of ductile metals. *Int. J. Fracture* **13**, 267–77.

Maji, A.K., Schreyer, H.L., Donald, S., Zou, Q. and Satpathi, D. (1995) Mechanical properties of polyurethane foam impact limiters. *ASCE J. Engng. Mech.* **121**, 528–40.

Mamalis, A.G. and Johnson, W. (1983) The quasi-static crumpling of thin-walled circular cylinders and frusta under axial compression. *Int. J. Mech. Sci.* **25**, 713–32.

Mamalis, A.G., Manolakos, D.E, Baldoukas, A.K. and Viegelahn, G.L. (1989) Deformation characteristics of crashworthy thin-walled steel tubes subjected to bending. *Proc. Instn. Mech. Engrs. Part C*, **203**, 411–17.

Mamalis, A.G., Manolakos, D.E., Baldoukas, A.K. and Viegelahn, G.L. (1991a) Energy dissipation and associated failure modes when axially loading polygonal thin-walled cylinders. *Thin-Walled Structures* **12**, 17–34.

Mamalis, A.G., Manolakos, D.E., Demosthenous, G.A. and Ioannidis, M.B. (1994) On the bending of automotive fibre-reinforced composite thin-walled structures. *Composites* **25**, 47.

Mamalis, A.G., Manolakos, D.E., Demosthenous, G.A. and Ioannidis, M.B. (1996a) The static and dynamic collapse of fibreglass composite automotive frame rails. *Composite Structures* **34**, 77.

Mamalis, A.G., Manolakos, D.E., Demosthenous, G.A. and Ioannidis, M.B. (1996b) Energy absorption capability of fibreglass composite square frustra subjected to static and dynamic axial collapse. *Thin-Walled Structures* **25**, 269.

Mamalis, A.G., Manolakos, D.E., Demosthenous, G.A. and Ioannidis, M.B. (1998) *Crashworthiness of Composite Thin-Walled Structural Components.* Technomic Publishing, Lancaster, PA.

Mamalis, A.G., Manolakos, D.E., Demosthenous, G.A. and Johnson, W. (1991b) Axial plastic collapse of thin bi-material tubes as energy dissipating systems. *Int. J. Impact Engng.* **11**, 185–96.

Mamalis, A.G., Manolakos, D.E., Viegelahn, G.L., Demosthenous, G.A. and Yap, S.M. (1991c) On the axial crumpling of fibre-reinforced composite thin walled conical shells. *Int. J. Vehicle Design* **12**, 450.

Mamalis, A.G., Robinson, M., Manolakos, D.E., Demosthenous, G.A., Ioannidis, M.B. and Carruthers, J. (1997) Crashworthy capability of composite material structures. *Composite Structures* **37**, 109–34.

Mamalis, A.G., Yuan, Y.B. and Viegelahn, G.L. (1992) Collapse of thin-wall composite sections subject to high speed axial loading. *Int. J. Vehicle Design* **13**, 564–79.

Martin, J.B. (1964) Impulsive loading theorems for rigid-plastic continua. *ASCE J. Engng Mech. Div.* **90**, 27–42.

Martin, J.B. (1975) *Plasticity: Fundamentals and General Problems*. M.I.T. Press Cambridge, MA.

Martin, J.B. and Symonds, P.S. (1966) Mode approximations for impulsively-loaded rigid-plastic structures. *ASCE J. Engng. Mech. Div. Proc.* **92**(EM5), 43–66.

Meng, Q., Al-Hassani, S.T.S. and Soden, P.D. (1983) Axial crushing of square tubes. *Int. J. Mech. Sci.* **25**(9/10), 747–73.

Merchant, W. (1965) On equivalent structures. *Int. J. Mech. Sci.* **7**, 613–19.

Meyers, M.A. (1994) *Dynamic Behaviour of Materials*. John Wiley & Sons, New York.

Mills, N.J. (1994) Impact response. Chapter 9 in *Low Density Cellular Plastics* (eds H.C. Hilyard and C. Cunningham). Chapman & Hall, London.

Mines, R.A.W., Worrall, C.M. and Gibson, A.G. (1998) Low velocity perforation behaviour of polymer composite sandwich panels. *Int. J. Impact Engng* **21**, 855–79.

Miscow, P.C.F. and Al-Qureshi, H.A. (1997) Mechanics of static and dynamic inversion processes. *Int. J. Mech. Sci.* **39**, 147–61.

Mohammadi, R. and Mahmudi, R. (2001) Ductile tearing energy of sheet metals determined by the multiple tensile testing (MTT) method. *Int. J. Plasticity* **17**, 1551–62.

Morris, A.J. (1971) Experimental investigation into the effects of indenting a cylindrical shell by a load applied though a rigid boss. *J. Mech. Engng Sci.* **13**, 36–46.

Morris, A.J. and Calladine, C.R. (1969) The local strength of a thin spherical vessel loaded radially through a rigid boss, in *Pressure Vessel Technology*, *Proceedings 1st International Conference on Pressure Vessel Technology*, Vol. 1 (ed. I. Berman). ASME, New York, 35–44.

Morris, A.J. and Calladine, C.R. (1971) Simple upper-bound calculations for the indentation of cylindrical shells. *Int. J. Mech. Sci.* **13**, 331–43.

Mukai, T., Kanahashi, H., Higashi, K., Yamada, Y., Shimojima, K., Mabuchi, M., Miyoshi, T. and Nieh, T.G. (1999) Energy absorption of light weight metallic foams under dynamic loading, in *Metal Foams and Porous Metal Structures* (eds J. Banhart, M.F. Ashby and N.A. Fleck), Verlag MIT Publishing, Bremen, Germany, 353–8.

Murray, N.W. (1973) Das aufrehmbare moment in einem zur richtung der normalkraft schrä gliegended plastischen gelenk, *Die Bautechnik* **50**, 57.

Murray, N.W. (1983) The static approach to plastic collapse and energy dissipation in some thin-walled steel structures. Chapter 2 in *Structural Crashworthiness* (eds N. Jones and T. Wierzbicki), Butterworths, London.

Murray, N.W. and Khoo, P.J. (1981) Some basic plastic mechanisms in the local buckling of thin-walled steel structures. *Int. J. Mech. Sci.* **23**, 703.

Nonaka, T. (1967) Some interaction effects in a problem of plastic beam dynamics, Parts 1–3. *ASME J. Appl. Mech.* **34**, 623–43.

Ohkubo, Y., Akamatsu, T. and Shirasawa, K. (1974) Mean crushing strength of closed-hat section members. Society of Automotive Engineers (SAE) Paper No. 740040.

Omar, T.A., Kan, C. and Bedewi, N.E. (1996) Crush behaviour of spot welded hat section components with material comparison. *Proceeding of the Crashworthiness and Occupant in Transportation Systems*, ASME, WMD, **218**, 65–78.

Ong, L.S. and Lu, G. (1996) Collapse of tubular beams loaded by an indenter. *Experimental Mechanics* **36**(4), 374–8.

Paine, F.A. (1991). *The Packaging Users' Handbook*. Blackie, Glasgow.

Palmer, A.C. and Martin, J.H. (1975) Buckle propagation in submarine pipelines. *Nature* **254**, 46–9.

Papka, S.D. and Kyriakides, S. (1994) In-plane compressive response and crushing of honeycomb. *J. Mech. Phys. Solids* **42**, 1499–1532.

Papka, S.D. and Kyriakides, S. (1998) Experiments and full-scale numerical simulations of in-plane crushing of a honeycomb. *Acta Mat.* **46**, 2765–76.

Parkes, E.W. (1955) The permanent deformation of a cantilever struck transversely at its tip. *Proc. Roy. Soc. London* **A228**, 462–76.

Perrone, N. (1972) Biomechanical problems related to vehicle impact, in *Biomechanics: Its Foundations and Objectives* (eds Y. C. Fung, *et al.*), Prentice Hall, Englewood Cliffs, NJ.

Prager, W. and Hodge, P.G., Jr (1951) *Theory of Perfectly Plastic Solids*. John Wiley, New York.

Prentice, J. (1986) Wedge drop tests to investigate plate tearing characteristics. Part II Project Report, Engineering Department, Cambridge University.

Pugsley, Sir A. and Macaulay, M. (1960) The large scale crumpling of thin cylindrical columns. *Quart. J. Mech. Appl. Math.* **13**, 1–9.

Ramakrishna, S. and Hamada, H. (1998) Energy absorption characteristics of crash worthy structural composite materials. *Key Engng. Mat.* **141–3**, 585–620.

Reddy, T.Y. (1989) Tube inversion–an experiment in plasticity. *Int. J. Mech. Engng. Educ.* **17**, 277–91.

Reddy, T.Y. (1992) Guist and Marble revisited – on the natural knuckle radius in tube inversion. *Int. J. Mech. Sci.* **34**, 761–8.

Reddy, T.Y. and Al-Hassani, S.T.S. (1993) Axial crushing of wood-filled square metal tubes. *Int. J. Mech. Sci.* **35**, 213–46.

Reddy, T.Y. and Reid, S.R. (1979) Lateral compression of tubes and tube-systems with side constraints. *Int. J. Mech. Sci.* **21**, 187.

Reddy, T.Y. and Reid, S.R. (1986) Axial splitting of circular metal tubes. *Int. J. Mech. Sci.* **28**, 1111–31.

Reddy, T.Y., Reid, S.R. and Barr, R. (1991) Experimental investigation of inertia effects in one-dimensional metal rings systems subjected to impact-II. Free ended systems. *Int. J. Impact Engng.* **2**, 463–80.

Reddy, T.Y., Reid, S.R., Carney III, J.F. and Veillette, J.R. (1987) Crushing analysis of braced metal ring using the equivalent structure technique. *Int. J. Mech. Sci.* **29**, 655–68.

Reddy, T.Y. and Wall, R.J. (1988) Axial compression of foam-filled thin-walled circular tubes. *Int. J. Impact Engng.* **7**, 151–60.

Reddy, T.Y., Yu, T.X. and El-Dufani, A.I. (1996) Plastic collapse of circular tubes under combined bending and torsion. *The Third Asia-Pacific Symposium on Advances in Engineering Plasticity and its Applications (AEPA '96)*, Hiroshima, Japan, August, 1996.

Redwood, R.G. (1964) Discussion of (DeRuntz JA & Hodge PG 1963). *ASME J. Appl. Mech.* **31**, 357.

Reid, S.R. (1983) Laterally compressed metal tubes as impact energy absorbers. Chapter 1 in *Structural Crashworthiness* (eds N. Jones and T. Wierzbicki), Butterworths, London.

Reid, S.R. and Bell, W.W. (1984) Response of 1-D metal ring systems to end impact, in *Mechanical Properties at High Rates of Strain* (ed. J. Harding), Institute of Physics Coference Series No. 70, Bristol, 471–8.

Reid, S.R., Bell, W.W. and Barr, R. (1983a) Structural plastic model for one-dimensional ring systems. *Int. J. Impact Engng.* **1**, 185–91.

Reid, S.R., Drew, S.L.K. and Carney, III J.F. (1983b) Energy absorbing capacities of braced metal tubes. *Int. J. Mech. Sci.* **25**, 659–67.

Reid, S.R. and Goudie, K. (1989) Denting and bending of tubular beams under local loads, in *Structural Failure* (eds T. Wierzbicki and N. Jones), John Wiley & Sons, New York.

Reid, S.R. and Gui, X.G. (1987) On the elastic-plastic deformation of cantilever beams subjected to tip impact. *Int. J. Impact Engng.* **6**, 109–27.

Reid, S.R. and Harrigan, J.J. (1998) Transient effects in the quasi-static and dynamic internal inversion and nosing of metal tubes. *Int. J. Mech. Sci.* **40**, 263–80.

Reid, S.R., Johnson, W. and Reddy, T.Y. (1980) Pipe whip restraint systems. *Chartered Mechanical Engineer*, June 1980, 55–60.

Reid, S.R. and Peng, C. (1997) Dynamic uniaxial crushing of wood. *Int. J. Impact Engng.* **19**, 531–70.

Reid, S.R. and Prinja, N. K. (1989) The mechanics of pipe whip. *Proceedings Conference Pipework Engineering and Operation, I. Mech. E.*, ASCE, 315–26, London, UK.

Reid, S.R. and Reddy, T.Y. (1978) Effects of strain hardening on the lateral compression of tubes between rigid plates. *Int. J. Solids Structures* **14**, 213.

Reid, S.R. and Reddy, T.Y. (1983) Experimental investigation of inertia effects in one-dimensional metal rings systems subjected to impact-I. Fixed ended systems. *Int. J. Impact Engng.* **1**, 85–106.

Reid, S.R. and Reddy, T.Y. (1986a) Static and dynamic crushing of tapered sheet metal tubes of rectangular cross-section. *Int. J. Mech. Sci.* **28**, 295–322.

Reid, S.R. and Reddy, T.Y. (1986b) Axial crushing of foam-filled tapered sheet metal tubes. *Int. J. Mech. Sci.* **28**, 643–56.

Reid, S.R., Reddy, T.Y. and Gray, M.D. (1986) Static and dynamic axial crushing of foam-filled sheet metal tubes. *Int. J. Mech. Sci.* **28**, 295–322.

Reid, S.R., Reddy, T.Y. and Peng, C. (1993) Dynamic compression of cellular structures and materials. Chapter 8 in *Structural Crashworthiness and Failure* (eds N. Jones and T. Wierzbicki), Elsevier, London, 295–340.

Reid, S.R., Tan, P.J. and Harrigan, J.J. (2001) The crushing strength of aluminium alloy foam at high rates of strain, in *Impact Engineering and Applications* (eds A. Chiba, S. Tanimura and K. Hokamoto). Elsevier, Amsterdam, 15–22.

Reid, S.R. and Wang, B. (1995) Large deflection analysis of whipping pipes: Part I – Rigid, perfectly-plastic model. *ASCE J. Engng. Mech.* **121**, 881–7.

Reid, S.R. and Wen, H.M. (2000) Perforation of FRP laminates and sandwich panels subjected to missile impact. Chapter 8 in *Impact Behaviour of Fibre-Reinforced Composite Materials and Structures* (eds S.R. Reid and G. Zhou), Woodhead Publishing, Cambridge, UK.

Reid, S.R., Yu, T.X. and Yang, J.L. (1995) Response of an elastic, plastic tubular cantilever beam subjected to force pulse at its tip – small deflection analysis. *Int. J. Solids Structures* **32**, 3407–21.

Reid, S.R., Yu, T.X. and Yang, J.L. (1998) An elastic-plastic hardening-softening cantilever beam subjected to a force pulse at its tip: a model for pipe whip. *Proc. Roy. Soc. London* **A454**, 997–1029.

Reid, S.R., Yu, T.X., Yang, J.L. and Corbett, G.G. (1996) Dynamic elastic-plastic behaviour of whipping pipes: experiments and theoretical model. *Int. J. Impact Engng.* **18**, 703–35.

Roach, A.M., Jones, N. and Evans, K.E. (1998) The penetration energy of sandwich panel elements under static and dynamic loading. Part II. *Composite Structures* **42**, 135–52.

Ruan, D., Lu, G., Chen, L. and Siores, E. (2002) Compressive behaviour of aluminium foams at low and medium strain rates. *Composite Structures* **57**, 331–6.

Ruan, D., Lu, G. and Wang, B. (2001). A study of dynamic deformation modes of intact and damaged aluminium honeycombs, in *Impact Engineering and Applications* (eds A. Chiba, S. Tanimura and K. Hokamoto). Elsevier, Amsterdam, 719–24.

Ruan, D., Lu, G., Wang, B. and Yu, T.X. (2003) In-plane dynamic crushing of honeycombs – a finite element study. *Int. J. Impact Engng.* **28**, 161–82.

Ruan, H.H. and Yu, T.X. (2003) Local deformation models in analysing beam-on-beam collisions. To appear in *Int. J. Mech. Sci.*

Ruan, H.H., Yu, T.X. and Hua, Y.L. (2003) Plastic modal approximations in analyzing beam-on-beam collision. To appear in *Int. J. Solids Structures.*

Santosa, S. and Wierzbicki, T. (1997) Effect of an ultralight metal filler on the torsional crushing behaviour of thin-walled prismatic columns. *Int. J. Crashworthiness* **2**, 305–31.

Santosa, S. and Wierzbicki, T. (1998a) On the modelling of crush behaviour of a closed-cell aluminium foam structure. *J. Mech. Phys. Solids* **46**, 645–69.

Santosa, S. and Wierzbicki, T. (1998b) Crash behavior of box columns filled with aluminium honeycomb or foam. *Computers and Structures* **68**, 343–67.

Sedov, L.I. (1993) *Similarity and Dimensional Analysis in Mechanics*. Tenth edition. CRC Press, Boca Raton, FL.

Shen, W.Q. and Chen, K.S. (1998) An investigation on the impact performance of pipelines. *Int. J. Crashworthiness* **3**, 191–209.

Shim, V.P.-W. and Stronge, W.J. (1986a) Lateral crushing of thin-walled tubes between cylindrical indenters. *Int. J. Mech. Sci.* **28**, 683–707.

Shim, V.P.-W. and Stronge, W.J. (1986b) Lateral crushing in tightly packed arrays of thin-walled metal tubes. *Int. J. Mech. Sci.* **28**, 709–28.

Shim, V.P.-W., Tay, B.Y. and Stronge, W.J. (1990) Dynamic crushing of strain-softening cellular structures – a one-dimensional analysis. *ASME J. Engng. Mat. & Tech.* **112**, 398–405.

Silva-Gomes, J.F., Al-Hassani, S.T.S. and Johnson, W. (1978). The plastic extension of a chain of rings due to an axial impact load. *Int. J. Mech. Sci.* **20**, 529.

Singace, A.A. (1999) Axial crushing analysis of tubes deforming in the multi-lobe mode. *Int. J. Mech. Sci.* **41**, 865–90.

Singace, A.A. (2000) Collapse behaviour of plastic tubes filled with wood sawdust. *Thin-Walled Structures* **37**, 163–87.

Singace, A.A. and El-Sobky, H. (1996) Further experimental investigation on the eccentricity factor in the progressive crushing of tubes. *Int. J. Solids Structures* **33**, 3517–38.

Singace, A.A. and El-Sobky, H. (1997). Behaviour of axially crushed corrugated tubes. *Int. J. Mech. Sci.* **39**, 249–68.

Singace, A.A. and El-Sobky, H. (2001) Uniaxial crushing of constrained tubes. *Proc. Instn. Mech. Engrs.* **21**, 353–64.

Singace, A.A., El-Sobky, H. and Reddy, T.Y. (1995) On the eccentricity factor in the progressive crushing of tubes. *Int. J. Solids Structures* **32**, 3589–602.

Song, H., Wan, Z., Xie, Z. and Du, X. (2000) Axial impact behaviour and energy absorption efficiency of composite wrapped metal tubes. *Int. J. Impact Engng.* **24**, 385–401.

Soroka, W. (1999) *Fundamentals of Packaging Technology*. Second edition. Institute of Packaging Professionals, Horndon, Va.

Steel, W.J.M. and Spence, J. (1983) On propagating buckles and their arrest in sub-sea pipelines. *Proc. Instn. Mech. Engrs.* **197A**, 139–47.

Stronge, W.J. (1985) Impact and penetration of cylindrical shells by blunt middiles, in *Metal Forming and Impact Mechanics* (ed. S.R. Reid). Pergamon Press, Oxford, UK, 289–302.

Stronge, W.J. (1993) Impact on metal tubes: indentation and perforation, in *Structural Crashworthiness and Failure* (eds N. Jones and T. Wierzbicki). Elsevier Applied Science, London.

Stronge, W.J. (2000). *Impact Mechanics*. Cambridge University Press, Cambridge, UK.

Stronge, W.J. and Shim, V.P.-W. (1987) Dynamic crushing of a ductile cellular array. *Int. J. Mech. Sci.* **29**, 381–406.

Stronge, W.J. and Shim, V.P.-W. (1988) Microdynamics of crushing in cellular solids. *ASME J. Engng. Mat. & Tech.* **110**, 185–90.

Stronge, W.J. and Yu, T.X. (1993) *Dynamic Models for Structural Plasticity*. Springer-Verlag, London.

Stronge, W.J., Yu, T.X. and Johnson, W. (1983) Long stroke energy dissipation in splitting tubes. *Int. J. Mech. Sci.* **25**, 637–47.

Stronge, W.J., Yu, T.X. and Johnson, W. (1984) Energy dissipation by splitting and curling tubes, *International Conference on Structural Impact and Crashworthiness*, London, July, 1984; in *Structural Impact and Crashworthiness*, Vol. 2 (ed. J. Morton), 576–87, Elsevier, London, 1984.

Su, X.Y., Yu, T.X. and Reid, S.R. (1995a) Inertia-sensitive impact energy-absorbing structures, Part I: Effects of inertia and elasticity. *Int. J. Impact Engng.* **16**, 651–72.

Su, X.Y., Yu, T.X. and Reid, S.R. (1995b) Inertia-sensitive impact energy-absorbing structures, Part II: Effect of strain rate. *Int. J. Impact Engng.* **16**, 673–89.

Symonds, P.S. (1965) Viscoplastic behavior in response of structures to dynamic loading, in *Behavior of Materials under Dynamic Loading* (ed. N. Huffington) ASME, 106–24.

Symonds, P.S. and Fleming, W.T. Jr (1984) Parkes revisited: on rigid-plastic and elastic plastic dynamic structural analysis. *Int. J. Impact Engng.* **2**, 1–36.

Symonds, P.S. and Frye, C.W.G. (1988) On the relation between rigid–plastic and elastic–plastic predictions of response to pulse loading. *Int. J. Impact. Engng.* **7**, 139–49.

Tabor, D. (1948) A simple theory of static and dynamic hardness. *Proc. Roy. Soc. London* **A192**, 247.

Tam, L.L. and Calladine, C.R. (1991) Inertia and strain-rate effects in a simple plate-structure under impact loading. *Int. J. Impact Engng.* **11**, 349–77.

Tani, M. and Funahashi, A. (1978) Energy absorption by the plastic deformation of body structural members. Paper 780368 Presented at SAT Annual Meeting, Detroit, February.

Tao, X.M., Xue, P. and Yu, T.X. (2003) New development of cellular textile composites for energy absorption applications. To appear in *J. Industrial Textile*.

Taylor, G.I. (1948) The use of flat-ended projectiles for determining dynamic yield stress I: Theoretical considerations. *Proc. Roy. Soc. London* **A194**, 289–99.

Thomas, S.G., Reid, S.R. and Johnson, W. (1976) Large deformation of thin-walled circular tubes under transverse loading. Part I. *Int. J. Mech. Sci.* **18**, 325–33.

Thornton, P.H. (1979) Energy absorption in composite structures. *J. Composite Mat.* **13**, 247–62.

Thornton, P.H. (1980) Energy absorption by foam-filled structures. SAE Paper 800081, SAE Congress, Detroit.

Thornton, P.H. and Edwards, P.J. (1982) Energy absorption in composite tubes. *J. Composite Mat.* **16**, 521–45.

Thornton, P.H., Harwood, J.J. and Beardmore, P. (1985) Fiber-reinforced plastic composites for energy absorption purposes. *Composite Sci. and Tech.* **24**, 275–98.

Timoshenko, S.P. and Gere, J.M. (1961) *Theory of Elastic Stability*. Second edition. McGraw-Hill, Tokyo.

Timoshenko, S.P. and Woinowsky-Krieger, S. (1959) *Theory of Plates and Shells*. Second edition. McGraw-Hill, Tokyo.

Ting, T.C.T. (1964) Plastic deformation of a cantilever beam with strain-rate sensitivity under impulsive loading. *ASME J. Appl. Mech.* **31**, 38–42.

Updike, D.P. (1972) On the large deformation of rigid plastic spherical shell compressed by a rigid plate. *ASME J. Engng. for Industry* **94**, 949–55.

Viano, D.C. and Lau, I.V. (1988) A viscous tolerance criterion for soft tissue injury assessment. *J. Biomechanics* **21**, 387–99.

Wang, B., Klepaczko, J.R., Lu, G. and Kong, L.X. (2001) Viscoplastic behavior of porous bronzes and irons. *J. Mat. Proc. Tech.* **113**, 574–80.

Wang, B. and Lu, G. (2002) Mushrooming of cylindrical shells under dynamic axial loading. *Thin-Walled Structures* **40**, 167–82.

Wang, B., Zhang, J. and Lu, G. (2003) Taylor impact tests for ductile-porous materials Part 2: experiments. *Int. J. Impact Engng.* **28**, 499–511.

Wang, X. and Lu, G. (2002) Axial crushing force of externally fibre reinforced metal tubes. *Proc. Instn. Mech. Engrs. Part C* **216**, 863–74.

Wang, X.G., Bloch, J.A. and Cesari, D. (1992) Static and dynamic axial crushing of externally reinforced tubes. *Proc. Instn. Mech. Engrs. Part C* **206**, 355–60.

Wasti, S.T. (1964) Finite deformation of spherical shells. *PhD Thesis*, University of Cambridge.

Watson, A.R., Reid, S.R., Johnson, W. and Thomas, S.G. (1976) Large deformations of thin-walled circular tubes under transverse loading – II. *Int. J. Mech. Sci.* **18**, 387–97.

White, M.D. and Jones, N. (1999) Experimental quasi-static axial crushing of top-hat and double-hat thin-walled sections. *Int. J. Mech. Sci.* **41**, 179–208.

White, M.D., Jones, N. and Abramowicz, W. (1999) A theoretical analysis for the quasi-static axial crushing of top-hat and double-hat thin-walled sections. *Int. J. Mech. Sci.* **41**, 209–33.

Wierzbicki, T. (1983) Crushing analysis of metal honeycombs. *Int. J. Impact Engng.* **1**, 157–74.

Wierzbicki, T. and Abramowicz, W. (1983) On the crushing mechanics of thin-walled structures. *ASME J. Appl. Mech.* **50**, 727–34.

Wierzbicki, T. and Bhat, S.U. (1986) On the initiation and propagation of buckles in pipelines. *Int. J. Solids Structures* **22**, 985–1005.

Wierzbicki, T., Bhat, S.U., Abramowicz, W. and Brodkin, D. (1992) Alexander revisited – a two folding elements model for progressive crushing of tubes. *Int. J. Solids. Structures* **29**, 3269–88.

Wierzbicki, T., Molnar, C. and Matolscy, M. (1978) Experimental-theoretical correlation of dynamically crushed components of bus frame structures. *Proceedings of the Seventeenth International FISITA Congress*, Budapest, June.

Wierzbicki, T., Recke, L., Abramowicz, W. and Gholami, T. (1994a) Stress profiles in thin-walled prismatic columns subjected to crush loading. I. Compression. *Computers and Structures* **51**, 611–23.

Wierzbicki, T., Recke, L., Abramowicz, W. and Gholami, T. (1994b) Stress profiles in thin-walled prismatic columns subjected to crush loading. II. Bending. *Computers and Structures* **51**, 625–41.

Wierzbicki, T. and Suh, M.S. (1988) Indentation of tubes under combined loading. *Int. J. Mech. Sci.* **30**, 229–48.

Wierzbicki, T. and Thomas, P. (1993) Closed-form solution for wedge cutting force through thin metal sheets. *Int. J. Mech. Sci.* **35**, 209–29.

Wu, E. and Jiang, W.-S. (1997) Axial crush of metallic honeycombs. *Int. J. Impact Engng.* **19**, 439–56.

Wu, K.Q. and Yu, T.X. (2001). Simple dynamic models of elastic–plastic structures under impact. *Int. J. Impact Engng.* **25**, 735–54.

Wu, L. and Carney, III J.F. (1997) Initial collapse of braced elliptical tubes under lateral compression. *Int. J. Mech. Sci.* **39**, 1023–36.

Wu, L. and Carney, III J.F. (1998) Experimental analysis of collapse behaviour of braced elliptical tubes under lateral compression. *Int. J. Mech. Sci.* **40**, 761–77.

Xue, P., Tao, X.M. and Yu, T.X. (2000a) Progress in the studies of energy absorption behaviour of textile composites (in Chinese), *Adv. in Mech.* **30**(2), 227–38.

Xue, P., Tao, X.M. and Yu, T.X. (2001a) Cellular textile composite: configuration and energy absorption mechanisms. Chapter 3 in *Engineering Plasticity and Impact Dynamics* (ed. L.C. Zhang), World Scientific, Singapore.

Xue, P., Yu, T.X. and Tao, X.M. (2000b) Effect of cell geometry on the energy-absorbing capacity of grid-domed textile composites. *Composites Part A – Appl. Sci. and Manufacturing* **31**, 861–8.

Xue, P., Yu, T.X. and Tao, X.M. (2001b) Flat-topped conical shell under axial compression. *Int. J. Mech. Sci.* **43**, 2125–45.

Yang, J.L., Yu, T.X. and Reid, S.R. (1998) Dynamic behaviour of a rigid, perfectly plastic free-free beam subjected to step-loading at any cross-section along its span. *Int. J. Impact Engng* **21**, 165–75.

Yu, T.X. (1979) Large plastic deformation of a circular ring pulled diametrically (in Chinese). *Acta Mechanica Sinica* **11**(1), 88–91.

Yu, T.X. (2002) Dynamic response and failure of free-free beams under impact or impulsive loading. Chapter 10 in *Advances in Dynamics and Impact Mechanics* (eds G.N. Nurick and C.A. Brebbia), WIT Press, Ashurst, Southampton, UK.

Yu, T.X., Hui, T.Y.J. and Ruan, H.H. (2003) Dynamic behavior of scaled-down corrugated guardrail beams under impact, IMPLAST2003, Delhi, India, 16–19 March 2003. In *IMPLAST 2003: Plasticity and Impact Mechanics* (ed. N.K. Gupta), Phoenix Publishing House, New Delhi, 444–53.

Yu, T.X. and Johnson, W. (1982) Influence of axial force on the elastic-plastic bending and springback of a beam. *J. Mechanical Working Technology* **6**, 5–21.

Yu, T.X., Reid, S.R. and Wang, B. (1993) Hardening-softening behaviour of tubular cantilever beams. *Int. J. Mech. Sci.* **35**, 1021–33.

Yu, T.X., Tao, X.M. and Xue, P. (2000) Energy-absorbing capacity of grid-domed textile composites. *Composites Sci. and Tech.* **60**, 785–800.

Yu, T.X. and Teh, L.S. (1997) Large plastic deformation of beams of angle-section under symmetric bending. *Int. J. Mech. Sci.* **39**, 829–39.

Yu, T.X., Xue, P. and Tao, X.M. (2001) Design and construction of cellular textile composites for energy absorption. *ICCM-13*, Beijing, China.

Yu, T.X., Yang, J.L. and Reid, S.R. (1997) Interaction between reflected elastic flexural waves and a plastic 'hinge' in the dynamic response of pulse loaded beams. *Int. J. Impact Engng.* **19**, 457–75.

Yu, T.X, Zhang, D.J., Zhang, Y. and Zhou, Q. (1988) A study of the quasi-static tearing of thin metal sheets. *Int. J. Mech. Sci.* **30**, 193–202.

Yu, T.X. and Zhang, L.C. (1996) *Plastic Bending: Theory and Applications*. World Scientific, Singapore.

Zhang, T.G. and Yu, T.X. (1989). A note on a 'velocity sensitive' energy-absorbing structure. *Int. J. Impact Engng* **8**, 43–51.

Zhao, H. (1997) Testing of polymeric foams at high and medium strain rates. *Polymer Testing* **16**, 507.

Zhao, H. and Gary, G. (1998) Crushing behaviour of aluminium honeycombs under impact loading. *Int. J. Impact Engng.* **21**, 827–36.

Zhao, X.L. and Hancock, G.J. (1993) Experimental verification of the theory of plastic moment capacity of an inclined yield line under axial force. *Thin-Walled Structures* **15**, 209–33.

Zhou, Q. (2001) Applications of cellular materials and structure in vehicle crashworthiness and occupant protection, in *Engineering Plasticity and Impact Mechanics* (ed. L.C. Zhang), World Scientific, Singapore.

Zhou, Q., Thomas, M. and Stibich, A.M. (1998) An analytical study of system variables for meeting FMVSS 201 head impact requirements, *Proceedings of Crashworthiness, Occupant Protection and Biomechanics in Transportation Systems*, 1998 ASME Congress, Anaheim, California.

Zhou, Q., Wang, D.M. and Yu, T.X. (1990) Energy absorption capacity of box structures under impact (in Chinese), *Chinese J. Appl. Mech.* **7**(3), 51–7.

Zhu, G., Goldsmith, W. and Dharan, C.K.H. (1992a) Penetration of laminated Kevlar by projectiles – I. Experimental investigation. *Int. J. Solids Structures* **29**, 399–420.

Zhu, G., Goldsmith, W. and Dharan, C.K.H. (1992b) Penetration of laminated Kevlar by projectiles – II. Analytical model. *Int. J. Solids Structures* **29**, 421–36.

Index

acceleration pulse
in car crash test, 6
asymmetric haversine, 368
axial force effect, 35, 91

ballistic limit
of composites, 344, 350
of metal tubes, 115
beam-on-foundation model, 261
bending collapse
of channel beams, 137
of circular tubes, 130
of composite tubes, 333, 334
of rectangular/square metal tubes, 124
biaxial bending, 130
Buckingham theorem, 70

Chest Injury Criterion, 10
coefficient of restitution, 177
energetic, 180
kinematic, 179
kinetic, 177
collapse mechanism, 34
incipient, 43
collapse mode
diamond, 144
global, 238
mixed, 144
ring, 6, 144
collision
complete elastic, 180
complete inelastic, 181
compression phase of, 175
direct central, 174
energy loss in, 56, 203
restitution phase of, 175
combined torsion and bending, 143
composite
bending of composite tubes, 333
cellular textile composite, 304
circular tubes, 318
conical shells, 332
metal tubes, 337
rail beam, 332
sandwich panels, 344
grid-domed textile composite, 305
impact of, 341
conical shell, 307, 332
contact spring, 194
Cowper-Symonds relation, 58
crashworthiness, 11
critical separation, 226
crooked plates, 199

densification strain *see also* locking strain, 271
DENT, (deep edge notched tension), 215
dimensional homogeneity, 68
dimensionless groups (or π groups), 70, 243, 348
double-hat structure, 164
drop height, 360
drop hammer, 80
ductile tearing, 215
dynamic enhancement factor, 382

eccentricity factor, 148
effective crush length,
of circular tube, 148
of square tube, 163
elastic, linear hardening material, 26
elastic, perfectly plastic material, 26
elastic, power hardening material, 26
elliptical tubes, 102, 331
energy absorption diagram, 366
energy ratio, 33
impact energy ratio, 111
equivalent structure technique, 89
essential work, 216

failure mechanism, 313
falling object protective structures (FOPS), 14

foam
  closed-cell, 278
  cushion curves of, 360
  filling, 143, 333
  master curve of, 362
  open-cell, 278
fracture strain, 26, 222,318, 349
fragility factor, 361
fragmentation mode, 322, 326
free–free beam, 54
free tube inversion, 249
fully plastic bending moment, 29
  of section of bi-materials, 339
fundamental dimensions, 68

Gadd Severity Index (GSI), 8
gas guns, 87
generalised force, 35
generalised plastic hinge, 37

half-length of a fold, 147, 162
hardening modulus, 27
Head Injury Criteria (HIC), 8, 368
Hertz theory, 181

impact test methods, 87
impulse, 53
  compression, 176
  over-loading, 53
  restitution, 177
indentation, 181
  elastic, 183
  elastic-plastic, 186
  energy dissipation due to, 190
inertia-sensitive energy-absorbing structures, 197

kinematically admissible velocity/displacement field, 40
knuckle radius, 251, 258

large deformation, 46
limit load, 34
limit state, 34
limit surface, 35
line of thrust, 88
load space, 35
local penetrating mode, 237
locking strain *see also* densification strain
  of foams, 169, 280
  of honeycombs, 271
  of wood, 287
lower bound theorem, 38

mass-spring model, 192, 195, 380
maximum bending moment, 125
maximum elastic bending moment, 29
maximum elastic curvature, 29
moment-curvature relationship, 29
moving hinge (see also plastic hinge), 90

non-dimensional boss size parameter, 114

oblique compression, 112

packaging, 17, 359
pendulum, 80
physical variables, 68
pipe-whip, 15
plane strain, 98
plastica, 99
plastic dent depth, 120
plastic hinge, 30, 124
  effective length of, 36, 99, 151
  travelling, 124, 158
plateau stress
  of foams, 280, 282
  of honeycombs, 271, 273, 276
  of woods, 286
pivot point, 92
propagating collapse, 261
prototype, 75

rigid, perfectly plastic material, 27
rigid-plastic idealisation, 32
rigid, strain-hardening material, 27
ring net, 351
rockfall protection, 351
roll-over protective structures (ROPS), 14

safety helmet, 16, 316
scaling factor, 75
shock wave, 50, 106, 294, 289
size effect, 79
sled, 80
small-scale model, 75
solidity ratio, 151, 152, 163, 165
spalling, 51
specific energy absorption capacity, 23, 284, 318, 328
specific work, 216
splaying progressive crushing, 321, 324
split Hopkinson pressure bar (SHPB), 82
square or hexagonally packed, 109
statically admissible stress field, 38
stopping distance, 369
strain-hardening, 25
strain-rate sensitivity of materials, 57
stress wave
  elastic, 48
  impedance of, 48
  plastic, 48
strip method, 133
structural effectiveness, 151, 152, 163, 165

Taylor theory, 87
Thoracic Trauma Index (TTI), 10
top-hat structure, 164
toroidal shell, 157
torsional crushing, 143
transition zone, 266
trousers type setup, 217
tube nosing, 255
type II structures, 197

ultimate stress, 26
upper bound
  method, 61
  of final displacement, 66
  theorem, 41

vehicle
  accident statistics, 2
  interior trim, 367

Wayne State Tolerance Curve, 7
W-beam guardrail system, 12, 373
Winkler foundation, 184
wrap-around, 106

yield
  strain, 27
  stress, 25
  velocity, 48